AF696144

Complex Networks and Dynamic Systems

Volume 1

Series Editor:
Terry L. Friesz
Pennsylvania State University, University Park, PA, USA

For further volumes:
http://www.springer.com/series/8854

Maria Giaoutzi · Bartolomeo Sapio
Editors

Recent Developments in Foresight Methodologies

Springer

Editors
Maria Giaoutzi
Department of Geography
and Regional Planning
National Technical University of Athens
Athens, Greece

Bartolomeo Sapio
Fondazione Ugo Bordoni
Roma, Italy

ISBN 978-1-4614-5214-0 ISBN 978-1-4614-5215-7 (eBook)
DOI 10.1007/978-1-4614-5215-7
Springer New York Heidelberg Dordrecht London

Library of Congress Control Number: 2012951671

© Springer Science+Business Media New York 2013
This work is subject to copyright. All rights are reserved by the Publisher, whether the whole or part of the material is concerned, specifically the rights of translation, reprinting, reuse of illustrations, recitation, broadcasting, reproduction on microfilms or in any other physical way, and transmission or information storage and retrieval, electronic adaptation, computer software, or by similar or dissimilar methodology now known or hereafter developed. Exempted from this legal reservation are brief excerpts in connection with reviews or scholarly analysis or material supplied specifically for the purpose of being entered and executed on a computer system, for exclusive use by the purchaser of the work. Duplication of this publication or parts thereof is permitted only under the provisions of the Copyright Law of the Publisher's location, in its current version, and permission for use must always be obtained from Springer. Permissions for use may be obtained through RightsLink at the Copyright Clearance Center. Violations are liable to prosecution under the respective Copyright Law.
The use of general descriptive names, registered names, trademarks, service marks, etc. in this publication does not imply, even in the absence of a specific statement, that such names are exempt from the relevant protective laws and regulations and therefore free for general use.
While the advice and information in this book are believed to be true and accurate at the date of publication, neither the authors nor the editors nor the publisher can accept any legal responsibility for any errors or omissions that may be made. The publisher makes no warranty, express or implied, with respect to the material contained herein.

Printed on acid-free paper

Springer is part of Springer Science+Business Media (www.springer.com)

To our children:
Dimitris, Edoardo, Francesco, and Stella for making our lives brighter.

"It's just the light … shadows of the night"
"May be the night that opened up, a blue pomegranate,
a dark chest, and filled you with stars,
cleaving time.
And yet the statues …"

George Seferis
from "Sensual Elpenor",
Collected Poems 1924–1955

About the Authors

Dr. Toni Ahlqvist is a senior scientist at VTT Technical Research Centre of Finland. Currently, he works as a postdoctoral researcher at the Academy of Finland. His current research focuses on socio-spatial transformations induced by science, technology and innovation policies and political economy of national and regional innovation systems. He has also published widely on the field of foresight, on topics such as roadmapping, emerging technologies and infrastructures and socio-technical change.

Dr. Aliye Ahu Akgün is a research assistant at the Istanbul Technical University (ITU) Department of Urban and Regional Planning. She received her M.Sc. in the field of regional planning from Istanbul Technical University in 2005 and her Ph.D. in the field of urban and regional planning in 2009. Her main research interests are urban and regional planning, rural development, suburbanisation, entrepreneurship and diversity. In recent years, she has focused her research on rural entrepreneurship and sustainable rural development.

Péter Alács (1976) is a Ph.D. student at the Futures Studies Department of the Corvinus University of Budapest (CUB). His current research focus is on mathematical-statistical methods of interactively obtained information in a foresight process. He has a degree in physics (2001) from Eötvös Lóránd University (ELTE) and in economics (2003) from Corvinus University of Budapest (CUB). He was a research fellow at CUB research programme, sponsored by the Hungarian Academy of Sciences (2003–2006). He has published extensively on demographics, evolutionary algorithms, optimisation of pension systems, stochastic processes and the different philosophical and methodological approaches of uncertainty and complexity.

Hedwig van Delden is a director of the Research Institute for Knowledge Systems (RIKS), Maastricht, since 2005. She works among others on model integration, land use modelling and scenario studies. Her research on the integration of spatially explicit models frequently leads to teaching and speaking assignments in universities and scientific and policy-oriented congresses and events worldwide. She is managing and scientific director of the RIKS, and manages and provides scientific leadership to national and international projects that focus on the design, development and use of integrated models for policy support.

Annele Eerola works as a principal scientist in the team 'Foresight and Socio-technical Change' at VTT. Her research focuses on the links between foresight knowledge, corporate strategy and innovation policy. She holds a Ph.D. from Helsinki Swedish School of Economics and Business Administration and a Lic. Tech. degree from Helsinki University of Technology.

Professor Dr. Jan Erik Karlsen, has 40 years of experience as a research scientist. Presently, he is serving as professor of industrial economics and change management at the University of Stavanger, Norway, and as senior research scientist at the International Research Institute of Stavanger. He is vice president of the Norwegian chapter of ILERA (International Labour and Employment Relations Association), board member of the ILERA International Executive Committee, vice president of the Academy of Sciences (in Stavanger), forum chair for EAIR 2012 (The European Higher Education Society) and member of a series of other academic and professional organisations within sciences, economics and sociology, including World Futures Studies Federation.

Dr. Masood Gheasi has been working in the Department of Spatial Economics at the VU University Amsterdam since 2009. He has a master's degree and Ph.D. in law and economics from Bologna University. He worked for a short period at the Sapienza University of Rome, Italy. His research focus is on international migration, especially knowledge migrants, and has a large number of publications in the field.

Maria Giaoutzi is a professor of regional planning and economic geography at the National Technical University of Athens (NTUA); a fellow of the Netherlands Institute of Advanced Studies (NIAS), the Royal Dutch Academy, since 1991; member of various International Research Networks (GREMI, NECTAR,URBINO, STELLA, etc.); elected chancellor at the Steering Committee of the International Regional Science Association (1990–1994); elected member of the Executive Committee of I.G.U-C18 (Telecommunications)/CNFG, (1992–1995); president of the Board of Directors of the Hellenic Tourism Organization Land Development Company (1998–2000); and consultant at international organisations, such as OECD, EEC, ERT, and ESF. She has long research experience from EU and national projects on foresight, e-Planning, participatory methods and modelling, technology assessment, etc. She published rather extensively (books and journals, among others) on the above topics.

Ruud van der Helm was trained as an engineer and holds a Ph.D. in environmental sciences with a thesis on 'the vision phenomenon'. He is working on the fundamentals of future studies, including the use of language as an instrument for studying the future. Together with Ted Fuller and Laurent Mermet, he coedited the special edition of Futures on 'Re-examining and renewing theoretical underpinnings of the Futures field' (Futures 41 (2009)). Currently, he is the first secretary to the Embassy of the Kingdom of the Netherlands to Ghana.

Hanne Karlsen is a doctoral candidate in Leadership and Organisation at BI – Norwegian School of Management. Her current research focus is on organisational change processes and organisational identity dynamics when organisations respond to and voluntarily adopt rules of standards that explicitly pertain to a wider set of organisations. She has previously published articles on future studies and knowledge creation for the future. She also lectures in change management, organisation theory and organisational communication.

Eric Kemp-Benedict is a senior scientist at the Stockholm Environment Institute and lead of SEI's Rethinking Development theme. His work focuses on sustainable development planning and scenario analysis. Kemp-Benedict actively develops and applies tools and methods for participatory and study-specific sustainability analyses. Other areas of interest include water, livestock and land use; rural livelihoods; poverty and income distribution; and social dynamics. He received his Ph.D. in theoretical physics from Boston University in 1997 and his MAT for secondary physics education from Tufts University in 2003.

Dr. Sirkku Kivisaari works as senior researcher in the team 'Foresight and Socio-technical Change' at VTT. Her educational background is in business management. Her research focuses on future-oriented technology assessment and innovation studies. Her special interest lies in enhancing the diffusion of innovations provoked by societal concerns for the well-being of the ageing society.

Prof. Raija Koivisto has over 25 years' experience in research, development, teaching and consulting in safety and foresight-related topics. Her current research is focused on integrating safety and security issues in foresight and future research studies. Current activities include national, European and international research, development, and consulting activities concerning security, safety and foresight – special areas of activity are mass transport and the economical evaluation of safety. She is a member of several related networks.

Dr. Totti Könnölä is a CEO of Impetu Solutions, a consultancy specialised in strategic innovation and ICT. He is also adjunct professor on operations in IE Business School, Madrid, and associate professor in industrial policy and technology in ICAI UPComillas, Madrid. He is a frequent contributor to international conferences and a reviewer, editor and author for a number of scientific publishers.

Carolien Kroeze received an M.Sc. degree in biology from the University in Groningen, the Netherlands, and a Ph.D. degree in environmental sciences from the University of Amsterdam, the Netherlands. She has worked at the National Institute for Public Health and the Environment (RIVM) in the Netherlands. Currently, she is professor of environmental pollution at Wageningen University. In addition, she is professor of environmental outlook studies at the Open University of the Netherlands.

Dr. Tuomo Kuosa is an expert on strategic foresight, scanning and visualising futures knowledge into alternative futures and developing futures research methodologies. He is interested in nonlinearities, triggering points and path dependencies in transformation. He is founding partner of AlternativeFutures, Finland, that merges strategic foresight with codesign, ideation, concept design and fast prototyping. Kuosa is a postdoctoral fellow at Singapore's Nanyang Technological University, Centre of Excellence for National Security (CENS), since 2010.

Dr. Matthias K.B. Lüdeke is a senior scientist at the Potsdam Institute for Climate Impact Research (PIK) where he works on climate change impacts on urban functions and global patterns of climate and global change vulnerabilities. Recently, he contributed to various urban modelling–related projects at PIK, including Hyderabad (BMBF), URBS PANDENS (EU) and Vulnerability Archetypes (PBL). He is lead author of UNEP's Global Environmental Outlook, lecturer in environmental sciences at the University of Potsdam, reviewer for Global Environmental Change and Environmental Modelling and Assessment and EU's research directorate general, the Swiss National Research Foundation and the German Federal Environmental Agency.

Erik Laes is a chemical engineer and holds several degrees in philosophy and environmental sciences. In 2006, he obtained a Ph.D. on 'The role of nuclear energy in a sustainable energy supply'. He worked (1999–2009) as a researcher at SCK•CEN (the Belgian nuclear research centre) on several projects involving sustainability appraisal, participative technology assessment and scenario development. In July 2009, he joined VITO in the unit 'Transition, Energy and Environment' where he is currently involved in projects concerning scenario analysis, sustainability evaluation of energy technology and transition management.

Dr. Eveline van Leeuwen is an assistant professor in the Department of Spatial Economics at the VU University, Amsterdam. Her background is in spatial planning and spatial economics. In the past years, her research focused on retail and rural development, urban–rural interactions, tourism and sustainable development issues. Apart from participating in national and EU-funded projects, she also teaches in the field of earth sciences and economics.

Fre Maes was educated in biology, environmental protection and microeconomics in Brussels. He has experience in participative policymaking both as a stakeholder and policymaker. He is former member of the cabinet of the minister vice president of Flanders, the federal council for sustainable development and the Flemish Environmental Impacts Reporting team (MIRA). He is currently working as climate expert for the federal Belgian environment administration and responsible for European emission trading scheme policy and implementation matters as well as issues related to the transition of Belgium towards a low-carbon society.

Peter Nijkamp is a professor of regional and urban economics and economic geography at the VU University, Amsterdam. In the past years, he has focused on new quantitative methods for policy analysis and on spatial-behavioural analysis of economic agents. He is past president of the Netherlands Research Council (NWO). In 1996, he was awarded the most prestigious scientific prize in the Netherlands, the Spinoza award.

Enrico Nicolò has a Doctor Laurea degree in electronic engineering, summa cum laude, from the Università degli Studi di Roma 'La Sapienza'. He is a senior engineering research scientist at Fondazione Ugo Bordoni (FUB), where he is focusing on scenario methodologies, project-network simulations, telecommunication networks, electromagnetic field biohazard and communication ethics. He deals with user experience of communication technologies, with focus on young users. His research interests include the language of photographic communication. Prior to joining FUB in 1986, he was for 3 years with SESA Italia, where he worked on telephone and packet-switched data networks.

Arturs Puga received his Dr. Hist. from the University of Latvia in 1992. He was in the diplomatic service during 1990–2002 and an ambassador during 1999–2002. His work relates to political and social sciences and international relations published in European countries. Since 2003, he is head of the Forward Studies Unit (independent, think–do body for European foresight research) located in Latvia. Dr. Puga has expertise in foresight, innovation and knowledge management. His recent research and publications focus on theoretical issues of knowledge development, dealing with futures at the organisational and the national level. He works with European policymaking bodies and science and technology institutions.

Victor van Rij is scientific staff member of the Netherlands Council for Science and Technology. He is an expert in the field of foresight and horizon scanning and identification of early warning signals. He has been involved in the national foresight activities of the Netherlands since 2001, in his capacity both as senior advisor in the unit of Foresight and Horizon Scanning of the Ministry of Education, Culture and Science and as assistant of the science foresight coordinator at the Royal Academy of Science and Arts in the Netherlands. Dr. van Rij was the Netherlands coordinator of the ERANet Forsociety and is member of the national network of future experts (NTV) within the Netherlands. He has been involved in many national and international foresight projects as consultant and contractor and had been member of the Science in Society Advisory group of the EC. He has been involved in the Dutch national horizon scan and has led a pilot project joining the horizon scans of the UK, Denmark and the Netherlands.

Da Ruan began his career at the Belgian Research Centre SCK•CEN in 1991, after obtaining his Ph.D. in mathematics from Ghent University (Belgium). Da completed the TRANSFUSION project successfully, as a postdoc, by developing a fast algorithm for the analysis of well-logging signals for the oil industry. Subsequently, he took the initiative in the FLINS (Fuzzy Logic in Nuclear Science) project and the FLINS conferences which grew to become the leading conference in this specialised field. Da Ruan searched tirelessly for applications of fuzzy logic and related theories, for example, in reactor control, cost estimates including uncertainties of large projects, energy policy decision support systems and the analysis of large data sets for safety culture and the nonproliferation of nuclear weapons. He published more than 90 peer-reviewed journal articles, 2 textbooks and 20 research books. His international recognition is also shown by numerous invitations to act as a keynote speaker at international conferences as well as by the award of an honorary doctorate from the Nuclear Power Institute of China. Da Ruan served as member of the editorial board of Fuzzy Sets and Systems, as regional editor for Europe of the *International Journal of Intelligent Automation and Soft Computing*, co-editor-in-chief of the *International Journal of Nuclear Knowledge Management*, editor-in-chief of the *International Journal of Computational Intelligence Systems*, editor of the book series Intelligent Information Systems and editor of the proceedings series Computer Engineering and Information Science. Da Ruan was guest professor at the Department of Applied Mathematics and Computer Science at Ghent University and at the Department of Applied Economics at Hasselt University and assistant professor at the Faculty of Information Technology at the University of Technology in Sydney, Australia *(in memoriam).*

Bartolomeo Sapio has a Doctor Laurea degree in electronic engineering, summa cum laude, at the University of Rome 'La Sapienza'; he is senior researcher and project manager at Fondazione Ugo Bordoni. He has participated in several international projects. He was chairman of COST (COoperation in the field of Scientific and Technical research) Action 298 'Participation in the Broadband Society' and member in the Management Committees of COST Action IS0605 'A Telecommunications Economics COST Network – Econ@Tel' and COST Action A22 'Foresight methodologies – Exploring new ways to explore the future'. From 2010 to 2011, he chaired the technical working groups of the TETRA Association, representing more than 160 organisations in the TETRA (Terrestrial Trunked Radio) business from all continents of the world.

Serge Stalpers is a doctoral candidate in Environmental Systems Analysis at Wageningen University. His research focus is on the science-policy interface in climate change policy, particularly the co-production of model results and scenarios in integrated assessments. Currently, he is education coordinator at the SENSE Research School for Socio-Economic and Natural Sciences of the Environment.

Dr. Anastasia Stratigea is a Dipl. Engineer and assistant professor at the Department of Geography and Regional Planning, National Technical University Athens (NTUA). She has a long experience in EU and national projects. Her recent research interests focus on foresight methodologies, scenario planning, participatory planning, ICTs, regional development, renewable energy, etc.

Dr. Erik Terk has been director of the Estonian Institute for Futures Studies for the last 20 years. He is specialised in subjects like transition economy and transition society studies, innovation studies as well as geo-economic scenarios and strategic planning. In the beginning of the 1990s, he coordinated, as the deputy minister of economy, the introduction of Western type of entrepreneurship forms and business support policy in the Republic of Estonia. Parallel to scientific and consulting work, he held membership positions on different boards and councils of institutions for the planning and/or executing of economic policy in Estonia: the Bank of Estonia

(National Bank), the Estonian Development Fund, the Financing Committee for Technology Development and the Port of Tallinn. Erik Terk was a council member of the International Institute for Applied Systems Analysis (IIASA).

Aviel Verbruggen was educated in engineering and applied economics at Louvain, Antwerp and Stanford University. He is full professor at the University of Antwerp, in energy and environmental economics and policies. He has published more than 450 journal articles, book chapters and advisory reports. Salient topics are cogeneration, electricity supply and demand, energy efficiency, renewable energy and policy planning. He is founder of research groups and advisory companies. He conceived, supervised and edited the State of the Environment Reports {MIRA} in Flanders (1993–1998). He was the first president of Flanders' Environmental Advisory Council {MiNa-Raad} (1991–1995) and contributor to the IPCC, Working Group III: Third, Fourth and Fifth Assessment Reports (1998–2014), and Special Report on Renewable Energy (2008–2011).

Acknowledgements

The authors wish to acknowledge COST Action A22 "Foresight Methodologies: Exploring new ways to explore the future" and its participants.

We offer special recognition to Ms. Patricia Ellman, whose language skills helped improve this book.

Contents

Part I
Introduction

Chapter 1
In Search of Foresight Methodologies: Riddle or Necessity

Maria Giaoutzi and Bartolomeo Sapio

> *I also realized along the way that the goal is not everything. Going through the process all together is important.*
>
> Michel Godet (2012)

To study the future is to study potential change – unveiling what is likely to make a systemic or fundamental difference over the next 10–25 years or more. Studying the future is not simply economic projection or sociological analysis or technological forecasting, but a multidisciplinary examination of change in all major areas of life in order to find the interacting dynamics that are creating the next age.

Futurists have not yet reached consensus on the name or definition of their activity. Some prefer the term 'futures research', meaning the use of methods to identify systematically both the consequences of policy options and alternative futures with policy implications for decision makers. Others prefer the term 'future studies', meaning the study of what might happen, and what we might want to become. Still others, apparently in Europe and francophone Africa, prefer 'prospective studies', meaning the study of the future in order to develop a strategic attitude of mind with a long-range view of creating a desirable future.

Foresight is a professional practice that supports significant decisions, and as such it needs to be more assured of its claims to knowledge (methodology). Foresight

M. Giaoutzi (✉)
Department of Geography and Regional Planning,
National Technical University of Athens (NTUA),
Heroon Polytechniou str. 9, Zographou Campus, 15780, Athens, Greece
e-mail: giaoutsi@central.ntua.gr

B. Sapio
Fondazione Ugo Bordoni,
Viale del Policlinico 147, Roma, 00161, Italy
e-mail: bsapio@fub.it

M. Giaoutzi and B. Sapio (eds.), *Recent Developments in Foresight Methodologies*, Complex Networks and Dynamic Systems 1, DOI 10.1007/978-1-4614-5215-7_1,
© Springer Science+Business Media New York 2013

is practised across many domains and is not the preserve of specialised 'futurists', or indeed of foresight specialists. However, the disciplines of foresight are not well articulated or disseminated across domains, leading to reinventions and practice that does not always make best use of experience in other domains.

Futures research can be directed to large- or small-scale issues, in the near or distant future, and can project possible or desired conditions. It is not a science; the outcome of studies depends on the methods used and the skills of the practitioners. Its methods can be highly quantitative or qualitative. It helps to provide a framework to better understand the present and expand mental horizons.

The value of futures research lies less in forecasting accuracy than in its usefulness in planning and opening minds to consider new possibilities, and thus change the policy agenda. Its purpose is not to know the future, but to help us make better decisions today via its methods that force us to anticipate opportunities and threats, and consider how to address them – strategically it is better to anticipate, rather than just respond to change.

The purpose of futures methodology is to systematically explore, create, and test both possible and desirable futures to improve decisions. It includes analysis of how those decisions might change as a result of the implementation of policies and actions, and the consequences of these policies and actions.

The use of futures methods enhances anticipatory consciousness, which in turn improves the foresight, thus making it possible to act faster or earlier, and making the organisation or individual more effective in dealing with change. The ability to anticipate gives extra time to better understand threats and opportunities, develop more creative strategies, create new product opportunities, and create and share vision for organisational change.

The methodological development of foresight is an important task that aims, firstly, to strengthen the pool of the tools available for application, thus empowering the actors involved in the realm of foresight practice. Secondly, elaborating further on methodological issues, such as those presented in this book, enables the actors involved in foresight to begin to critique current practice from this perspective and, thirdly, to begin to design foresight practice with greater reflexivity.

The present trends towards methodological concerns indicate a move from 'given' expert-predicted futures towards a situation in which futures are nurtured through the dialogue between 'stakeholders', i.e. those with a stake in the future of the particular issue under study.

The focus of this book is on recent developments in foresight methodologies and relates in its greater part to the work done in the context of the COST A22 network of the EU on foresight methodologies.

This book has five parts elaborating on a set of aspects of foresight methodologies.

In Part I, Chap. 1, 'In Search of Foresight Methodologies: Riddle or Necessity', by Maria Giaoutzi and Bartolomeo Sapio, elaborates on the insights and perspectives of the foresight practice and presents the organisation of this book.

In Part II, a set of three chapters focuses on 'Theorizing About Foresight Methodologies'.

More precisely in Chap. 2, 'Defining the Future: Concepts and Definitions as Linguistic Fundamentals of Foresight', Ruud van der Helm stresses the importance of semantics that warrants an analysis of how language functions and what it contributes to a better understanding of the future. This chapter focuses on one particular dimension of this wide domain of semantic research: concepts and definitions. The objective of this contribution is to provide more insights into how definitions of future relevant terminology are shaped and laid down in semantic research. Finally, the chapter draws some conclusions on how to proceed with semantic research and the enrichment of the futures domain that it may yield.

In Chap. 3, 'Classification of Tools and Approaches Applicable in Foresight Studies', Jan E. Karlsen investigates the inherent ontological and epistemic premises embedded in the application of quantitative and qualitative foresight methods and tools and provides a taxonomy for the classification of such approaches. His conclusion challenges researchers in the field by maintaining that an optimal combination of qualitative and quantitative approaches is only limited by the ingenuity of the researchers themselves, not by the intrinsic characteristics of the approaches.

In Chap. 4, 'Bridging Qualitative and Quantitative Methods in Foresight', Matthias Lüdeke focuses on the shortcomings of epistemological approaches to prediction, oriented to the case of classical mechanics, which appears as a more or less singular stroke of luck in the history of science. First, the role of quantitative models in foresight studies is presented. Then, after a short overview of the four main approaches to foresight according to Kreibich (2006), the chapter proceeds with a discussion of qualitative, as compared with quantitative, concepts in science and concludes with a range of approaches that bridge the gap between the two traditions.

In Chap. 5, 'New Emerging Issues and Wild Cards as Future Shakers and Shapers', by Victor van Rij, *the* focus is on wild cards which in his view represent the occurrence of singular (idiosyncratic, historically original), sudden (abrupt, fast), surprising (unexpected, startling), and high-impact (severe) events. Thus, wild cards circumscribe the following: (1) one-of-a-kind discrete incidents (2) that arise rapidly (3) in a way not fully recognisable ex ante from past information and (4) that lead to profound perturbations and alterations in the known state of affairs. Hence, wild cards not only have an impact in quantitative terms on the absolute level of a pre-existing trend or alter its rate of growth but also are likely to transform the qualitative attributes of the phenomenon, possibly setting new directions for future evolution. This chapter explains how the wild cards are identified and can be used to safeguard policies against their unwanted effects.

In Part III, six chapters are included which elaborate on 'System Content Issues'.

In Chap. 6, 'Forms of Reasoning in Pattern Management and in Strategic Intelligence', Tuomo Kuosa deals with the issue of strategic intelligence, an emerging field of business consulting that presents large, complex, or complicated issues of transformation in a more understandable form. Pattern management is seen as one field of strategic intelligence based on empirical data and formal structures of strategic intelligence, but also on a heuristic approach enabling inter alia, quantitative data, reasoning, and narratives to be integrated. The main aim of this chapter is to show the most commonly used ways of managing, finding, drawing, reasoning, or anticipating patterns in our environment, but also to locate how the concept of *pattern* can be perceived in different ways.

Péter Alács, in Chap. 7, 'Micro-meso-macro: From the Heritage of the Oracle to Foresight', provides a methodological framework for designing and better understanding the foresight process. Towards this end, he presents a three-level approach, where *uncertainty*, *complexity,* and *time* are dealt with differently at each level. At the meso-level, both narratives and numbers are used to assess the knowledge on the foresight issues, while at the micro-level, this is prevented by not integrating narratives and numbers. Finally, integration is possible only at the macro-level by using narrative tools. The chapter shows that the above methodological approach to foresight not only expands the potential of the foresight activity but also clarifies its methods.

In Chap. 8, 'Going from Narrative to Number: Indicator-Driven Scenario Quantification', by Eric Kemp-Benedict, the focus is on the limitations of the predictive mathematical models, conventionally used in policy analysis, in their potential capacity as exclusive tools in futures studies, since they cannot integrate the sudden changes seen in real societies. As an alternative, the field of complex systems has successfully produced similar changes in simplified model systems but has been less successful in practical futures work.

Some recent scenario exercises are presented, such as the IPCC, UNEP's GEO-3 scenarios, the work of the Global Scenario Group, and the European VISIONS project, which have all addressed this issue by combining wide-ranging narratives with quantitative models, demonstrating that a synthesis between qualitative and quantitative approaches is possible.

The characteristics of computer models, appropriate for use in foresight, are presented, and examples of appropriate models are described.

In Chap. 9, 'On Foresight Design and Management: A Classification Framework for Foresight Exercises', Totti Könnölä, Toni Ahlqvist, Annele Eerola, Sirkku Kivisaari, and Raija Koivisto claim that while the expansion of foresight scope to include systemic processes and societal considerations has provided significant opportunities for learning and synchronised action between different business units and/or policy fields, it may also have caused digression and ambiguity in the theory and practice of the management of foresight processes. This is true, in particular, in contract research organisations that have faced major challenges to reorganise their foresight activities, as part of the changes in their innovation practices.

This chapter examines this shift and consequent methodological responses at the VTT Technical Research Centre of Finland. In particular, it considers the design and implementation of recent VTT road mapping and other foresight processes which typically apply and link diverse methods in order to best respond to case-specific expectations. This chapter also develops a coherent classification framework in support of the design and management of foresight processes. The experience of VTT is geared towards the coherent and modular application of foresight methods and the responsive engagement of stakeholders.

In Chap. 10, Arturs Puga elaborates on the question: 'Will Entrepreneurship, Knowledge Management and Foresight Emerge in a System?' by focusing on the results and experience gained by foresight projects and workshops in Latvia and by making the assumption that an understanding of the knowledge management (KM) terminology and processes enables the active and effective participation of researchers in foresight projects. It also facilitates the development of the foresight culture at both the individual and the organisational levels and presents examples of how organisational and personal KM models operate in the development of project activities.

In Chap. 11 on 'Scenario Transfer Methodology and Technology', Bartolomeo Sapio and Enrico Nicolò are concerned with facilitating the understanding and exploitation of the results obtained through the application of foresight methodologies by decision makers and strategic planners.

The availability of complex mathematical outputs has often discouraged top managers and decision makers from adopting suggestions derived from the utilisation of these methods and has limited the potential of their conceptual frameworks and computerised tools.

The authors propose that by exploiting the available technological tools, scenario modellers should be able to integrate the capabilities offered by technology to *transfer* effectively and efficiently the acquired results to strategic end-users, so that the comprehension, interpretation, acceptability, and usability of scenario methods and their results can be facilitated and increased. To this end, this chapter introduces some fundamentals of a *scenario transfer methodology and technology*, which are developed within the logical framework of *scenario engineering.*

Part IV, 'Foresight Tools and Approaches', consists of seven chapters presenting a broad range of applications of foresight tools and approaches which are included in EU and other projects.

Chapter 12, 'Willingness of Stakeholders to Use Models for Climate Policy: The Delft Process', by Serge Stalpers and Carolien Kroeze, focuses on participatory integrated assessment (PIA) approaches which ensure consideration of multiple perspectives on climate change through science-stakeholder dialogues, while simultaneously respecting decision stakes. More precisely, the authors have investigated the information needs of the participants of the Delft Process in order to assess their willingness-to-use (WTU) model results. Building on the Delft Process case, a conceptual model of WTU assessments is presented, assuming that participants often assess WTU implicitly based on their own expectations, and on scattered information provided by scientists.

The Delft Process illustrates how in a PIA, WTU is not always explicitly assessed by the participants. An explicit WTU assessment, such as that suggested by the conceptual model presented in this chapter, could improve the communication between scientists and stakeholders by ensuring that information provided by the scientists better matches the information needs of participants for assessing their WTU models.

In Chap. 13, 'Linking Narrative Storylines and Quantitative Models to Combat Desertification in the Guadalentín Watershed (Spain)', Kasper Kok and Hedwig van Delden elaborate on a foresight approach, developed in the framework of an EU-financed project, dealing with desertification in the Mediterranean region, where multi-scale scenarios were developed for Europe, the Northern Mediterranean, and four other local areas. This approach involves the participation of stakeholders in the scenario development process, which links all these narrative storylines with an integrated quantitative model. A Policy Support System (PSS) is also presented. Developed in the same project, this PSS has as its main objective to establish a link between the qualitative scenarios and the PSS for the watershed of the Guadalentín River in Spain. From the results of two scenario workshops, three scenarios were selected, each linked to the same Mediterranean scenario. The purpose of this selection was to maximise both the variety in the narrative storylines and the expected output of the PSS. The chapter illustrates the practical potential and pitfalls of linking qualitative storylines and quantitative models. Future research should, however, also focus on the more fundamental theoretical obstacles that can be easily overlooked.

In Chap. 14, 'Scenario Planning as a Tool in Foresight Exercises: The LIPSOR approach', Anastasia Stratigea and Maria Giaoutzi focus on the potential of scenario planning for regional future studies with the support of scenario planning tools. The chapter first discusses the role of scenario planning in foresight, and then considers an application of the scenario planning participatory model LIPSOR, in the region of Crete. Finally, some conclusions are drawn as to the strong and weak points of the suggested approach.

Chapter 15, 'Foresights, Scenarios and Sustainable Development: A Pluriformity Perspective' by Eveline van Leeuwen, Peter Nijkamp, Aliye Ahu Akgun, and Masood Gheasi, concerns future sustainable development strategies from a stakeholders' perspective. The chapter reviews various sustainable development contributions and also addresses various methodological issues pertaining to sustainable development. The literature review lays the foundation for the operational analysis in the chapter. Based on a multidimensional indicator system, reflecting a pluriformity in approaches and viewpoints, a systemic perspective based on a multicriteria model is proposed against the background of an 'amoeba' diagram. By means of this model, a set of local or regional empirical case studies is presented originating from five European countries: namely Italy, Spain, Romania, Finland, and Scotland. To map out and analyse sustainable development of the areas under consideration, four scenarios are developed (Competitiveness, Continuity, Capacity, and Coherence) for each of these five European cases. These cases are evaluated on the basis of the viewpoints of relevant stakeholders regarding future sustainable development. The scenarios are then systematically assessed with a view to the identification of the most preferred future. Their results indicate that, in general, the most preferred sustainable future is that of the Coherence scenario, in which ecological and social factors are the most influential sustainability factors.

In Chap. 16, 'Methodological Challenges in Combining Quantitative and Qualitative Foresight Methods for Sustainable Energy Futures: The SEPIA Project', by Erik Laes, Da Ruan, Fre Maes, and Aviel Verbruggen, the focus is on the merits and challenges of combining participatory fuzzy-set multicriteria analysis (MCA) with narrative scenario building, supported by (quantitative) energy system modelling within the context of the SEPIA project ('Sustainable Energy Policy Integrated Assessment'). SEPIA aims to provide tools and methods to support deliberations on a sustainable energy future for Belgium. The project set-up presented includes the following phases: methodological reflections on sustainability assessment; participatory construction of long-term sustainable energy futures; participatory construction of a value tree, including sustainability criteria; and a deliberation on these futures with the aid of a fuzzy-set MCA decision support tool that is both methodologically sound and legitimate from a stakeholder point of view.

Chapter 17, 'Building Strategic Policy Scenarios for EU Agriculture: AG2020 ', by Maria Giaoutzi and Anastasia Stratigea, first focuses on the development of a methodological framework, in support of the decision-making process for reforms in the Common EU Agricultural Policy (CAP). The use of this framework will support the structuring of a range of backcasting policy scenarios, based on quantitative and qualitative analysis of the future. The chapter then describes the methodological approach for building strategic policy scenarios for AG2020. Following on from this, the process of building the Images of the Future in EU agriculture in 2020 is presented. This is put into practice in building the policy framework that will support the target achievement in the Images of 2020. The chapter closes with some conclusions, and future prospects are discussed.

In Chap. 18, 'Opportunities for Combining Quantitative and Qualitative Approaches in Scenario-Building: The Experience of the "Estonia 2010" Project', Erik Terk claims that the qualitative and quantitative approaches, or narratives and numbers, are one of the most exciting problems in the development of the methodology of foresight/futures studies. The building of user-oriented scenarios is not just one of the futures studies methods but rather a broader methodological construction, providing a discussion platform on how the qualitative and quantitative methods in scenario building could be combined so as to ensure both the consistency and cohesion of the created constructs, and also their user-friendliness for the decision makers. The chapter also considers the delicate issue of integrating numbers or narratives with alternating weighting procedures throughout the scenario-building period and policy implementation.

References

Godet, M. (2012). To predict or to built the future? Reflections on the field and differences between foresight and 'la prospective'. *The Futurist, 46*(3).

Kreibich, R. (2006). Future issues and future science. Research report No26 Berlin: IZT- Institute for future studies and technology assessment.

Part II
Theorizing About Foresight Methodologies

Chapter 2
Defining the Future: Concepts and Definitions as Linguistic Fundamentals of Foresight

Ruud van der Helm

> *Once having recognized that the dictionary is not a stable and univocal image of a semantic universe, one is free to use it when one needs it.*
>
> Umberto Eco

2.1 Introduction

The future is explored and created through language. Terminology, concepts and definitions form fundamental ingredients for foresight, leading into inferences, conjectures, narratives and stories. Many futures methods rely on a specific lingo, some of which has even been trademarked. Although not always duly recognised, the importance of language as an instrument for foresight cannot be overstated. Or, in the words of Richard Slaughter (1996), introducing his Advanced Futures Glossary[1]: "It's well known that concepts and words are bearers of thoughts and ideas. What's less well known is that the language of Futures Studies is a rich and powerful symbolic resource in its own right that opens up new worlds of understanding and possibility".

The importance of semantics warrants an analysis of how language functions and what it contributes to a better understanding of the future. We will focus here on one

This chapter was inspired by the work of Working Group Zero on Concepts and Definitions of the European COST Network on Foresight Methodologies (COST A22).

[1] See: www.foresightinternational.com.au

R. van der Helm (✉)
Embassy of the Kingdom of the Netherlands to Benin,
Ministry of Foreign Affairs HMA COTONOU / Loc. 134, P.O. Box 12200,
2500 DD The Hague, Netherlands
e-mail: ruudvanderhelm@gmail.com

M. Giaoutzi and B. Sapio (eds.), *Recent Developments in Foresight Methodologies*, Complex Networks and Dynamic Systems 1, DOI 10.1007/978-1-4614-5215-7_2,
© Springer Science+Business Media New York 2013

particular dimension of this wide domain of semantic research: concepts and definitions. The objective of this chapter is to provide more insight into how definitions of future-relevant terminology are shaped and laid down in semantic resources.

We proceed by taking the following five steps. First, we develop the reasons and the relevance of our undertaking. There are differing views on the importance of semantic research, especially in cases where futures studies are seen as a practical exercise. Then we look into how the futures domain has dealt with concepts and definitions, focusing on existing glossaries, dictionaries and encyclopaedias. After that, we move back into theory by discussing four approaches to establishing a definition. We will deepen the obtained insights by analysing how semantic issues have been dealt with by two recent Ph.D. theses that clearly position themselves in the futures domain. Finally, we come up with some conclusions on how to proceed with semantic research and the enrichment of the futures domain that it may yield.

2.2 Concepts, Vocabulary and Definitions

Since futures knowledge is to a large extent captured in language, semantic research is a necessary part of the futures domain (see Mermet et al. 2008). Since methodological and applied approaches dominate the futures domain, this dimension of maturing foresight has largely been overlooked in recent years. This is, in contrast to the pioneering years of futures studies, where several authors – like Gaston Berger, Bertrand de Jouvenel and Fred Polak – have done important philosophical groundwork, in which semantics plays an important role (in particular for De Jouvenel and Polak, the future field could only be developed with an appropriate and well-founded futures vocabulary. De Jouvenel remains the spiritual father of the *futurible*; the attempts by Polak to establish his *prognostics* were less successful[2]).

This does not mean that no attention at all has been paid to concepts, definitions, vocabulary and wording. Quite the contrary. Many authors are aware that key terminology needs to be clarified. For example, most – perhaps all – scenario studies include at least some notions on what is meant by a "scenario", leading to a myriad of definitions and pseudo-definitions. A recent experiment with content analysis by Sam Cole provides another example of how concepts can offer particular insights (Cole 2008).[3] But the practical or applied nature of many foresight activities and

[2]For a complete overview of Polak's contributions to the futures field, see Van der Helm (2005).

[3]Cole's work on content analysis is based on word count within the body of futures literature (in particular, the journal *Futures*). Unfortunately for the scope of our work, his approach largely overlooks how these concepts are defined and interpreted by their authors and how the meanings of concepts (and different concepts with similar meanings) evolve over time. Two of his cases that would warrant more in-depth research are the distinctive use of "future" (singular) and "futures" (plural) and of "futures studies" and "foresight" over the years.

studies hardly ever leads to more in-depth reflection on the meaning and interpretation of key terminology.

Getting concepts and vocabulary right remains a challenge for the futures domain for three reasons. First, the futures domain heavily borrows from other domains, and coming from different angles, the overall futures lexicon is a blend rather than an autonomous entity. The grounding of futures terminology can only be achieved by increasing the awareness of its (theoretical) sources.[4] François Hetman (1969), author of the first comprehensive futures dictionary (which we more amply discuss in the next section), was very well aware that by 1969 the terminology had several very distinct sources but that an attempt could be made to work towards one linguistic body. Arguably, we have not moved much further ahead, the futures domain still being a patchwork of different universes, with partially shared but also partially very distinct and even competing lexicons (the latter addressed by Laurent Mermet (2008) as the "metonymical hustle").

Second, not everybody is convinced of the relevance of in-depth semantic research. This uneasiness with semantic research may result from the risk that the search for precision will lose its practical meaning and that it becomes an end in itself (and hence a sterile sophism). But Pero Mićić (2006), whose thesis we will discuss more in-depth below, rightfully refers to authors like Gottlob Frege and Bertrand Russell, who have extensively argued that, with imprecise concepts, there cannot be precise knowledge. They put forward this thesis notwithstanding the fact that they are also fully aware that concepts are always imperfect and imprecise, in that they can never describe all the aspects of features of the thing in itself. In practice, few would argue that the formal bearers of semantics – glossaries, (etymological) dictionaries and encyclopaedias – are not useful, at least sometimes. They may provide some anchor points for starting a research study or help in increasing the precision of our thinking and communication. As long as we appreciate the value of their contribution (see Eco 1984), there still is much to build on. And even though introductions such as "According to the Oxford Dictionary..." or "The concept was used by the Greeks..." may seem obligatory or cliché-like at times, they point towards the importance that semantic reflexivity may have in work that does not intend to be semantic in itself.

Last, precision should not lead to rigidity or claims of universal value. Following Umberto Eco's reflection on dictionaries and encyclopaedias, it is important to recall that a "dictionary is not a stable and univocal image of a semantic universe" (Eco 1984: p. 86). This exactly underscores the importance of semantic research: since wording and meaning are not codified once written down in a glossary or dictionary, we need to keep on reflecting on how the instrument of language does, or could, advance on thinking about the future.

---

[4]For this reason the Working Group on Concepts and Vocabulary of the COST A22 which inspired this chapter abandoned the idea to work on vocabulary in favour of working on the theoretical underpinning. Some of the work has been assembled in a special edition of *Futures* (2008). See Mermet et al. (2008).

2.3 Futures Dictionaries, Glossaries and Encyclopaedias

Although semantic research is relatively scarce, the futures field is certainly not devoid of glossaries, dictionaries and encyclopaedias.[5] But to our best knowledge, very few of them are systematic attempts to bring together a lexicon for futures studies.

To date, the most comprehensive semantic work in this regard remains *The Language of Forecasting* by François Hetman (1969).[6] It contains 383 entries flowing from the work of 215 different authors, of whom Bertrand de Jouvenel (20 references) and Herman Kahn (12 references) are best represented. Published by the French *Futuribles*, it is also one of the very few bilingual French-English futures vocabularies (even including the translations of entries into German) and no doubt the only one with such a broad scope. This dictionary was conceived in a period in which futures research was growing rapidly into many different domains, from the philosophical domain (inspired by Gaston Berger) into the technological domain, economics, (global) environmental forecasts, etc. Strictly speaking, Hetman's dictionary is more an encyclopaedia and is based on a selection and compilation (rather than a reworking) of, by then, existing futures literature. Often, no definition in the strict sense of the word is provided; entries are circumscribed rather than defined. Most entries are extracted from a single source, and this source is subsequently used as basis for its definition.

Hetman saw his dictionary clearly as "an attempt at an analytical and systematical analysis of the main elements of 'forecasting' ('futures research'). It [has been] devoted in particular, and almost exclusively to their ontological, conceptual, praxeological, methodological, operational and normative aspects" (p. XIII). Furthermore, he stated that "[t]his work calls for further discussion. It is offered as a tool – the first of its kind. Ulterior refinements will be necessary to meet the wishes, experience, and research of all its readers" (p. XIV). And to position the dictionary even better, the editorial note states that "[i]t is necessary at times for someone to have the courage to offer for criticism, and thus improve, a work whose shortcomings will bring out the problems that need to be solved" (p. IX).

Hetman's work remains an interesting reference work, although mainly in a historical sense. In many respects, the work would need a refurbishment to become valuable for current futurists, in two, main ways. First, it would have to be updated in terms of lexicon (adding new entries, removing obsolete ones) and adding

---

[5] Much can be said about the differences between glossaries, dictionaries and encyclopaedias, but this debate goes beyond the scope of our undertaking (for an overview, see Eco 1984). For the purposes of our discussion, we follow in general the name that the authors have given to their work. In some cases, we may add some remarks regarding the justification of the name with the intention to focus the reader on what can actually be found in the text concerned.

[6] The book has been out of print for a long time. Thanks to LIPSOR at the French Centre national des arts et des métiers (CNAM), the complete volume has been made available via the Internet and can be downloaded from http://www.cnam.fr/lipsor/eng/data/langageprevision.pdf.

relevant reference material since 1969. This in itself would be a titanic job. Second, the work would clearly benefit from stronger analytical, descriptive and ontological work, not only reciting what others have said but digging into entries in order to bring out the essence, evolution and semantic diversity or overlap (entries having different meanings, or different entries having similar meanings). This would require even more perseverance, since each entry may require quite some in-depth research (e.g. see Van der Helm (2006) on *probability, possibility, and plausibility* and Van der Helm (2009) on *visions*).

The second important reference work has been the futures glossary that has been prepared by Richard Slaughter as part of the Knowledge Base of Futures Studies (Slaughter 1996).[7] It contains roughly 190 entries, most of them explained or defined in one or two phrases. In contrast to Hetman's encyclopaedic approach to move towards a futures lexicon as part of the shaping of the futures field, Slaughter's objective resides essentially in broadening the audience of futures studies. His attempt is deliberately communicative, as he is "interested in seeing the symbolic resources of the field, i.e. concepts, metaphors, language and ideas, taken up and used more widely". Therefore, his purpose is "to de-mystify some of the esoteric-sounding terms that are used in futures". For that reason references are relatively scarce and mainly used as illustrations and not as underpinning.

For the purposes of semantic research, however, Slaughter's glossary is of rather limited value. In that sense, he rightly suggests that it "is a personal, and perhaps idiosyncratic overview of futures terms". Many of the latter are subject to wide interpretation, so my attempts at clarity should be seen as providing numerous starting points for enquiry and understanding. But it is indeed in the nature of a glossary to avoid lengthy discussions on origins, evolution and multiple interpretations. A similar observation could be made about many other glossaries. Slaughter's is as far as we know currently the most comprehensive.

The third and last reference that cannot be left unmentioned is Wikipedia. Unimaginable in 1969 (Hetman) and unavailable in 1996 (Slaughter) – Wikipedia was officially launched on 15 January 2001, according to Wikipedia (!) – it may well have become the most advanced semantic base for futures studies. This is partly thanks to the proactive role that several futurists have taken, (inter alia, with a World Futures Studies Federation (WFSF)-inspired Wikipedia raid), in order to enter and refresh future-relevant entries. A special domain has been set up for future-relevant articles: Future Wikia.[8] The number of relevant entries is obviously unknown, not least because the domain does not have clear boundaries. The Future Wikia alone already contains 784 entries,[9] but this by no means includes all relevant

[7] We refer here to the glossary which is part of the CDROM: *Knowledge Base of Futures Studies*, Millennium Edition CD ROM, since it has been part of a milestone project to establish the foundations of futures studies. This glossary is dated in 1996. An updated version of the glossary (Advanced Futures Glossary 2005) can be obtained from Foresight International (www.foresightinternational.com.au). The glossary can also be consulted online.

[8] http://future.wikia.com/wiki/Main_Page

[9] Reference date: 2 January 2009

entries. Most terminology can be found in the encyclopaedia, and key terminologies like "scenarios", "vision", "futures studies" and "foresight" are amply worked out, often completed by references and cross-linkages. The quality of entries is nevertheless uneven and often idiosyncratic, notwithstanding the extensive quality controls that Wikipedia has put in place, turning the Wikipedia itself into an interesting source material for semantic research. Even though it is not necessarily the exact update that Hetman's work needed, Wikipedia provides at least a contemporary answer to his "first draft" (Hetman 1969: p. IX).

Having identified these three (contrasting) semantic sources, we need to deepen our understanding one level further. These and many other resources make a claim on relevant concepts and their meaning (any glossary, dictionary or encyclopaedia could in this sense be described as a set of concepts (entries) and their claim on meaning and/or interpretations). Often the meaning is derived from, or framed as, a definition of what the particular concept is or does. But definitions may differ considerably, in terms of substance, historic sensitivity and completeness, as well as in terms of tolerance for alternative meaning. For that reason we propose to move towards a reflection on how definitions come about and how we can better appreciate them.

2.4 Defining Definitions

A definition is often defined by "a statement of what a thing is" or "a statement which states the essential properties of the things to which a given concept applies" (Mautner 1996). But language philosophy distinguishes more categories that establish that different approaches to definitions are possible. For his own research purposes, the Dutch philosopher Hans Achterhuis has translated these categories into distinct ways of dealing with definitions beyond linguistics (Achterhuis 2008), which we will translate into four approaches that are relevant for our current undertaking (the essentialistic, descriptive, stipulative and normative definitions):

- The *essentialistic definition* is based on the expression of the essential properties of the object or concept to be defined. Such a definition tries to focus as much as possible on the distinguishing properties. In itself this approach is ontological and aims to bring objects/concepts back to the core, to their unalterable essence. But a more pragmatic approach is possible here and more common in everyday life. Obviously, a chair cannot fully be defined as "an object with four legs on which one can sit", because it unduly excludes chairs with three legs and includes non-chair objects with four legs but on which one can sit, like a table. But in practice, such an essentialistic definition is often sufficient to make one's statements understood. In the case of an essentialistic definition, the characteristics of the object/concept determine whether a specific term applies or not. Its weaknesses, however, are that the approach is insensitive to historical evolution, and contextual and normative interpretation.

- A *descriptive definition* is based on an analysis of how a term is used in practice. It is based on observations of how other "authors" define or use an object/concept. It leads in general to an array of definitions rather than to one single definition. Dictionaries testify to this diversity. In the social sciences, the observation that a concept has a different meaning for different people has become commonplace. This often leads to outcries that a term with so many meanings does not have any meaning at all. However, the best one can say is that such a term is at least not devoid of meaning, even though it may lack the necessary precision.[10] A descriptive approach is sensitive for the historic evolution of meaning, contextual and even idiosyncratic meaning. It refrains from normative claims (i.e. one definition cannot be said to be better than another, even though this does not exclude reflections on precision and appropriateness).
- A *stipulative definition* determines how a specific linguistic expression is to have a certain meaning. This definition demarcates what is in and what is out. If a chair is defined as an object with four legs, then a similar object with three legs does not qualify under the definition: a three-legged chair would be an oxymoron. It is the author who is responsible for this determination. Therefore, stipulative definitions cannot be said to be correct or incorrect. However, stipulative definitions may or may not be generally accepted. Since this approach is to a large extent arbitrary, attributed meaning may be open for discussion. This does not exclude that in many cases stipulative definitions may be so well accepted that there will be little argument. The way the colour red is defined is fully arbitrary, yet generally accepted. The usefulness of stipulative definitions resides clearly in their potential for precision (compare with the descriptive definition). The most rigorous approach in this regard comes from mathematics, where unique definitions are attributed to symbols and permutations. The danger of stipulative definitions, however, lies in their claim for universality of a specific meaning, whereas most stipulative definitions only have meaning within the boundaries of a specific domain (be it a book, a method, a field of research, etc.).
- A *normative definition* is a variant of the stipulative definition. In this case characteristics are projected on the object/concept that, strictly speaking, goes beyond the semantic domain by putting a value-laden claim on its purpose. This includes, in particular, performative qualities, in which case the author defines a concept in such a way that it does not only assign meaning to the term (stipulative definition) but also drives at a particular behaviour. Defining injustice by "a state of inequality that has to be eradicated" transfers not only a meaning but also a value-laden action that is strictly spoken outside the semantic domain. Where a stipulative definition has an analytical purpose, a normative definition can be said to be action oriented. Obviously, this may be seen as either a strength (in particular

[10] A clear example from the futures field has been delivered by Wendell Bell, who argues that the term "paradigm" should no longer be used, due to the wide array of different meanings. Instead, he proposes to use "transdisciplinary matrix" for one of the submeanings attributed to "paradigm" (Bell 2003: pp. 184–187).

when the behaviour is intentional and desired) or a weakness (when it triggers undesired behaviour). Normative definitions risk being insensitive to essentialistic thinking and descriptive diversity.

As is the case with many categorizations, they tend to describe ideal and therefore identifiable cases. In practice, they become mingled again. As such, a dictionary may contain under one entry all forms of definitions (although this is often done from a descriptive perspective). Glossaries tend to use stipulative definitions when they have to establish a coherent working vocabulary for a precise purpose but may include dictionary-like descriptive definitions if the purpose (and target group) is only loosely defined. So how do futurists reach their definitions? In the absence of a commonly accepted "Oxford Futures Dictionary" – the Future Wikia being still too eclectic – they tend to establish their definitions themselves. In the next section, we look into some examples of how futures definitions are forged.

2.5 Two or Three Attempts to Get Concepts and Definitions Right

Although all futurists are to a certain extent aware of the importance of concepts and definitions, academic work stands out as an opportunity to explore the issue. For that reason, we have chosen to look into two recent Ph.D. theses in the domain of foresight/futures studies and to analyse what attempts have been undertaken to get the definitions right.

The two theses are as follows:

(a) Philip van Notten's *Writing on the Wall* (2005), dealing with discontinuity and scenario development, defended at the University of Maastricht (Netherlands)
(b) Pero Mićić's *Phenomenology of Future Management in Top Management Teams* (2006), dealing with corporate foresight, defended at Leeds University (UK)

Van Notten's thesis reveals two key areas of semantic analysis, which cover the main elements of the work: scenarios (Chap. 2) and discontinuity (Chap. 3). The chapter on *scenarios* starts with a descriptive approach, listing "the numerous definitions of scenarios" (p. 17). The list covers roughly 20 definitions. From these definitions, a shortlist of essential characteristics is deducted. This shortlist includes nine characteristics, of which one (i.e. plausibility) is immediately dismissed since it "is not a manageable characteristic for our definition due to its normative nature" (p. 20). The remaining eight characteristics are transformed into an "inclusive working definition", which states that "scenarios are coherent descriptions of alternative hypothetical futures that reflect different perspectives on past, present, and future developments, which can serve as a basis for action" (p. 20). What we see happening in this case is a mix of three different approaches in three subsequent steps: (1) descriptive listing of existing definitions, (2) essentialist approach by listing essential characteristics, leading, finally, to (3) a stipulative (working) definition, which is used to demarcate the boundaries of the research. The author explicitly refrains from a normative definition by ruling out the plausibility characteristic.[11]

A critical examination of this approach reveals a number of weaknesses. First, the descriptive approach remains underdeveloped. The author includes neither the historical perspective (Does the definition change over time?) nor the contexts in which these definitions were made. The extent to which the definitions represent the purposes of their authors would clearly deserve much more attention. When Kahn and Wiener (1967) define scenarios as having to be "causally coherent", from a descriptive point of view, this can clearly be placed in the emergence of systems dynamics by that time. When scenarios become referred to more as "perceptions", it leads us to the influence of the constructivists. Second, the essentialistic second step does not lead to a reflection on the reasons for taking the selected characteristics and not any others. Hence, this phase remains rather arbitrary and becomes little more than the intermediary step from the descriptive to the stipulative phase. Third, the stipulative definition seems to suffer from a need for all-inclusiveness. The author's concern is clearly to cover the diverse array of scenario studies, but one might wonder if a simpler definition would not have led to the same potential. It is not our objective to redesign the definition, but our guess would be that a stipulative definition, like "scenarios are descriptions of alternative future developments", would have been largely sufficient. But since the author has clearly demarcated the validity of the definition – it is a working definition for the purposes of the research that is to follow – one could also leave it at that.

When tackling *discontinuity*, Van Notten follows a different and more in-depth approach. In contrast to scenarios, the author argues, through citing H. Brooks, that "our understanding of discontinuities is [...] poorly formulated" (p. 47). Furthermore, he observes "some confusion regarding the concept of discontinuity" (p. 44). His response to this confusion is a more elaborate descriptive approach, which includes the study of related concepts (in particular "surprise" and "wild card") and a return to theory (in this case, the three time scales of the French historian Fernand Braudel). The stipulative definition that emerges from this elaboration is therefore much better grounded than the one on scenarios. A discontinuity is defined as "a temporary or permanent, sometimes unexpected, break in a dominant condition in society caused by the interaction of events and long-term processes" (p. 55). The author is well aware of the wider scope of this definition and suggests that this should be seen as a proposal for a definition of discontinuity in the context of scenario development (hence, beyond the domain of the thesis itself). Interestingly, the definition is immediately completed by a list of six characteristics, which seems a redundant completion after the groundwork that was done to reach the definition in the first place.

[11] Even though one could question whether this is done based on the correct underpinning, plausibility may indeed be a subjective criterion, but this does not necessarily lead to a normative definition. Furthermore, the author does not explain why coherence would not lead to a similar dismissal. For a more in-depth reflection on the concept of plausibility, see Van der Helm (2006).

When we move to the second thesis, Pero Mićić's *Phenomenology of Future Management in Top Management Teams* (2006), we encounter a contrasting approach to concepts and definitions. Whereas Van Notten's work needs definitions for the two major aspects of the thesis work, Mićić has deliberately and extensively dug into the way managers and authors use definitions. These definitions form the semantic foundation of the "integrated model of futures management" that forms the motif and the objective of the thesis. The main research question is precisely: "How shall an integrated model for future management, in top management teams, be designed to provide a *semantic framework*?" (p. 40; our highlighting), and the first sub-question is "What are the phenomena that managers use to describe and analyse the future?" (idem). If one were still in doubt about the linguistic perspective, Mićić clarifies later on that "building and improving the model of future management is an activity with linguistic character" (p. 87). Here, language acts as both instrument and object.

Mićić's approach claims to be strongly essentialistic. He applies phenomenological concepts and tools with the intention to identify and describe the "essential thingness" of futures concepts as autonomous objects and in relation to one another (pp. 85–88). He distinguishes two different phases. First, he follows a "semasiological" approach, taking the terms and concepts used by managers and the literature as starting points and asking for their meaning. The second phase consists of an "onomasiological" approach that departs from the "things-in-themselves" and asks for their names. The focus here was "on identifying and analysing the words that participants [of workshops and seminars] used to describe certain phenomena in the future and in the part of the present that is relevant for the future" (p. 88). By triangulating the findings and combining them with the perceptions and observations of the researcher, it is claimed that "relatively solid definitions and understandings were achieved" (idem).

Chapter 4 (pp. 161–290) is dedicated to the results of the effort, being an annotated list of objects. An object in this approach is defined as "a phenomenon used by managers and experts to describe an entity in their thinking about the future" (p. 164). The list contains ten main entries (like trend, surprise, vision, or strategy) and 11 sub-entries (like eventuality, goal, or task). Each object is described and defined according to the same logic: (1) morphology of characteristics, (2) definition, (3) elements (i.e. elements to improve understanding), (4) reasons (i.e. explaining inclusion in the list) and (5) criteria of quality.

Having gone through this theoretical rigour, the object model becomes surprisingly straightforward when is finally put into practice. Here we recognise a similar approach to the one Van Notten used for his scenario definition, albeit in a very well documented and extremely rigorous way. Hence, Mićić, as well, passes through the three earlier mentioned subsequent steps: (1) descriptive listing of existing definitions, (2) essentialist approach by listing essential characteristics, leading, finally, to (3) a stipulative (working) definition. Step 2 is in general absorbed by Step 1 as part of the morphology of characteristics. Step 3 is very dominant, since it is a requirement for the model to have focused and clear definitions, which together form a coherent semantic framework. One could say that the ultimate object model consists of a list of stipulative definitions, which are grounded in a descriptive approach. It is unclear whether the author considers his definitions to have validity beyond the model itself. Where elements are mentioned, they specify the definition, as was done by Van Notten in the case of the discontinuity definition.

In contrast to Van Notten, Mićić's definitions are sharper and more concise. However, one could question whether Mićić's approach is genuinely essentialistic as he suggests. In other words, do his definitions capture the essence of the objects? If that had been the case, much more effort should have been put in establishing the ontology of the concepts. Mićić dismisses such an approach by claiming that none of the objects has a physical nature but again admits that phenomenology and ontology do not exclude one another. Uncomfortably limping from one leg to the other, the essentialistic approach is largely abandoned, as we have seen, for a combined descriptive and stipulative approach. This can be illustrated by two examples from the set of object definitions: (1) a *surprise* is an event or development of low probability and high impact (p. 232) and (2) a *strategic vision* is a concrete picture of an ambitious, jointly desired and feasible future (p. 249). These definitions do not get close to the essence of the objects, even though they may describe some essential elements. Many ontological questions remain unasked, e.g. in case (1) there is no clear underpinning of why surprise could not exist in a case where two high-impact events with 0.5 probability coexist, or, inversely, there is no reflection about low probability events which are clearly unsurprising (like throwing a six (probability 0.17 against 0.83 for not-six) when rolling the dice). In case (2) there is no clear underpinning of why a vision should be desired (dystopia as a vision to avoid) and feasible (since this characteristic can only be determined in retrospect, when the vision has ceased to exist).[12] In view of Mićić's objective to use them in the model for future management, these concise and underpinned stipulative definitions seem clearly sufficient. But, in the end, these definitions represent rather common ground to the extent that one could wonder why Mićić put so much effort in getting to existing definitions, notwithstanding the fact that the effort to document his approach is so rigorously laudable.

What our discussion reveals is that the shaping of definitions itself is a fascinating task (in the case of Mićić even a very laborious process), which helps us understand how authors sharpen the language instrument for the purposes of saying something sensible about the future and also where this could be improved.

2.6 Conclusion

Suggestions for working with concepts and definitions are as follows. The objective of this chapter was to provide more insight into how definitions of future-relevant terminology are shaped and laid down in semantic resources. Under the assumption

[12] For an elaborate research study into the nature and functioning of visions of the future, see Van der Helm (2005, 2009).

that key vocabulary is an important (perhaps even the most important) instrument to create future knowledge, the way we define concepts determines to a large extent the way in which we understand the future.

It was argued that precision – but not rigidity – of definitions is important (*with imprecise concepts, there cannot be precise knowledge*). However, this precision can be obtained in several ways. Definitions can be established using an essentialistic, descriptive, stipulative or normative approach, or a combination of these. The two cases we discussed (Van Notten and Mićić) combined a descriptive and a stipulative approach, although in rather contrasting ways. When attempting to understand how other authors use key concepts, it may therefore be useful to both understand what definition they use *and* how they have arrived at this definition. Semantic research can add much reflexivity to the use, diversity and evolution of definitions in this way.

Whether this should lead to a more commonly shared futures lexicon, like François Hetman attempted with his "first draft" of *The Language of Forecasting*, is a question that we left floating. Wikipedia has somehow taken over this role, at least in terms of comprehensiveness, even though the quality remains uneven. We would argue, however, that quality semantic research that addresses key concepts could very well replace the drive towards comprehensiveness. Evidently, this would be under the assumption that futurists would be willing to inform their work by this type of more fundamental research. A glossary is clearly not sufficient in this regard, even though it may facilitate understanding and coherence within the well-defined scope of a method or a deliberation, or help for communicative purposes.

To conclude, it remains important to reiterate that language has an enormous potential to be tapped by futurists. This clearly goes beyond getting the concepts and definitions right. But it cannot do without this. That is why we have to keep on defining the future.

References

Achterhuis, H. (2008). *Met Alle Geweld, Een Filosofische Zoektocht*. Netherlands: Lemniscaat.

Bell, W. (2003). *Foundations of futures studies, history, purposes, and knowledge, human science for a new era* (Vol. 1). USA: Transactions.

Cole, S. (2008). The Zeitgeist of futures? *Futures, 40*, 894–902.

Eco, U. (1984). *Semiotics and the philosophy of language*. Bloomington: Indiana University Press.

Hetman, F. (1969). *Le Langage de la Prévision/The language of forecasting*. Paris: Futuribles/S.E.D.E.I.S.

Kahn, H., & Weiner, A. J. (1967). *The year 2000: A framework for speculation on the next thirty-three years*. New York: Macmillan.

Mautner, T. (Ed.). (1996). *The Penguin dictionary of philosophy*. London: Penguin.

Mermet, L. (2008). Extending the perimeter of reflexive debate on futures research: An open framework. *Future,* Available through ScienceDirect.

Mermet, L., Fuller, T., & Van der Helm, R. (2008). Re-examining and renewing theoretical underpinnings of the futures field: A pressing and long-term challenge. *Future,* Available through ScienceDirect.

Mićić, P. (2006, December). *Phenomenology of future management in top management teams*. A thesis submitted in partial fulfillment of the requirements of Leeds Metropolitan University for the degree of Doctor of Philosophy.

Slaughter, R. (1996). Futures glossary. In: *Knowledge base of futures studies* (Millennium Edition). CD ROM.

Van der Helm, R. (2005). The future according to Frederik Lodewijk Polak: Finding the roots of contemporary futures studies. *Futures, 37*, 505–519.

Van der Helm, R. (2006). Towards a clarification of probability, possibility and plausibility: How semantics could help futures practice to improve. *Foresight, 8*(3), 17–27.

Van der Helm, R. (2009). The vision phenomenon: Towards a theoretical underpinning of visions of the future and the process of envisioning. *Futures, 41*, 96–104.

Van Notten, P. (2005). *Writing on the wall: Scenario development in times of discontinuity*. Dissertation.com, USA.

Chapter 3
Classification of Tools and Approaches Applicable in Foresight Studies

Jan Erik Karlsen and Hanne Karlsen

3.1 Theoretical Background of Foresight Methodologies

Are predictive quantitative methods too limited to serve as tools in foresight studies? This concern has recently been met by the emerging application of qualitative methods as a means to complement and compensate for the perceived weaknesses of quantitative methods. It is particularly in terms of reflecting sudden changes or detecting incremental and weak signals of change in real societies that quantitative methods are deemed too static. A productive foresight analysis will need a more differentiated sense-making and robust repertoire (Rossel 2010, 2012). Krawczyk and Slaughter (2010: p. 75) state:

> The development of futures studies and the continuing advancement of its methodological base is a consequence of changing human needs regarding present and future as well as its cultural and social foundations. As in recent decades the world has been dramatically transformed, people's ways of approaching, considering and addressing these transformations have been constantly evolving thereby also stimulating the development of new futures methods.

Some recent foresight programmes and projects (e.g. COST A22) have challenged this issue by combining narratives and thick descriptions with games, simulations, or computer-based models and calculations.

J.E. Karlsen (✉)
Faculty of Social Sciences University of Stavanger,
Stavanger 4036, Norway
e-mail: jan.e.karlsen@uis.no

H. Karlsen, M.Phil.
Department of Leadership and Organizational Management at BI,
Norwegian School of Management, Nydalsveien 37,
0484 Oslo, Norway
e-mail: hanne.karlsen@bi.no

M. Giaoutzi and B. Sapio (eds.), *Recent Developments in Foresight Methodologies*, Complex Networks and Dynamic Systems 1, DOI 10.1007/978-1-4614-5215-7_3,
© Springer Science+Business Media New York 2013

The range of approaches to foresight research clearly indicates that the dichotomization between quantitative and qualitative methods is an unbalanced and oversimplified one. Qualitative research represents an approach rather than a dedication to a specific set of methods and techniques. Its application and appropriateness is contingent on the kind of foresight phenomena to be studied, discussed, and described. The same can be said about quantitative research: It is directed by the phenomena under scrutiny.

It is evident that there is a gap between the complexity of future options and pathways which is addressed in foresight studies and the analytical tools applied to map this complexity. And there is no consensus on an appropriate methodology balance between the qualitative and quantitative approaches. The lack of a common and approved methodology originates, at least partly, from the fact that the inherent ontological and epistemological characteristics of qualitative and quantitative methods differ when it comes to capturing the complexity of issues addressed in foresight exercises (Karlsen et al. 2010). Also, the choice of appropriate methodologies depends on the context, so it is natural that a general consensus view is missing.

Quantitative approaches most often investigate concepts, constructs, and variables; apply numerical values; investigate cause-and-effect or functional relationships; and focus on reliability, validity, generalizability, and objectivity as indicators of the quality of the evidence given by the analysis. The qualitative approach addresses phenomena and concepts, applies narrative descriptions, seeks patterns of association as connections between concepts, and uses dependability, credibility, transferability, and confirmability as core concepts to substantiate the quality of the evidence produced.

A classical orientation has been either to use quantitative methods and data to substantiate or underpin qualitative descriptions, or vice versa to apply brief narratives to frame the quantitative analysis. Previously, it has been suggested that when applying 'forecasting methods' we find that certain combinations of futures research methods fit together, while others do not, depending on the thematic focus of the analysis.

Our approach addresses how to find a balance between the two major approaches when analysing foresight issues. We offer a taxonomic matrix in which one might compose an optimal (in the sense of obtaining vigorous explanations) combination of approaches appropriate for telling robust stories of the future.

3.1.1 Material

Foresight expert group exercises are especially intriguing since they are purely mental processes (Karlsen and Karlsen 2007). They are intellectual thought processes about some aspects of the future (e.g. foresight tools). During the period 2004–2006, a series of 33 foresight methods were selected by a group of COST A22 foresight research scientists and listed as candidates to represent tools in which there are assumed to be a mixture of quantitative and qualitative data and approaches.

In a workshop meeting, the selection of foresight tools was subsequently refereed on a judgemental basis; candidate tools included on their face validity had to comply with both the number and narrative criterion.[1]

The material analysed in this chapter consists of data assessed by 12 foresight experts[2] (all members of work group 2, carried out and collected in two successive COST A22 sessions during 2006). The various tools were ranked on a scale from 1 to 10 regarding six different criteria inherent in the application of such methodological approaches. The batch of approaches assessed requires different scales of mobilization of domain expertise; may be applied on various foresight domains; and demonstrates different levels of complexity as to data input, process, and output. In addition to the ranking of the tools, qualitative comments as to the applicability and prerequisites of the approaches have been collected from the foresight methods experts.

3.2 Ontological and Epistemological Assumptions

Contemporary foresight activities are dominated by commitment to research methods, almost as an end in itself, resulting in abstracting modes of futures empiricism based on both quantitative and qualitative methods. Arguably, there has been a general failure to examine and explicate the relationship between theory and method. The application of a particular method has been seen as a sufficient requirement or justification of a foresight study, seemingly loosely coupled to the wider issues the study is designed to highlight. However, qualitative and quantitative methods pertain to theory differently depending on which ontological and epistemic assumptions guide their application. The point can be split in two: We cannot speak about *a theory of foresight* as a foundation for scientific inference in this field. Foresight studies seem to be dominated by the application of various tools and techniques. But, applying scientific techniques is assumed to legitimize the foresights. However, neither qualitative nor quantitative techniques are value-free or generic. They initiate from specific philosophical reflections about what the reality really is and how we can obtain knowledge about this world. Ergo, the techniques are not value-free when it comes to conceptualizing the future.

Ontology is the theory of the conceptions of reality, providing criteria for distinguishing between various types of objects and their perceived relationships. Objects can be concrete or abstract, existent or non-existent, real or ideal, as can their ties be

[1] All candidate tools are described in this book. According to a general template, some common characteristics such as scope, first appearance, mathematical/formal/logical representation, type of mobilization, existing software, sector/scale of application, type of output, strengths and weaknesses, and, finally, a bibliography are all used to describe the characteristics of the foresight tools.

[2] All experts were members of the COST A22, work group 2, which focused on the methodological integration of narratives and numbers in foresight exercises.

relations, dependences, or predications. Foresight studies deal with the 'future', and perhaps the only meaningful activity concerning the future is to think about it. As a mental object, the future is abstract, non-existent, and ideal.

The link to the future may be relational, dependent, and predictive. So, when we deal with future, we have to ask the basic question: 'What is there, or what will be there?' How do we know that future is or will be, and how do we know the characteristics of the future itself, if it exists? If future just refers to a collection of mental events experienced by a person, how do we recognize its very being, its entities, and their interrelationships? Furthermore, how can we make a catalogue of elements that will constitute the future in some specific domain, e.g. in a particular industrial sector, in a certain society, or pertaining to a pathway of a novel technology? The answer is that we most often revert to some basic ontological assumptions about the future, e.g. as a reflection of human imagination, or as a projection of today's situation. And we select and apply foresight methods accordingly.

The social sciences, which often offer their methods to foresights, adopt different ontological world views. One approach is the *realism school*, advocating the idea that the future is out there just waiting to be described and discovered. Another school of thought is the *empiricists*, who claim that we can observe the future (e.g. by weak and early signals) and evaluate those observations in relation to former experiences and facts. On the other hand, *positivists* put emphasis on the observations themselves, perhaps being more attentive to the claims about the facts than to the facts themselves. *Postmodernists* state that facts are fluid and elusive; consequently, we should focus only on our observational claims.

Next, how do we establish knowledge of the future, i.e. statements that are both supposed to be true and believed, thus supporting us with a firm epistemic ground for our justified true beliefs? What is measurable? Is the future really measurable, or do we have to distinguish between the knowledge about the past and the present as opposed to future events? How should we deal with 'time'? Foresight implies the explication of the concept of future time, what is the proper conceptualization?

Knowledge implies belief, and foresights often require some (often strongly normative) belief about a future state. But the future is a different country; they do things differently there. So differently, in fact, that only the most fabulous fortune teller can ever hope of getting more than a fraction of any scenario right. The future is not transparent and lucid; if it was, the meaning of life would wither and the science would fade away. Arguably, to think about the future is probably both a most stimulating intellectual exercise and a way of helping us to understand the present more accurately. We may even say, this is the most valuable contribution from foresight exercises.

3.3 So, What Is Foresight Methodology?

As discussed, *methodology* reveals both ontological and epistemological implications. But methodology does not come without some kind of philosophical or theoretical baggage (Hatch and Yanow 2008). A major objective for research is to

Table 3.1 Basic assumptions which characterize foresight methods

	Subjectivist approach	Objectivist approach
Core ontological assumptions	Future as a projection of human imagination	Future as a concrete structure
Basic epistemological stance	Obtain phenomenological (unique) insight of the future	Construct a positivist (general) science of the future
	Revelation of the future	Future sciences
	Knowledge is local and unique	Knowledge is cumulative
Foresight methods	Exploration of pure subjectivity about the future:	Extrapolation and simulation of objective characteristics of a perceived future:
	Inductive logic	Deductive logic
	Proximity	Distance
	Favourably disposed towards the values and interests of the principal players	Neutral and objective

Source: Adapted from Morgan and Smircich (1980) and Jacobsen (2005)

document facts, and methodology can be seen as approaches that give explanations concerning how we know (the practical reason of) a certain fact and what the (theoretical reason of a) certain fact is. In such a manner, methodology contains puzzle-solving devices and instruments, which require us (qua research scientists) to make assumptions about what the world (including the future world) really is, i.e. ontological assumptions, and what stands for knowledge, i.e. epistemological assumptions.

Such distinctions, illustrated as opposite poles of scientific interference, are depicted in Table 3.1 (Morgan and Smircich 1980).

The three levels: ontology, epistemology, and methodology are closely connected. Foresight methodology is not a means in itself; it is applied to give direction to approaches appropriate to map the reality, i.e. the future. However, with different and conflicting ontological and epistemological starting points, we should not be surprised that there exists strong discord about which methods are the most suitable when mapping the future.

For instance, when researchers define the future as a social construct but then attempt to measure global attributes and dimensions, they are most probably mixing philosophies of science. They apply points of view from both the positivist and the subjectivist school in conflicting (and maybe confusing) ways, likely violating the subjectivist ontology and interpretive epistemology underlying their social constructivist approach to defining 'the future'. As we know from standard scientific practice, appropriate operationalization comes from a clear understanding of the theoretical, ontological, and epistemological positions through which we approach a particular construct, e.g. 'the future'.

So, is 'the future' real (objective) or imaginary (subjective)? Although we neither desire nor expect to achieve consensus about the concept of future, we believe that progress can be made in foresight by anchoring our theorizing and our empirical explorations in transparent and explicit methods and by stating the ontological and epistemic assumptions underpinning these approaches.

Table 3.1 illustrates how we may interpret the use of foresight methods in a contrasting image of a subjectivist versus an objectivist approach to (social) sciences. According to Morgan and Smircich (1980: p. 492), both reality and knowledge can be understood and classified on a continuum between these two opposites. Between the pure subjectivist ontological views that reality is just a mental and imaginary picture and the opposite objectivist views that reality is something robust and concrete, they position reality as a 'social construction', as a 'realm of symbolic discourse', as a 'contextual field of information', or as a 'concrete process'. The corresponding epistemological stance will focus on: how the future as a 'social reality' is created, how the future as 'patterns of symbolic discourse' is to be understood, how the future as 'contexts' is mapped, or how its 'systems, processes, and changes' are studied. Related to these different ontological and epistemological views, there will be corresponding categories of research methods: 'hermeneutics', 'symbolic analysis', 'contextual analysis of Gestalten', and 'historical analysis' of processes.

In Table 3.1, we just apply the outliers of the typology described by Morgan and Smircich, implying that we either explore the future as 'pure subjectivity' or as 'pure objectivity'. As such, subsets of a foresight methodology pertain to both quantitative and qualitative *approaches*, and subsets of each approach comprise various *tools/techniques* that can be applied within the chosen foresight methodology.

The *objectivist (or positivist) school* is an approach within the philosophy of science resting on the ontological stance whereby the world (including the future) can be described by laws and structures. This stance legitimates the epistemic claim that we can develop a general science of the futures and build an accumulated body of knowledge about this future reality. Furthermore, we may collect objective data about an objective reality by applying a particular approach, i.e. quantitative (numerical) data with a strong emphasis on deductive strategies and appropriate distance to the subject investigated.

The *subjectivist (or interpretationist) school* does not talk about an objective reality, where everyone will perceive the future homogeneously. Neither does it expect to find nor to rely on strict laws and logical structures; this school also often applies inductive strategies to collect information (words, images, etc.) about a phenomenon from a short distance.

Arguably, the framework of a subjectivist and an objectivist approach for studying the future resembles the distinction between foresight-as-invention and foresight-as-prediction (see Cunha 2002). The basic ontological and epistemic assumptions come within each of the approaches, implying different views of the kind of research undertaken, as well as distinguishing between the major schools of methods applied.

The basic challenge for foresight studies relates to the fact that events, which have not yet taken place, cannot be described in objective terms, simply because they have not yet happened. To be strictly logical, the most reasonable action is to *think* about the future, implying that we treat the future as a purely subjective entity to be conceptualized, measured, and debated via a qualitative approach. However, when we apply quantitative methods we mostly treat the future as if it is an objective entity, measurable in real and objective quantities, although it only exists in our minds. On the other hand, quantitative models and measurements can also be treated as social constructions by those involved in a foresight exercise. So, in the end, they also are based on subjective assumptions.

One way to deal with this paradox is to deconstruct both the past and the present objective circumstances, thus inferring the future will turn out likewise. The past has once been a future, and so has the present. Thus, the future will always (sooner or later) present itself as the present, and later as the past. What are subjective and mental exercises today will be objective and real tomorrow. In such a way, we may validate our conceptions of the future tomorrow. This assumption makes us more confident that we also can apply quantitative methods to describe measure and understand the future as such. Although we may be totally misconceptualizing the content of the coming future, this specific future will not be subjective forever. However, the future as a generic entity will in epistemological terms always be subjective. It is not something that has really happened except in our minds.

3.4 The Conception of 'Time' in Foresight

This subjective-objective perspective illuminates our conception of time. Time travel may never be possible, at least not in physical terms (Holden 2005). We can only travel in time in our minds. Suddendorf and Corballis (1997, 2007) coin the term 'mental time travel' to refer to the faculty that allow us to mentally project ourselves backwards to relive or forwards to pre-live events. They claim that past and future travels share phenomenological characteristics and activate similar parts of the brain. They also indicate that, since humans have been extremely clever in foreseeing, planning, and shaping the future, and since they cannot trace similar characteristics in non-human beings, man is most probably the only species which has the capacity for effective mental time travel, including travel into the future.

Cunha (2002) proposes that time travelling is viewed differently according to what kind of paradigm applied: the foresight-as-prediction or foresight-as-invention view. When operating inside the prediction paradigm, we treat 'time' as objective and hence predictable. Consequently, we seek to develop (almost only quantitative) techniques to increase the potency of prediction. Future is an extrapolation of the past, and based on adequate tools, it could be produced with reasonable accuracy. Within the invention paradigm, foresight is a never-ending interaction between past experiences, today's realities, and possible trajectories. The future is unpredictable but is still an outcome of complex interactions. The foresight is thus some kind of temporal (operations of fantasy) or self-conscious reflexivity, in which meaning is reconstructed from the process of interpreted feedback (Weick and Sutcliffe 2001;

Bell 2002), or even time-stretching (Tsoukas and Hatch 2001). Cunha indicates that invented futures are the outcomes of this interaction or a circular dance between multiple time horizons, lessons from history, options of today, and visions of the future. Such time horizons are not easy to separate and distinguish.

However, in our culture as in others, we still find the idea that the future is predetermined. Hawking (1988) distinguishes between imaginary (as in imaginary numbers) and real time. The *imaginary time* cannot be separated from different directions in space, i.e. one may move both North and South, forwards and backwards. In this respect, there is no principal difference between the directions forwards and backwards. When we deal with 'real' time, there is, as we all know, a large difference between the directions forwards and backwards. So, where does this distinction between past and future originate? Why do we remember the past, but not the future?

Hawking argues that the development of nature moves in the direction of increasing disorder, and consequently we need more than one explanation for one and the same phenomenon. The laws of science do not really distinguish between movements forwards or backwards in time. However, Hawking is offering us three 'time arrows' which give time a direction: a first thermodynamic arrow where disorder or entropy increases; a second cosmological arrow where the universe expands; and a third, psychological arrow of time. The latter conveys the direction in which we feel time is running, the direction in which we recall the past, but not the future. The psychological time arrow is determined in our brains by the thermodynamic arrow and it runs concurrently with the latter. Disorder increases over time, just because we measure time in the same direction as disorder increases. In a situation of increasing disorder, people would have a time arrow pointing backwards, almost like playing a film backwards.[3]

If we follow the reasoning of Hawking, the universe will end up in a situation of increasing disorder, and people will remember the past like we do now, not the future. If this also goes for the socially constructed realities and for social change, a social forecaster should eventually unveil this future disorder (i.e. the change of the present order) to us. The foresighter must penetrate the veiling of present time, unveil and reduce the uncertainty as to what comes tomorrow, and make it possible to behave strategically in relation to the future.

If the future were predetermined, what would then be the meaning of foresight? Two motivations may be important: human curiosity and the potential of personal gain by knowing what will happen tomorrow. To write off curiosity as motivation would not be wise. To think in a structured and creative way about the future can

[3]The reader should be cautioned: We probably cannot take the second law of thermodynamics (increasing disorder with time in closed systems) to define a more psychological time arrow: Arguably, we experience a rather cyclical development, alternating between increasing and decreasing order (birth-death, destruction-construction of institutions, etc.) which is typical for open systems (life is the export of entropy, overcoming the increasing disorder trend). The only closed system is the cosmos which we definitely do not experience as a whole.

give anyone a strategic advantage when it comes to adapt to and exploit the options embedded in a change process. The potential for personal gain, by believing that the future for everyone else other than oneself is predetermined, is however, extremely illogical. When applying foresight methods, time is the most fundamental and indispensable, however implicit, characteristic. To disregard this would be disastrous to the improvement of a solid foresight methodology.

3.5 Some Classification Schemes of Foresight Methods

Many foresight methods and tools are rather flexible in their application. However, most classifications of the methods applied in foresight studies do not explicate their underlying epistemological and ontological assumptions; the typologies often seem ad hoc, pragmatically or a posteriori-generated. Popper (2008: p. 62) claims that, until recently, the selection of foresight methods has been 'dominated by the intuition, insight, impulsiveness and – sometimes – inexperience or irresponsibility of practitioners and organizers'. In this section, we adapt four different typologies offered in the foresight literature to classify examples drawn from our list of 33 foresight methods (i.e. tools and approaches: see Appendix A). These candidate methods are characterized by a mix of qualitative and quantitative characteristics, thus bridging the pillars between a subjectivist and an objectivist approach.

The objective is to see whether such classifications are sufficiently extensive and inclusive with respect to our spectrum of candidate tools and applications.[4] The test is whether these commonly used classifications are robust and exhaustive enough to encompass all our 33 candidate methods in one classification scheme.

One way to classify our candidate foresight tools is according to the sorts of output generated, i.e. to distinguish between quantitative and qualitative approaches, and then add a secondary criterion characterizing the purpose of the foresight, e.g. normative versus exploratory design (Gordon 1994: p. 3). The reasoning behind this classification scheme (applied in the Millennium project) is that all foresight tools, beyond scoring on the qualitative/quantitative dichotomy, also rest on some inherent and inevitable ontological and epistemic assumptions. These assumptions characterize how we perceive the future as something imaginary or concrete and at the same time as something that reveals or constructs the future. Using just a few of our candidate foresight tools (see the list in Appendix A) as an illustration, the matrix could appear like Table 3.2 below.

This illustration demonstrates that some frequently applied foresight tools may have different characteristics concerning upon which ontological and epistemic assumptions they (implicitly or explicitly) rely. *Scenario analyses* consist of visions of future states and courses of development, organized in a systematic way as texts,

[4] In this chapter, we use the terms typology, taxonomy, and classification interchangeably.

Table 3.2 Classification of foresight methods according to output

Tool	Quantitative	Qualitative	Normative	Exploratory
Scenario analysis	X	X	X	X
Delphi		X	X	X
Relevance trees		X	X	
Decision matrices	X			X

Source: The Millennium Project, Gordon (1994)

charts, etc. They are supposed to have a very broad application, often constructed either as a quantitative or a qualitative approach (or both), implying both a normative and an exploratory mission. The *Delphi* method has most of these characteristics, except that it is (mainly) qualitative. It is constructed as a method of obtaining a consensus of the opinions of a group of experts from their answers to a series of questionnaires, interspersed with controlled opinion feedback. The *relevance tree* is just qualitative and normative, constructed as a method of selecting the optimal pathway to a desired future state. Finally, *decision matrices* (and their mathematical representations) are labelled as both quantitative and exploratory. However, in our opinion, this method is constructed on normative assumptions of rationality.

The classification presented above surely highlights our search for quantitative and qualitative groupings. However, it may be objected that (nearly) any foresight technique can be applied to both normative and exploratory forecasting; it is just a matter of how the technique is applied. Hence, this taxonomy does not give us much more than the most obvious groupings where techniques are sorted on their face validity.

Another approach would be to group the tools according to their major perspective (or target) on thinking about the future. A classification, first presented by Miles and Keenan (2003) applied on a few of our other candidate tools, would look like Table 3.3 below.

From our list of candidate tools, we chose *SWOT analysis* as an illustration of a foresight tool applied when we wish to map a certain condition. SWOT is a technique, which is based on identifying the strengths (S), weaknesses (W), opportunities (O), and threats (T) in any situation, including a future state. *Multi-agent simulation* is most often applied to explore possible pathways into or in the future by using various quantitative modelling techniques. On our list of candidate foresight tools, we have several examples in the group 'creative' approaches, e.g. expert panels, cross impact analysis, and scenarios. Here, we have selected *brainstorming* as an illustration. This is a method used in groups to support creative problem-solving, the generation of new ideas and greater acceptance of proposed solutions. Finally, tools of prioritization can embrace techniques such as critical (and key) technologies, gap analysis, and input-output analysis, just to mention a few from our list. In Table 3.3, we have used (technology) *road mapping* as an illustration. This technique is a goal-oriented tool for supporting (technology) management and planning and for prioritizing between different pathways.

Table 3.3 Foresight methods classified according to perspective

Perspective	Method
Identifying issues	SWOT analysis
Extrapolative approaches	Multi-agent simulation
Creative approaches	Brainstorming
Prioritization	Roadmapping

Source: Miles and Keenan (2003)

Although this typology or grouping presented by Miles and Keenan (2003) may be useful in certain contexts (e.g. when applied on technology foresights), we immediately sense the weakness for the purpose of our study: It covers and camouflages the basic distinction between quantitative and qualitative approaches, and it does not offer any help as to whether the tools are normative or exploratory. In such a manner, it avoids analysing and discussing the ontological and epistemological challenges inherent in the tools and techniques themselves.

Still another way of classifying is by sorting the foresight tools according to the purpose of the foresight exercise itself (Gordon 1994: p. 10). Then we obtain a third taxonomy model, applied on some of our other candidate tools, illustrated in Table 3.4 below.

The *focus group* is a cost-effective tool to collect judgements by the explicit use of group interaction to produce data and insights that would be less accessible without the interaction found in a group (Morgan 1988: p. 12). While it is an approach that can be used as a stand-alone tool, focus group interviews are often employed with other methods, both quantitative and qualitative (Kreuger 1988: p. 7).

Forecasting is the process of estimation in unknown situations. Prediction is a similar, but more general term, and usually refers to estimation of time-series, cross-sectional, or longitudinal data. Forecasting is commonly used in the discussion of time-series data.

Micro-simulation models are computer models that operate at the level of the individual behavioural entity, such as a person, family, or firm. Such models simulate large representative populations of these low-level entities in order to draw conclusions that apply to higher levels of aggregation, such as an entire country.

Risk assessment is measuring two elements of any risk: the magnitude of the potential loss and the probability that the loss will occur. There is a distinct element of uncertainty in the measurement of the future, i.e. the future is risky.

Backcasting presupposes the identification of a particular scenario and traces its origins and pathways of development back to the origin, or the present.

Even this classification scheme, later reviewed by Miles and Keenan (2003), is seemingly generated on an ad hoc basis. Though a nice solution, since almost every foresight has a specific purpose, it does not offer any meta-explanation as to why we

Table 3.4 Foresight methods classified according to purpose

Purpose of the foresight	Appropriate tool/technique
Collect judgements	Focus groups
Forecast time series	Forecasting
Understand the linkages between events, trends, and actions	Micro-simulations
Determine a course of action in the presence of uncertainty	Risk assessment
Portray alternate plausible futures	Backcasting

Source: The Millennium Project, Gordon (1994)

should regard these five categories as sufficient for all kinds of foresight purposes. And it does not offer us any explicit distinction between the quantitative and qualitative dimension.

May (1996) proposes a simple classification according to the approaches that the foresight studies take to the future: foreseeing, managing, or creating. Applied to our candidate methods the fourth model (as summarized in Table 3.5) is depicted as follows.

Foreseeing (i.e. prediction, prophecy, forecast, foresight, foreknowledge) attempts to see the future in advance. We assume predictability and that we can obtain knowledge before an event occurs, e.g. the weather forecast. Nature will create and determine the outcome, but we may be able to prepare and adapt to a possible future situation. Methods in this category could either be extremely subjective, like statements from (genius) weather prophets, or highly formalized, analytical tools which emphasize possible and probable futures. Most foreseeing methods often regard the future as inevitable, or even as something objective which already exists. However, it is only in such cases where the future is 'foreshadowed in the discernible patterns of the past and present [that] rational prediction becomes possible' (Rescher 1998). Today we are having a debate about the causes and trends of climate change. The core of this debate is about whether to adapt to or reduce emissions. Is the future really unpredictable and the climate varying according to its own laws? Or do our current conflicting views originate from our limited understanding of the ecological system, or just from a limited ability to foresee the future climate, which we may be able to improve with enforced ICT and modelling capacity? Whatever the answer, climate projections may be well predicted, but sudden shifts, shocks, and discontinuities are not well predicted in such formalized models.

Methods for *managing* the future accept that it is unpredictable, it will not contain just one and only one outcome, and we will not be able to forecast it with certainty. Consequently, a smart approach is to focus on managing change. Change can be addressed reactively (e.g. crisis management), or proactively (e.g. strategic management). Present activity is concerned with preparing for uncertain future outcomes.

Methods for *creating* the future assume the future does not (yet) exist; it has to be created by human action, and it depends on what we plan to do. Such methods

Table 3.5 Foresight methods classified according to their approach to the future

Approach	Concept	Technique	Assumption
Foreseeing	Prediction	Surveying	Predictability/transparency
	Extrapolation	Linear input-output analysis	Stability/consistency
		Time-series analysis	
	Analytical forecasting	Integrated assessment	Modelling ingenuity
Managing	Judgmental	Delphi	Expert opinion
		Expert panels	
	Forecasting	Cross impact	Interactions
		Input-output analysis	
	Management	Integrated assessment scenarios	Analytical skills
		Cost-benefit analysis	
Creating	Policy making	Relevance tree	Logical steps
		Backcasting	
		Roadmapping	
	Speculation	Trend spotting	Imagination
	Imaging	Brainstorming	Creativity

Source: Unido Technology Foresight Manual (Module 3, 2005), adapted from May (1996)

are proactive, often focusing on the optimal options based on our normative assessment of a preferable future. We have to be creative to imagine what we would like to achieve or avoid. Many foresight programmes, often commissioned by governments, come into this category. The methods chosen strengthen the picture of the future in the directions considered favourable to the major players. Such pictures may become self-fulfilling since the decision makers are often in a position to make them happen. Making and publishing the foresight influences subsequent actions and pushes the social events (e.g. technology development, science) towards the prophecy itself (Karlsen 2002).

As for most of the typologies discussed so far, this one also camouflages the distinction between numbers and narratives in the tools used. Additionally, it is not exhaustive – it does not offer an adequate subcategory for all the tools on our candidate list.

Yet another approach by Popper et al. (2007) reports the frequency of methods used in recent foresights in Europe and the rest of the world. Out of a total of 1,405 entries in their database, 785 are mapped on 'methods'. Three main methods clusters are identified:

- Foresight and forecasting methods, e.g. scenario, Delphi, roadmapping, and backcasting
- Data analysis methods, e.g. trend analysis, SWOT, and MIC/MAC
- General approaches, e.g. expert workshops, interviews, and literature review

On a more specific level, 26 major techniques were mapped. The most frequent and popular techniques are also the most generic, e.g. *literature review*, applied in more than

50% of the reported cases. The next most frequently applied approach is *expert panel*, i.e. a technique belonging to the same cluster of general approaches as literature reviews. Keenan et al. (2006) comment that such generic approaches are most probably under-reported in the survey, i.e. since they are not considered to be 'pure' foresight tools.

Scenarios rank third, being the most popular approach within the cluster of 'foresight and forecasting' techniques. *Futures workshops* come next, belonging to the same cluster as the scenario technique. *Trend extrapolation* ranks fifth, as the most important candidate of the data analysis methods cluster.

Lagging behind these top five approaches, come *brainstorming, mega-trend analysis, surveys*, and *key technologies*, closely followed by *interviews* and *Delphi*. More rarely used techniques involve environmental scanning, modelling and simulation, backcasting, and relevance trees.

Although most of the techniques are used at all territorial levels (supranational, national, subnational), some techniques are considered to be more global than others. Amongst the top five, scenarios are more popular in transborder exercises than at the national level, and expert panels are more popular at the national level than at the other levels.

As for the time horizon set for the foresight exercise, creative approaches such as scenarios, workshops and essays, and methods offering long-range forecasting, such as trend extrapolation, modelling and simulation, and mega-trend analysis, were most frequently applied for the long-term and distant future (2025–2050). In the short-term (<2010) perspective, techniques like interviews and SWOT analysis are often applied. Medium-term (2015–2020) perspectives often correspond to Delphi, key technologies, and brainstorming.

Multiple techniques are often used in several foresight exercises. Keenan et al. (2006) mention some frequent pairs of techniques, such as technology roadmapping with key technologies, SWOT analysis and Delphi with brainstorming, and scenarios with both modelling and simulation and with trend extrapolation. Likewise, Delphi is the technique which normally engages the largest number of participants in the exercise. Other techniques do not distinguish themselves with regard to the mobilization of the experts engaged.

Although their study provides a lot of interesting information about the present use of various foresight methods and tools, it does not discuss the underlying ontological and epistemological assumptions connected to the various approaches. Neither does it offer any substantial taxonomy except the empirical overview of the tools sorted on various background criteria.

3.6 What Have We Learned So Far?

In summing up, we observe that the four existing groupings and taxonomies in the literature fail to offer a classification that is sufficiently exhaustive for our purpose. Given that foresight tools and techniques are not prescriptively rigid – rather, they can be altered and adapted according to field, subject, time horizon, and purpose of the study – we need a thorough understanding of the underlying logic of the foresight exercise to make a proper and corresponding taxonomy.

Most of the typologies presented in this section do not make explicit how they perceive the design and logic of the foresight exercise, and neither do they indicate what kind of ontological and epistemic characteristics relate to the tools applied. Most typologies also omit the distinction between qualitative and quantitative approaches.

3.7 An Alternative Meta-classification

We now return to the challenge of making the philosophical presuppositions embedded in foresight methods explicit. Amongst the most commonly applied foresight tools available, we have chosen tools and approaches where both the quantitative and qualitative elements are represented. State-of-the-art taxonomies, as illustrated in the previous section, are mostly ad hoc and context-specific. Qualitative and quantitative methods are interchangeably supportive and complementary in most foresight studies. Hence, ontological and epistemological presuppositions ought to be made explicit in foresight studies.

One part of our rationale for the following classification is to explore the potential embedded in triangulation of methods (both qualitative and quantitative) as part of foresight study designs. And we also describe the tools themselves that have incorporated quantitative and qualitative data elements in their application. There is a qualitative/quantitative divide embedded within most foresight study designs, i.e. in order to arrive at some kind of answers to research questions, and to be able to have a degree of confidence in them. And there is also a triangulation element embedded within the tools applied for generating and analysing data. These are called foresight tools and are built on qualitative and quantitative assumptions, without necessarily making them explicit. There is obviously room for including triangulation, i.e. use of both quantitative and qualitative methods, in the same foresight study. This is a slightly different point made with regard to the mix of qualitative and quantitative elements within the different tools used in foresight studies.

To arrive at an alternative classification, the foresight experts first selected and assessed 33 tools commonly used in foresight exercises (see Appendix A). These tools were primarily chosen to meet the meta-criterion of being applied and applicable both in a quantitative and a qualitative setting. Tools and approaches deemed purely quantitative (e.g. regression analysis, S-curve) or purely qualitative (e.g. genius forecasting, essays) were not eligible to be included in the list of candidate tools. The experts secondarily added five sub-criteria, thus presenting the following list:

- Qualitative/quantitative (meta-criterion)
  - Evaluates the presence of narratives/metrics in the representation of information (1 = purely qualitative, 10 = highly quantitative)

- Mobilization
  - Evaluates the numbers of experts required for the application of the tool (1 = lowest number of experts, 10 = highest number of experts)
- Scope
  - Evaluates the number of domains to which the tool is applied (1 = lowest number of domains, 10 = highest number of domains)
- Complexity: input
  - Evaluates the complexity (magnitude, interrelations, sources) of input data (1 = lowest complexity, 10 = highest complexity)
- Complexity: process
  - Evaluates the complexity (mathematical, philosophical, emergence, stakeholders) of the process (1 = lowest complexity, 10 = highest complexity)
- Complexity: output
  - Evaluates the complexity (representation, communication, transparency) of the output (1 = lowest complexity, 10 = highest complexity)

The list of assessment criteria addresses the mixture of the quantitative/qualitative balance but also includes the magnitude of human resources (i.e. the number of domain experts) which is required to obtain sufficient information, the breadth of application areas and themes where the tools normally are used, and the complexity (i.e. the interrelatedness of different components in the entity studied) of the input information, and the process of the information gathering itself, as well as the platform for interpretation and use of the data processed and disseminated.

Having such a set of assessment criteria, the experts felt confident to referee the inherent characteristics of the methods without being derailed by or having to take into account the perspective, purpose, or approach to the future itself in the foresight exercise. The assessment focus is on the tools themselves and their face validity characteristics.

All the 33 tools are assessed on a quantitative scale (1–10) for each of the five criteria. For each criterion, we have illustrated just the top, medium, and bottom pairs of tools, according to their mean values. The lowest mean value for any tool on any of the criteria measured is 1.5, and the highest mean value is 8.8. The overall picture of the initial assessment is presented in the six subsequent Figs. 3.1, 3.2, 3.3, 3.4, 3.5, and 3.6.

The qualitative versus quantitative mix ranges from mean values 2.5–2.6 (brainstorming/lateral thinking) to 8.2–8.3 (fuzzy logic/cost-benefit analysis). Typical middle range tools are key technologies (5.7) and integrated assessment (5.8).

The range of mobilized human resources varies largely between the candidate tools, ranking lateral thinking (1.5), together with neural networks (2.0) as the least costly in terms of expert manpower required. CBA (3.9) and backcasting (4.3) are in the middle, while surveying (6.8) and Delphi (8.8) ranks top (no surprise?!).

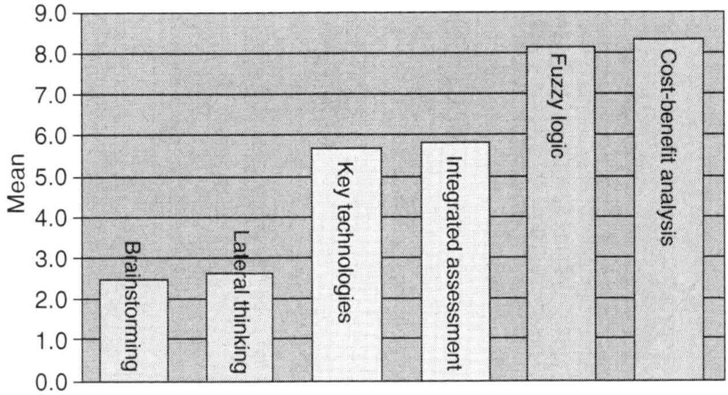

Fig. 3.1 Qualitative-quantitative mix

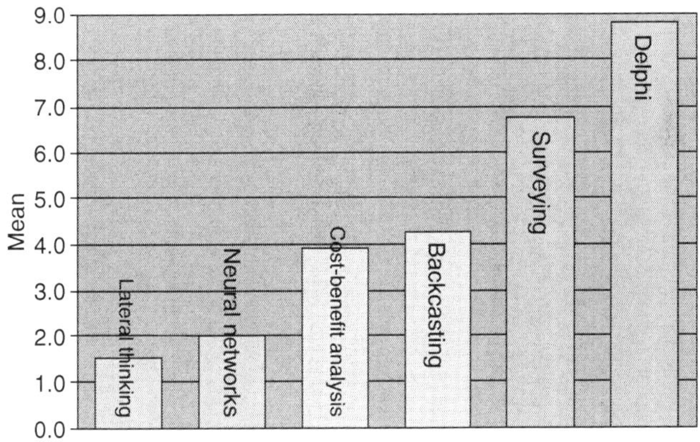

Fig. 3.2 Mobilization

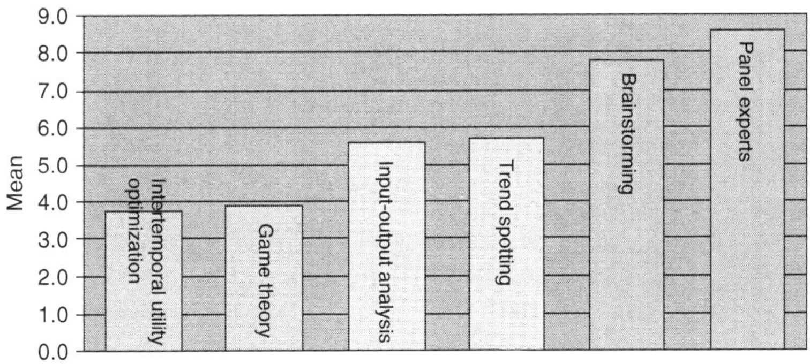

Fig. 3.3 Scope

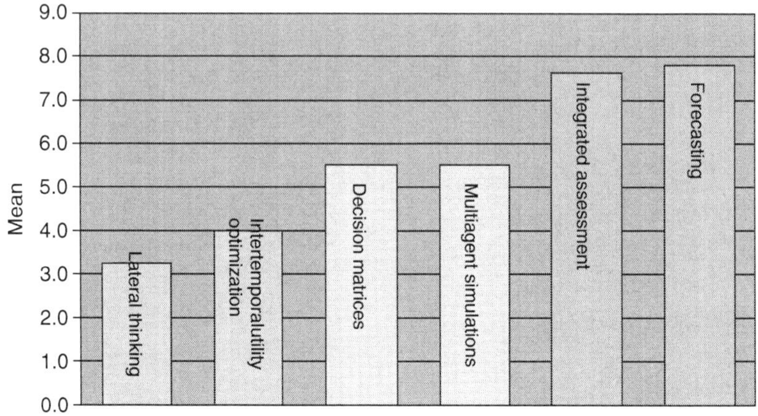

Fig. 3.4 Complexity input

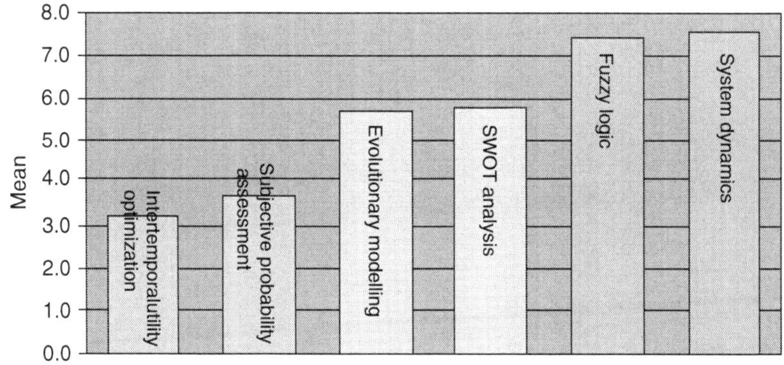

Fig. 3.5 Complexity process

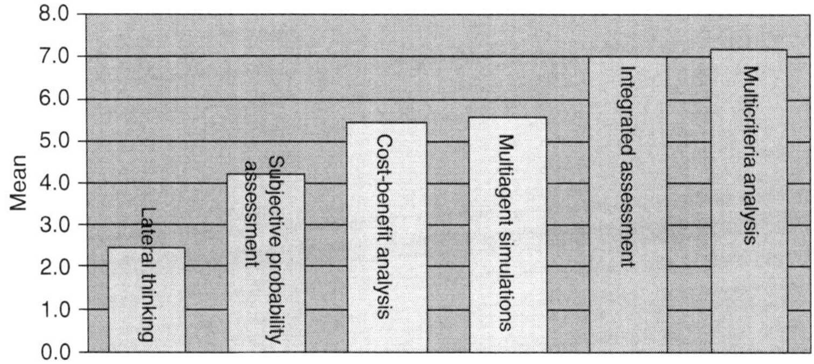

Fig. 3.6 Complexity output

Expert panels (8.5) and brainstorming (7.8) are assessed to be the most applicable tools, measured in the range of domains noted. Input-output analysis (5.6) and trend spotting (5.7) scores are in the middle, while game theory (3.9) and inter-temporal utility optimization (3.8) are deemed applicable to the smallest number of domains.

The complexity of input is deemed most demanding for forecasting (7.8) and integrated assessment (7.6). Lateral thinking (3.2) once again scores at the bottom end, together with inter-temporal utility optimization (4.0). Multi-agent simulations and decision matrices (both 5.5) rank as middle complex tools in this respect.

When it comes to process complexity, we once again find inter-temporal utility optimization scores low, together with subjective probability assessment (3.3 and 3.7, respectively). Fuzzy logic (7.4) and system dynamics (7.5) have top scores, while evolutionary modelling (5.7) and SWOT (5.8) are ranked as middle complex process tools.

Finally, lateral thinking (2.4) and subjective probability assessment (4.0) are ranked lowest when it comes to output complexity, i.e. are deemed the simplest tools in the box. Multi-criteria analysis (7.1) and integrated assessment (7.0) have top scores, while CBA and multi-agent simulations are typical tools scoring in the middle.

Summarizing these findings, we may say that most of the tools selected cluster around the median value for each of the criterion. Just a few candidates position themselves at the very extremes of the scale. When we look at the score summed over all the five sub-criteria, we find the five lowest-scoring tools to be inter-temporal utility optimization (18.0 points), lateral thinking, relevance tree, neural networks, and game theory. This group of low-scoring tools have a series of five to six neighbouring tools when it comes to total sum score. All these tools are assessed to be less demanding with respect to mobilization, less adaptable to all kinds of foresight exercises, and less complex than the rest of the candidate tools.

At the opposite end of the scale, we find (with increasing sum score) cross impact assessment, Delphi, multi-criteria analysis, expert panels, and at the top integrated assessment (34.4 points). This group has fewer neighbours: Just risk assessment and forecasting are tools in the vicinity of this high-ranked category.

Intuitively, one may think that the high-ranked category contains mainly quantitative tools and the low-ranked category mostly qualitative tools. Although we find examples which confirm this reasoning, it does not hold as an explanation for the whole list of tools. When analysing the underlying sum-score distribution, controlled for quantitative versus qualitative scores, the picture is as illustrated in Table 3.6 below.

In this table, the 33 candidate tools are analysed as to whether they belong to a group characterized mostly on the qualitative side or mostly on the quantitative side. The split between the two categories has been done according to the median value (5.7) in order to have two categories of equal size. The mean score for the qualitative category is 4.2, and for the quantitative category, it is 7.0. As can be seen from Table 3.6 above, we find no significant differences between the average sum scores for either of these categories. Hence, the bridging of quantitative and

Table 3.6 Foresight methods classified according to quantitative-qualitative score

Category	Sum score sub-criteria Low	High	Mean
Mostly qualitative tools	18.1	32.7	26.7
Mostly quantitative tools	18.0	34.4	25.8

qualitative approaches which was decided as a meta-criterion for inclusion does not give a clear-cut distinction between the tools assessed according to the sub-criteria applied.

Hence, our findings indicate that we find more practical guidance than substantial insight when classifying the inventory of candidate tools using such a set of criteria as demonstrated in this section. Still, the expert assessment demonstrates that all 33 tools and applications easily fit into this five-category classification scheme.

3.8 Discussion

In many ways, this study unveils a general problem or challenge within the foresight school as an intellectual endeavour: Though it emphasizes a critical dialogue about scientific vision and self-contained truth, foresight is mostly a conceptual framework and hardly a scientific discipline in itself. Yet, foresight exercises apply scientific methods and involve domain experts on a broad scale, perhaps to pretend to be more 'scientific' than there is supporting evidence for?

Some epistemological lessons, where the traditional application of research methods collides with the eagerness to adapt such methods to the vital endeavour of foresight exercises, seem to emerge from the study. Foresight is usually considered conducive to strategic reasoning, research processes, and policy development. The status of futures literacy, however, is both an undefined and under-communicated part of the scientific research process (Miller 2007). What, therefore, could be a fruitful relation between foresight and common research methodologies, in particular methods of sociological imagination? Could we talk about an epistemology of futures intelligence, and consider systematic futures orientations as research activities, and also as part of the theory of knowledge? Would it be helpful to require foresight studies to explicate the ontological assumptions about the future as a dependent variable and the corresponding methods applied? Would a bridging of qualitative and quantitative methods really help us overcome the pre-scientific barriers of the present state of foresight exercises?

Obviously there is space for a compromise in the debate about foresight methods, combining the strengths of quantitative and qualitative approaches and bridging the ontological and epistemological gap we find between the subjectivist and the objectivist approach. At present, we probably see quantitative and qualitative approaches not as competing but more as complementary ways of collecting empiri-

Table 3.7 Pragmatic approach to foresight methods application

Key elements in methods application	Pragmatic approach
Ontological assumptions	Future as (uncertain) regularities
Epistemic stance	There are both general and shared constructions of the future
	Intersubjectivity; common perceptions of the future
	Knowledge is partly cumulative but also context dependent
Foresight methods	Inductive/deductive logic adapted to the current problem
	Reciprocal action between individual and context
	Balance between distance and proximity to the problem
	Neutrality as an ideal
	Narratives and numbers supplement each other

Source: Adapted from Jacobsen (2005: p. 42)

cal evidence. Rarely will one approach be able to replace the other: Most often, they reciprocally supplement each other.

On the other hand, we regularly have to choose between qualitative and quantitative approaches, not because they are principally different but because the research strategy and design require us to do so. For scientific reasons, none of them will necessarily be deemed the better or superior. Patton (1994) states that because quantitative and qualitative methods have different strengths and weaknesses, they constitute alternatives to each other and, therefore, are not mutually exclusive approaches. Thus, both types of information may be collected and applied in the same foresight study.

So, let us try to propose a kind of pragmatic approach to the application of foresight methods, leaning on the ontological Popperian position of probability rather than certainty, i.e. on regularities rather than laws and strict systems. In epistemological terms, it is difficult to imagine pure objective knowledge about the future. What we think about the future is a result of our interests and of our socialization and education. Hence, we need just to be able to say anything at all about the future and introduce some kind of bridging concept. The concept of intersubjectivity may function as such a proxy, rather than the objectivity/subjectivity controversy. Having such a platform, we may propose a classification as listed in Table 3.7.

This pragmatic approach, which in fact embraces the comprehensive foresight toolbox, rejects the objectivist assumption that the future is pre-generated by laws (which we are bound to discover and apply) and the subjectivist assumption that the future is always unique. Popper (1957, 1990) claims that social systems may be subject to certain 'laws', but these are not absolute as in the natural sciences. Rather, we may be able to observe some regularities, events that repeat themselves (e.g. cohorts), but not causally determined chains of events. Probability replaces certainty as an ontological principle. In epistemological terms, knowledge about social systems will not be objective as in the physical world, but we may obtain knowledge which yields a higher consensus amongst actors. The subjective position may be replaced by some kind of 'intersubjectivity', like what we obtain when we perform a Delphi, an expert panel, or even a survey. Fuller and Loogma (2009) discuss the relationship between social constructivism (as an epistemological perspective) and foresight methodology. They hold out that a social constructivist perspective 'enables a methodological reflection on how, with what legitimacy, and to what social good, knowledge is produced' (ibid:p. 71). Foresights must engage 'intersubjective' meaning; otherwise, they do not anticipate nor produce futures. As such, foresight is a social construction produced by, and for the benefit of, the collectivity, meaning that what is regarded as valid knowledge about the future is collectively assessed and shared. 'Intersubjectivity' puts consensus amongst referees (or interviewees) in place of 'truth'. Then we may talk about a reality that transcends and exceeds the barriers set by a specific context. This epistemic stance implies that some kind of reality (the future) exists out there, different from the subjective sense of the foresight actor herself. This future may be addressed by both quantitative and qualitative methods using narratives as well as numbers.

3.9 Conclusion

Foresights as statements about the future can only be verified or falsified post priori, i.e. when the imaginary future has become the present, and as such revealed its secrets. Additionally, it cannot be deemed science, since it fails to meet crucial scientific requirements, such as here-and-now testability.

As stated previously in this chapter, foresight studies apply scientific reasoning and methods in an attempt to be scientifically legitimized. However, foresight studies are mainly a pragmatic way of solving anticipated (and uncertain) problems of the future. By investigating the ontological and epistemological assumptions underlying the various tools and techniques applied, we can also spot the contours of the kind of theory which supports, or is inherent in, foresight studies – theories about what the future is. Foresight studies are not theory-independent, even if they often present themselves as neutral assertions of the future. They are not inductive, even if we believe that since we do not know what the future will bring, we therefore have to construct our knowledge from the data. The pragmatic taxonomy proposed in Table 3.7 demonstrates in fact the opposite: The conception about what the future is (theory) will never be independent of the techniques applied to measure it. And this point is crucial; otherwise, foresight studies will risk being tautological, i.e. just confirm what we already know.

So far, everything that has been said here points to a neglected and omitted feature of foresight methodology: that it is often based on implicit, untested, and ad hoc basic assumptions. The most important methodological issues revolve around the problems of crucial testing and debating the grounds of the rival views presented.

For the most part, foresight researchers have been so concerned with producing applicable, useful, and relevant foresight studies that articulate a view of the future consistent with the inherent and implicit methodological assumptions that the more fundamental and critical view on the inclusion and use of the methods and tools has passed almost unobserved. To borrow a perspective from Morgan and Smircich (1980: p. 499): Here rests the main challenge of the analysis and the proposed taxonomy of foresight methods.

The basic assumptions, both ontological and epistemological, embedded in the present use of foresight methods should be challenged. It is necessary to transcend the current debate about foresight methodology with respect to the merits of its theoretical basis, and to its methods, both qualitative and quantitative, taken from empirical research, which dominates the contemporary scene of foresight studies.

The major challenge in foresight studies is not that we apply methods belonging to different traditions within the theory of science but rather that the future, i.e. our study object, slides outside the realm of the logic of scientific inference. In principle, there may be two approaches for such a blending of methods: either taking components of methods from both strands as input to the overall inquiry or integrating different method types at all phases of the foresight. Foresight researchers who apply primarily quantitative methods tend to be more inclined to see the value of qualitative approaches as an add-on to their design than vice versa. However, some researchers think that a true integration of the two strands is not feasible. To some, 'the quest for meaning and the quest for measurement are incommensurable' (Massé 2000: p. 411). This point of view is possibly the least fruitful way of reasoning on this topic. Arguably, the ways foresight researchers might choose to combine qualitative and quantitative approaches are only limited by the ingenuity of the researchers themselves, not by the intrinsic characteristics of the qualitative and quantitative approaches.

Appendix A

A.1 Tools/Approaches: Towards a Classification

On a scale of 1–10, evaluate the following aspects regarding the tools/approaches with which you are familiar.

A.2 Qualitative/Quantitative

Evaluates the presence of narratives/metrics in the representation of information (1 = purely qualitative, 10 = highly quantitative)

A.3 Mobilization

Evaluates the number of experts required for the application of the tool (1 = lowest number of experts, 10 = highest number of experts)

A.4 Scope

Evaluates the number of domains to which the tool is applied (1 = lowest number of domains, 10 = highest number of domains)

A.5 Complexity: Input

Evaluates the complexity (numerosity, interrelations, sources) of input data (1 = lowest complexity, 10 = highest complexity)

A.6 Complexity: Process

Evaluates the complexity (mathematical, philosophical, emergence, stakeholders) of the process (1 = lowest complexity, 10 = highest complexity)

A.7 Complexity: Output

Evaluates the complexity (representation, communication, transparency) of the output (1 = lowest complexity, 10 = highest complexity)

Appendix B

Table B1 Tools and approaches assessed by the experts

1. Analogy analysis	12. Game theory	23. Expert panels
2. Backcasting	13. Gap analysis	24. Qualitative dynamic modelling
3. Brainstorming	14. Input-output analysis	25. Relevance trees
4. Cost-benefit analysis	15. Integrated assessment	26. Risk assessment
5. Cross impact analysis	16. Inter-temporal utility optimization	27. Roadmapping
6. Decision matrices	17. Key technologies	28. Scenario analysis
7. Delphi	18. Lateral thinking	29. Subjective probability assessment
8. Evolutionary modelling	19. Micro-simulations	30. Surveying
9. Focus groups	20. Multi-agent simulations	31. SWOT analysis
10. Forecasting	21. Multi-criteria analysis	32. System dynamics
11. Fuzzy logic	22. Neural networks	33. Trend spotting

References

Bell, W. (2002). A community of futurists and the state of the futures field. *Futures, 34*(3/4), 235–247.

Cunha, M. P. (2002). *Time travelling: Organisational foresight as temporal reflexivity*. Paper presented at Probing the Future Conference, Glasgow, July 11–13, 2002.

Fuller, T., & Loogma, K. (2009). Constructing futures: A social constructionist perspective on foresight methodology. *Futures, 41*, 71–79.

Gordon, T. J. (1994). *Integration of forecasting methods and the frontiers of futures research*. AC/UNU Millennium Project Paper.

Hatch, M. J., & Yanow, D. (2008). Methodology by metaphor: Ways of seeing in painting and research. *Organization Studies, 29*(1), 23–44.

Hawking, S. (1988). *A brief history of time*. London: Bantam Books.

Holden, C. (2005). Time's up on time travel. *Science, 308*, 1110.

Jacobsen, D. I. (2005). *How to conduct research (Hvordan gjennomføre undersøkelser?)*. Kristiansand: Norwegian Academic Press.

Karlsen, J. E. (2002). The self-fulfilling prophecy. In S. U. Larsen (Ed.), *Theory and methods in the social sciences*. New York: Columbia University Press.

Karlsen, J. E., & Karlsen, H. (2007). Expert groups as production units for shared knowledge in energy foresights. *Foresight, 9*(1), 37–49.

Karlsen, J. E., Oeverland, E. F., & Karlsen, H. (2010). Sociological contributions to futures' theory building. *Foresight, 12*(3), 59–72.

Keenan, M., Butter, M., Sainz, G., & Popper, R. (2006). *Mapping foresight in Europe and other regions of the world: The 2006 annual mapping report of the EFMN* (Report to the European Commission). Delft: TNO.

Krawczyk, E., & Slaughter, R. (2010). New generations of futures methods. *Futures, 42*, 75–82.

Kreuger, R. A. (1988). *Focus groups: A practical guide for applied research*. London: Sage.

Massé, R. (2000). Qualitative and quantitative analyses of psychological distress: Methodological complementarity and ontological incommensurability. *Qualitative Health Research, 10*, 411–423.

May, G. H. (1996). *The future is ours: Foreseeing, managing and creating the future*. Westport/London: Praeger/Adamantine.

Miles, I., & Keenan, M. (2003). *Overview of methods used in foresight*. Paper presented at the Technology Foresight for Organizers Training Course, Ankara, December 8–12, 2003.
Miller, R. (2007). Futures literacy: A hybrid strategic scenario method. *Futures, 39*, 341–362.
Morgan, D. L. (1988). *Focus groups as qualitative research*. London: Sage.
Morgan, G., & Smircich, L. (1980). The case for qualitative research. *Academy of Management Review, 5*(4), 491–500.
Patton, M. Q. (1994). *Qualitative evaluation and research methods* (2nd ed.). Newbury Park, CA: Sage.
Popper, K. (1957). Propensities, probabilities, and the quantum theory. In D. Miller (Ed.) (1985), *Popper selections*. Princeton: Princeton University Press
Popper, K. (1990). *A world of propensities*. Bristol: Thoemmes Press.
Popper, R. (2008). How are foresight methods selected? *Foresight, 10*(6), 62–89.
Popper, R., Keenan, M., Miles, I., Butter, M., & Sainz, G. (2007). *Global foresight outlook 2007: Mapping foresight in Europe and the rest of the world* (Annual Mapping Report). European Commission, EFMN, pp. 66. Available at http://prest.mbs.ac.uk/efmn/gfo_2007.pdf
Rescher, N. (1998). *Predicting the future: An introduction to the theory of forecasting*. New York: State University of New York Press.
Rossel, P. (2010). Making anticipatory systems more robust. *Foresight, 12*(3), 73–86.
Rossel, P. (2012). Early detection, warnings, weak signals and seeds of change: A turbulent domain of futures studies. *Futures, 44*(3), 229–239.
Suddendorf, T., & Corballis, M. C. (1997). Mental time travel and the evolution of the human mind. *Genetic Social and General Psychology Monographs, 123*, 133–167. http://cogprints.org/725/.
Suddendorf, T., & Corballis, M. C. (2007). The evolution of foresight. What is mental time travel and is it unique to humans? *Behavioral and Brain Sciences, 30*, 299–351.
Tsoukas, M., & Hatch, M. J. (2001). Complex thinking, complex practice: The case for a narrative approach to organizational complexity. *Human Relations, 54*, 979–1013.
UNIDO Technology Foresight Manual (Module 3) (2005). Foresight Brief No 45.
Weick, K. E., & Sutcliffe, K. M. (2001). *Managing the unexpected*. San Francisco: Jossey Bass.

Chapter 4
Bridging Qualitative and Quantitative Methods in Foresight

Matthias K.B. Lüdeke

4.1 Introduction

There is a long-lasting and controversial discourse on the role of quantitative and qualitative data and methods in science, at least since the "Newtonian turn" in physics in the seventeenth century. After this successful step in the mathematical formalization of a large branch of physics, nowadays called "classical mechanics", it was used as a kind of paradigmatic case by many theorists of science. Thereby, standards for scientific processes and theory structures were imposed on realms of science dealing with dramatically different subjects and having different purposes than classical mechanics. This was controversially discussed within the debate on positivism, but it still has a strong influence on our understanding of science.

Why is this relevant for the discussion of quantitative and qualitative concepts in foresight?

Firstly, this paradigmatic case deals with the motion of objects in space (planets, cannonballs, cars), that is, it deals explicitly with the time dimension. Therefore, a new kind of mathematics was developed by Newton and Leibniz: the differential calculus. The general laws of motion could then be formulated as a set of differential equations which calculate the (observed or future) time courses of the object's location from given initial (and boundary) conditions. These laws of motion described a number of observations and experiments so well that at the beginning of the nineteenth century, a mechanistic world view was formulated, assuming that, once set in motion, the universe would work like clockwork, following eternally the

M.K.B. Lüdeke (✉)
Department for Climate Impacts and Vulnerabilities,
Potsdam Institute for Climate Impact Research,
P.O. Box 60 12 03, Potsdam, D-14412, Germany
e-mail: luedeke@pik-potsdam.de

M. Giaoutzi and B. Sapio (eds.), *Recent Developments in Foresight Methodologies*, Complex Networks and Dynamic Systems 1, DOI 10.1007/978-1-4614-5215-7_4,

© Springer Science+Business Media New York 2013

Newtonian laws of motion (theological complications could be resolved[1]). Although this extreme view was revised for several reasons,[2] the relation between the explanation of phenomena and their prediction is still a vital point for the controversial understandings of foresight.

Secondly, the cited paradigmatic case is a fully quantitative theory where each basic concept (like "length") is operationalized by a measurement procedure ("compare with the 'mètre des archives' in Paris") which assigns the respective variable ("s") a real number ("5.51 m"). This constitutes a clear-cut relation between the quantitative theory and its real-world subject and makes a variable-oriented approach to scientific explanation and prediction very appealing.

Thirdly, the Newtonian laws of motion are valid for a huge number of different experimental and observed situations (all macroscopic mechanical phenomena with relative velocities significantly less than the speed of light). This implies that science is able to find general laws with very wide ranges of applicability.

In the next section, we analyse the shortcomings of epistemological approaches to prediction which are oriented to the above paradigmatic case, while it appears as a more or less singular stroke of luck in the history of science. In the following section, we then discuss the role of quantitative models in foresight studies. After a short overview of the four main approaches to foresight according to Kreibich (2006), we proceed with a discussion of qualitative concepts compared with quantitative concepts in science and conclude with some approaches which bridge the gap between the two traditions.

4.2 Explanation and Prediction

As mentioned above, within the epistemological position of logical positivism, there is no difference in the inferential pattern between explanation and prediction. Following the argumentation of Aligica (2003), this can best be demonstrated by the position of one of the proponents of the "nomologico-deductive" school of explanation:

> An explanation (…) is not complete unless it might as well have functioned as a prediction; if the final event can be derived from the initial conditions and universal hypotheses stated in the explanation, then it might as well have been predicted, before it actually happened, on the basis of a knowledge of the initial conditions and general laws (Hempel 1963).

Thus, the explanation of a phenomenon includes the information about antecedent conditions and general causal laws. Hempel called these "covering-laws" or

[1] For example, by reformulating the laws of motion as a variational principle by Lagrange (eighteenth century), implying more a (divine) purpose of the whole trajectory/history than reducing the options of God to defining the initial condition and the Newtonian laws

[2] Even by inner physical reasons like macroscopic irreversibly or, later, deterministic chaos

"nomologico-deductive" argument, when an observed phenomenon can be reconstructed along these lines.

The main arguments against the above concept as a general structure of valid science (Rescher 1998; Aligica 2003) include historical but also purely logical aspects:

- There are generally accepted and important scientific explanations without predictive power (e.g. the mechanisms which generate earthquakes or the evolutionary explanations of the emergence of new species).
- There are successful predictive methods without any explanatory content like time series analysis and correlational or analogical approaches.
- The history of science shows many examples of successful predictions based on poor or even wrong explanations, as well as wrong predictions based on good explanations.

The latter emphasizes the role of explanatory scientific theories as steps in an ongoing process instead of being already "close to the truth" and reflects on the fact that empirical theories (in contrast to mathematical statements) cannot be proved in a strictly logical sense but, according to Popper (1934 [2004]), can only be falsified.

This makes clear that – at least for a large part of relevant scientific predictive endeavours – the nomologico-deductive approach is not the most promising one: possibly either no general laws exist or the number of observed instances is, for systematic reasons, too low to perform a significant formal falsification procedure.

Aligica (2003) summarizes: "In the prediction's domain even the best confirmed theories are no more than reasonable and provisional estimates of the truth". And he concludes by stressing the principal epistemic difference between explanation and prediction (or retrodiction[3]):

> Explanations try to reveal connections between events, phenomena and states and if possible to reveal the fact that they are part of larger patterns, regularities and laws. The primary function of predictions and retrodictions is to acquire and offer more knowledge of specific, concrete events and occurrences. The idea is to export from premises the necessary epistemic weight needed to gain credibility. The primary function of such arguments is simply to establish or prove the conclusion. Consequently in a prediction or retrodiction argument, the application of general laws is not essential. An argument that makes appeal to general laws is always welcome but still it is as good as any other argument; and thus in the last instance it is inessential. Using the covering-law model to make a prediction or retrodiction is sufficient, but not necessary. Statements of restricted regularities, quasi-laws, statistical laws, the so-called common sense generalizations or accidental generalizations can viably be employed in projective arguments.

Even if this argumentation seems self-evident to many practitioners in future studies, it becomes crucial in foresight projects which include researchers from those disciplines where a covering-law-type understanding of science is (still) dominant.

[3]Retrodiction means the prediction of an event in the past from initial situations and conditions even further in the past.

In climate change, economic or ecological theories quantitative (dynamic) modelling plays a widely accepted role, and these models are often assumed by their authors to operationalize Hempel's general laws, allowing for explanation and prediction at the same time – consequently, they are hardly inclined to accept that their model-based predictions play a comparable role to "common sense generalizations" in foresight.

4.3 The Role of Quantitative Modelling in Foresight

Indeed, the argument has to be handled with care: some of these predictive models are closer to the above-mentioned epistemological "stroke of luck" than others, in particular the atmospheric climate forecast models (known as Atmospheric General Circulation Models) which incorporate a great deal of Newtonian mechanics and can dispose of a large (and increasing) amount of standardized data for validation via retrodiction. Of course the future could in principle falsify the model, but it is anchored very deeply in systematically accumulated empirical evidence. But already the next step to answer the question what the global climate will look like under a given human impact scenario requires more complex physical Earth system models which integrate oceans, the kryosphere and bio-geochemical cycles. Although these additional components are still purely subject to the laws of nature, the data situation for validation becomes more critical and (consequently?) the underlying theories more controversial. Here, the argument certainly becomes more relevant: that our current theoretical understanding may be more a historical phase than already "close to the truth". The fact that the theory has the same form (a dynamic quantitative model) as others which are closer to Hempel's paradigm should not be of any relevance in this context – its role in foresight exercises becomes relativized and in this case: "An argument that makes appeal to general laws is always welcome but still it is as good as any other argument" (Aligica 2003).

To stay with the forecast example of global climate change assessments, the next step, using what are called Integrated Assessment Models (IAMs), is to incorporate human actions and reactions into the formal model on the basis of the plausible argument that (anthropogenic) changes in the physical environment will have feedbacks on human actions – a relation which questions the possibility of reasonable a priori definitions of, for example, scenarios of anthropogenic CO_2 emissions. This means that socio-economic theories enter the physical Earth system models and with them all specific problems like reflexivity and the related problem of the separation of the observer (modeller) from its subject (e.g. society). While quantitative modelling approaches are well established in economics, these are highly contested in sociology and policy science. But, even in economics – similar to the situation described for physical Earth system models – the quantified theory is far from

Hempel's paradigm: independent of the obviously poor quality of predictions,[4] the basic hypotheses of mainstream theory are still used to guide economic policies.

IAM modellers are well aware of these shortcomings in the predictive ability of their integrated models and make attempts to quantify the uncertainty of their forecasts. Meanwhile, there are classifications of the sources of uncertainty in quantitative models, ranging from numerical failures to uncertainties in the choice of relevant variables and their interactions (structural uncertainties[5]). This also spans the range of the possibility of a formal uncertainty assessment from "manageable" to "almost impossible".

To deal with this situation, Jan Rotmans, an experienced IAM modeller (Rotmans and de Vries 1997), originally called for a proper interpretation of quantitative forecasts of large integrated models: "Don't trust the numbers, just trust the trends". This seems to be a possibility for a more careful interpretation of quantitative prediction, although it is not clear under which conditions totally uncertain numbers produce trustworthy trends.

From our experience in predictive formal modelling activities – mostly for purposes of policy assessment (e.g. Petschel-Held and Lüdeke 2001; Eisenack et al. 2006) – we would argue that the whole modelling process – not only the resulting prediction – is the relevant input into an assessment or foresight exercise. If all assumptions underlying the model are made explicit and transparent, mathematics (supported by computers) is an unrivalled means for correct and comprehensive logical deduction. A model used in this manner in a foresight process provides more "food for thought" than a black box and contributes to reasonable projections.

This understanding of the role of quantitative modelling in foresight has far-reaching consequences as it demands that a model used in foresight:

- Can either be made fully transparent with respect to its underlying assumptions to everybody who interprets its predictions
- Or is close to the paradigmatic case of classical mechanics (see the preceding section) and has proved its predictive capacity in many instances under widely varying conditions

In the integration section, we present an approach to dynamic modelling which is intrinsically appropriate to fulfil the first requirement.

So far, we have discussed the role of the most complex quantitative concept, dynamic modelling based on assumptions on mechanisms and interactions. This is mathematically realized either in the form of deterministic/stochastic ordinary/partial differential equations or their discrete counterparts. Our starting point was the critique of the generalization of the epistemic identity of prediction and explanation, a position which is oriented at a quantitative theory of exactly this form.

[4]Poor predictions of the economic cycle, wrongly predicted convergence of developing and developed countries, etc.

[5]For an interesting approach to deal with structural uncertainty, see Van Asselt and Rotmans (2002). They suggest a systematic exploration of different combinations of modules of an IAM along ideas of cultural theory.

As mentioned already, there are quantitative methods relevant for foresight without explanatory pretensions, for example, correlational approaches and time series analysis. Particularly in situations where only poor mechanistic knowledge is available, these open the possibility for temporal extrapolation. But one should keep in mind that virtually all of these statistical extrapolation methods are implicitly related to classes of mechanistic assumptions. To take a simple example, to choose a linear extrapolation instead of a quadratic one, even if the first reproduces the observed time series a bit better, implies the assumption that there is no significant positive feedback and that this will be also the case in future. We would therefore argue that the mechanistic assumptions which underlie the predictions should be made transparent whenever possible.

4.4 Approaches to Foresight

From the practice of foresight, Kreibich (2006) identifies four different approaches, which show that the paradigm discussed above covers only a small part of relevant predictive abilities:

1. *The explorative empirical-analytical approach.* Based on available explicit knowledge and actual data, probable and possible future developments are systematized under explicit assumptions and boundary conditions. These developments are then analysed according to specific rules.
2. *The normative-intuitive approach.* Experience and, more generally, partly tacit knowledge are used in an imaginative and creative way to generate desirable visions of the future.
3. *The planning approach.* Here the focus is the process of shaping the future towards a desirable vision. Stocks of knowledge and experience are used creatively to suggest new communication, decision making, participation and implementation processes.
4. *The communicative-participative approach.* The integration of actors from different societal sectors increases the amount of knowledge on possible future developments. In particular, the aspects of shaping and implementation possibilities become substantiated. The same is valid for the normative aspect (desirability).

Practical foresight exercises show that usually a combination of the above approaches is applied. For example, in their future study on global sustainability, *The Great Transition*, Raskin et al. (2002) applied a combination of the first three approaches, which is nicely documented by the structure of their final report. It starts in an explorative empirical-analytical manner by analysing historical transitions and developing from these global scenarios (see Fig. 4.1) by applying a defined set of philosophies.[6]

---

[6]Smith, Keynes, Malthus, Hobbes, Morris, Mill

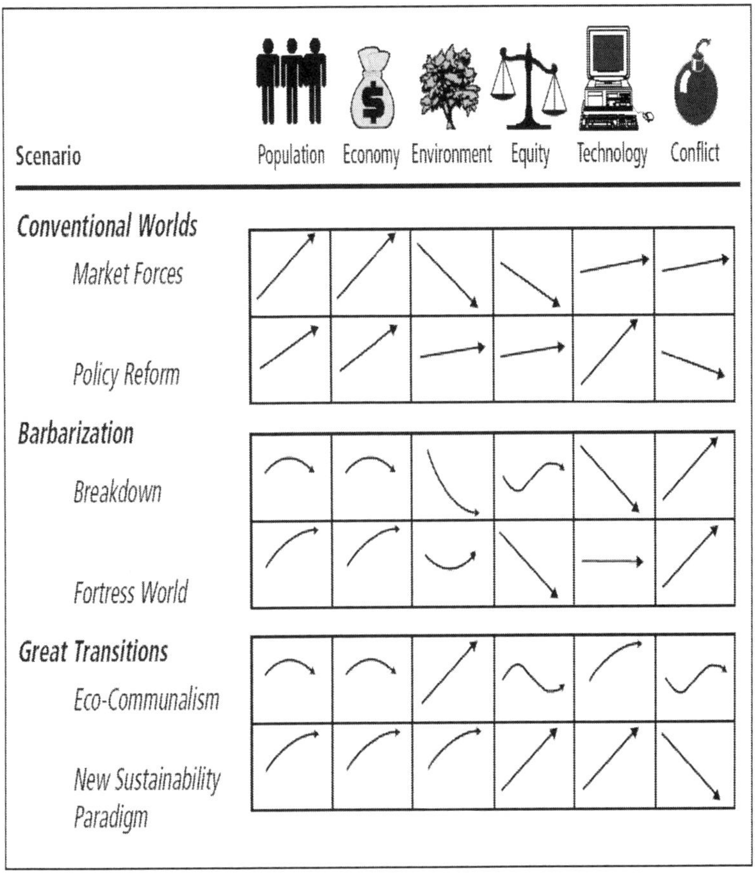

Fig. 4.1 Scenario structure with illustrative patterns (After Gallopín et al. 1997)

There is some overlap to the normative-intuitive approach as the set of applied philosophies is not sufficient to define the different future visions, and there is a strong normative component in imagining the "new sustainability paradigm".

After that, the "planning approach" is applied, asking how the desirable scenario could be implemented – the results are clearly represented in Figs. 4.2 and 4.3.

With respect to the choice of qualitative and quantitative methods in this future study – as in many others – a mix was applied, ranging from the citation of results from models as discussed in the preceding two sections to qualitative arguments, while the use of quantitative approaches is concentrated in approaches (1) and (3). However, it is instructive to have a brief look at some of the specifics of these two methodological schools in order to sharpen the view for the problem of the most satisfactory choice of method.

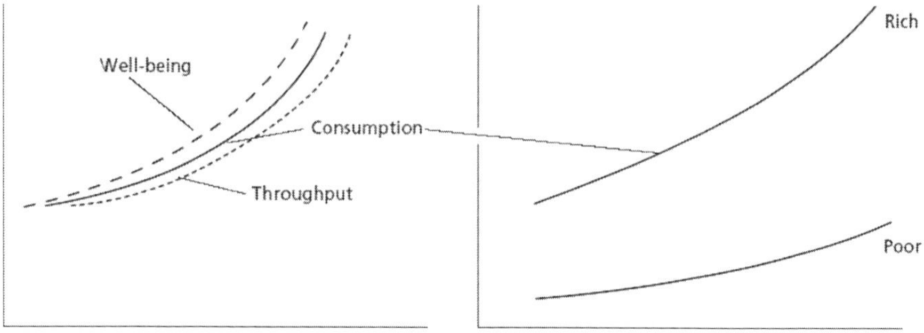

Fig. 4.2 "Market forces" scenario, where well-being, consumption and material/energetic throughput increase in parallel, while the consumption gap between the rich and the poor is increasing

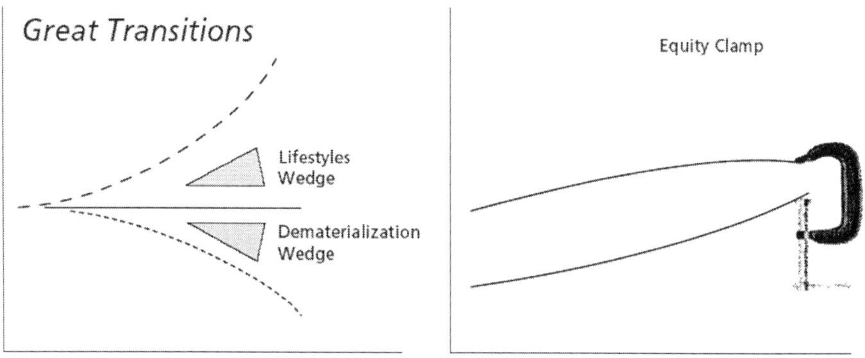

Fig. 4.3 "Instruments" for a transition to sustainability: the "lifestyles wedge", decoupling well-being from consumption; the "dematerialization wedge", decoupling consumption from material/energetic throughput; and the "equity clamp", forcing the redistribution of wealth on the globe (After Raskin et al. 2002)

4.5 Qualitative Versus Quantitative Methods

A common property of all quantitative methods in foresight – from complex dynamic modelling to simple statistical correlations – is that they are variable-oriented. This has far-reaching consequences. Firstly, all of these methods begin after the variables are defined – so the obviously crucial step of variable definition lies outside their scope.

Related to this, the concept of the quantitative variable is two-edged: on the one hand, it is clearly operationalized by a specific measuring process, thereby standardized and highly comparable, independent of the location and time of its measurement.

On the other hand, this has to be paid for by "context stripping", that is, it is abstracted from the original context in which it had a specific meaning.

However, when using quantitative variables, the systematic comparison of a large number of cases becomes feasible – which is important for the explorative empirical-analytical approach in foresight. With respect to the statistical evaluation of data intended to provide a basis for temporal extrapolation or to obtain relationships between variables which can, for example, be used in dynamic system models, one specific point has to be stressed: "outliers are no problem". This means that it is assumed to be irrelevant when the identified interrelation is invalid for some of the observed cases.

The statistical use of quantitative variables has to be distinguished from their use in system analytic models. Here, the time courses of the variables are deduced from their hypothesized interrelations which allow complex feedback nets to be evaluated. This is applicable in foresight if one can formulate explicitly and quantitatively the mechanisms which contribute to the process which has to be predicted. A further condition is that the assumed interrelationships stay valid, and the chosen variables stay relevant during the forecast period.

Quantitative variables may be measured on different scales, allowing for different mathematical operations:

(a) Ratio – all mathematical operations
(b) Interval – differences
(c) Ordinal – greater than, less than
(d) Nominal – discrete, no ordinal relationships

The less is the demand for measurement, the fewer the number of mathematical operations which are possible on the variables. System theoretical models need variables on a ratio scale while statistical evaluations are possible for all scales.

The characteristics of qualitative data and methods are significantly different. The form of the data is much richer – one can almost state that every type of information which is not a variable is qualitative data. Typical examples are a text, a photo, and a movie. The character is exactly the opposite of the "context-stripped" variable: which is a "meaningful but complex configuration of events and structures" or a "singular, whole entity purposefully selected".

Retrieval techniques for such qualitative data are, amongst others, interviews, observations, oral history, focus groups, and Delphi groups, which establish the link to the communicative-participative approach in foresight.

Data analysis techniques are hermeneutics (evaluating text and context), grounded theory (identify concepts across different texts), and others. One important aspect in qualitative methodology is the concentration on each single case. It may even be productive to look for the extreme cases rather than for the typical – in clear contrast to the treatment of "outliers" in quantitative statistical approaches.

The related process of thinking is a more circular one: during the process of foresight activities, definitions, and even aims, may be modified if appropriate – this, again, is in clear contrast to variable-oriented foresight which is more linear, in the sense that, after the initial variable definitions are made, the process has to stay with them – at least for a considerable time.

4.6 Integration of Qualitative and Quantitative Methods

One way to integrate the different methodological traditions is on the level of the organization of foresight projects, which allows the results of different quantitative and qualitative methods to be integrated. This is certainly a step forward, but does not guarantee the mutual understanding of the reasoning behind these results – which is a severe shortcoming in the communication process. Therefore, it seems to be valuable to look for existing methods at the interface between the qualitative and the quantitative tradition.

One class of these "interface methods" retains the variable orientation (including a more linear research process) but tries to deal with weaker scales (as far as is known, the following methods cover the main ideas in this realm).

Here, one possibility is statistics with multidimensional nominal data: as a two-valued nominal variable is already very close to a qualitative concept (something is either green or not), a, for example, cluster algorithm on multidimensional nominal data, yields qualitative constellations rather than quantitative cluster centroids. On the other hand, we still have the typical characteristics of ignorance with respect to single outliers which is unacceptable for important traditions of qualitative research.

This problem disappears if modified system analytic approaches on weaker scales are chosen. One example is the qualitative case study analysis (QCA) after Ragin (1994) on the basis of Boolean algebra: this uses Boolean variables (with the values true/false) to transparently deduce rules applicable for several cases. On the other hand, it is still variable-oriented with the typical consequences for the research process.

A more explicitly time-related approach in this class is systems analysis with ordinal variables (QDEs) after Kuipers (1994): this method allows possible future trend combinations to be deduced from very loosely characterized feedback structures – it makes the advantages of systems analysis available for only weakly quantified systems. As this approach is rather new, it will be described in some more detail in the following paragraphs.

QDEs are based on system theoretical process thinking, that is, the state of a system is related to its rate of change. In the realm of usual quantitative modelling, this is formalized by differential (since Leibnitz and Newton) or difference equations, where explicit numerical relationships between the variables and their rates of change are needed. In contrast, QDEs try to deduce the time development of the variables from a much weaker, namely, a "qualitative", understanding of the interactions of the system elements. This qualitative understanding can be characterized by the following hierarchy of determination:

1. Which elements are directly related (e.g. A and B are directly related, A and C are not: A – B)?
2. What is the direction of the influences (e.g. B influences A: A<–B)?
3. Is it a strengthening or diminishing influence (e.g. B diminishes A)?
4. Is it an influence on the variable or on its rate of change (e.g. B diminishes the change of A)?

Levels 3 and 4 above imply that it is possible to describe the elements of the system by ordinal scale variables, that is, a "greater/less than" relationship can be defined.

At level 4 of determination, QDEs can be applied and will generate the time course of the variables by their trends and trend changes. As QDEs are a generalized system analytic method, the boundaries of the system, its elements, their qualitative relationships, and exogenous drivers have to be identified. In all cases where this can be done, at least in parts, the method is applicable.

With respect to the mathematical representation, a QDE can be understood as a whole class of ordinary differential equations (ODEs), which are solved simultaneously. In its simplest form, the right-hand sides of the ODEs are only defined by their monotonicities, that is, only the signs of the Jacobi matrix elements are known. The results one can obtain from such a weak systems characterization (compared with a numerical model) are combinations of trend directions of the variables and sequences of such combinations. Depending on the input, branching and/or cyclic time developments may be the result, that is, different possible futures. Branching points identify critical stages in the development: depending on influences which are beyond the functional resolution of the assumed model, different paths may be entered.

QDEs can be considered as a kind of automatic phase space analysis which yields possible sequences of monotonicity cells. The algorithm works like a filter: starting with one trend combination, all possible successor combinations are generated. Then the algorithm filters all transitions which are not in accordance with the given system, that is, the given Jacobi matrix. For the remaining ones, again all valid successors are generated, and so on. This results in a "tree" where each branch represents a possible sequence of trend combinations or "qualitative trajectory".

To apply QDEs, it is necessary to construct an influence diagram which depicts the system's elements and their qualitative relationships. To obtain this, techniques of qualitative data collection (interviews, oral history, focus groups, Delphi groups) and data analysis (hermeneutics, discourse analysis, grounded theory) can be applied (for the potential role of these techniques in the different stages of model development and the interpretation of model results, see Luna-Reyes and Andersen 2003).

The method was originally applied by Kuipers and his group on qualitative physics and human physiology. In the realm of sustainability science, it was applied on smallholder agriculture in developing countries, urban development, fisheries management and industrial agriculture. In these cases, it was the aim: to calculate possible future developments from qualitative systems understanding; to choose from these a set of possible futures, that is, the desirable ones; to identify critical branching points; and to assess policy options to influence the development positively.

The strength of QDEs is that powerful mathematical system theoretical methods become available even if only qualitative knowledge of the interactions of the system's elements is available, for example, in the form of an influence diagram. This allows us to fulfil the requirement of the transparency of the model assumptions for the interpreter of the results as formulated earlier in the Sect. 4.3 on the role of quantitative modelling in foresight.

One disadvantage is that, in some cases, the result, that is, the qualitative trajectories, may be very ambiguous, in the sense that very many branching points occur. The extreme case would be that the filtering ability of the qualitative model is so weak that almost every future development is possible. But this simply means that the input – our knowledge of the system – is insufficient to make any forecasts.

Another class of "interface methods" deals with the systematization of a research or forecasting process that integrates quantitative and qualitative methods. A relevant example based on our own research experiences is the use of qualitative data retrieval and analysis to construct and validate/falsify system analytic models (Luna-Reyes and Andersen 2003): the purely deductive part of the whole forecast process is done via systems analysis (e.g. the above-mentioned QDE approach) while the – extremely important – remaining steps are done with qualitative methods. As the qualitative steps interact with the system analytic process at several points, the danger of insufficient mutual understanding of systems scientists and qualitative researchers is minimized. This method can be interpreted as an elaborated version of triangulation (Denzin 1970), which follows the idea of corroborating a result by obtaining it with different methods.

4.7 Conclusions

We started with a discussion of the most important paradigm of quantitative foresight: the concept of quantitative dynamic modelling. Its promises, limitations and chances were elaborated. From the limitations, the importance of alternative, inter alia qualitative, methods became clear. With respect to the chances, the transparency of the underlying assumptions and/or a long-standing, successful history of validation are identified.

Obviously the approaches to foresight are necessarily too diverse to be subsumed under the nomologico-deductive concept. As one more satisfactory possibility to frame the broad field of foresight activities, the systematization of Kreibich (2006) was adopted and – for illustration – applied to the "great transition" study of SEI. It occurred that – at first sight – in this study the explorative empirical-analytical (1) and the planning approach (3) were used. Closer inspection showed that also normative-intuitive aspects (2) played an important role. So, only one of the four approaches, the communicative-participative approach (4), was definitely not used in this study. Quantitative methods were mainly used in approach (1) and to some extent in (3). For approach (2), quantitative methods are clearly less appropriate, while they may have a role in the communicative-participative approach.

After a description of the important properties of quantitative and qualitative data and methods, a hierarchy of depths integration depth of the "two cultures" was identified: the most superficial way is the collection of qualitative and quantitative "black box" results gained by different members of the foresight activity – when the danger of unrecognized inconsistencies in the basic assumptions leading to the respective results is obvious. Then, for a somewhat deeper integration, two classes of "interface methods" were suggested: the very fast alternating application of qualitative and quantitative steps (e.g. Luna-Reyes and Andersen 2003) and the use of variable-oriented methods working with data on weaker than ratio scale.

As a very promising example for the latter interface methods, the systems analysis with ordinal variables was presented in more detail. It occurs that models (and projections) constructed with this method fulfil the above-mentioned precondition of transparency for all members of the foresight activity and allows us to map the uncertainty or ambiguity of assumptions, of course resulting in possibly very weak and ambiguous projections. In general, this example shows that there are current developments within mathematical systems theory which concentrate more on uncertainties in the system definition and, with respect to projections, more on corridors than on trajectories. This chapter has tried to show that this offers the chance of deeper integration of quantitative and qualitative methods in foresight activities.

References

Aligica, P. D. (2003). Prediction, explanation and the epistemology of future studies. *Futures, 35*, 1027–1040.

Denzin, N. (1970). *The research act*. Chicago: Aldine.

Eisenack, K., Lüdeke, M. K. B., Petschel-Held, G., Scheffran, J., & Kropp, J. (2006). Qualitative modelling techniques to assess patterns of global change. In J. Scheffran & J. Kropp (Eds.), *Advanced methods for decision making and risk management in sustainability science*. Hauppauge: Nova.

Gallopín, G., Hammond, A., Raskin, P., & Swart, R. (1997). *Branch points: Global scenarios and human choice* (PoleStar Series Report No. 7). Stockholm: Stockholm Environment Institute.

Hempel, C. G. (1963). Explanation and prediction by covering laws. In B. Baumin (Ed.), *Philosophy of science: The Delaware seminar* (Vol. 1). New York: Wiley.

Kreibich, R. (2006). *Zukunftsforschung*. Arbeitsbericht Nr. 23/2006. Institut für Zukunftsstudien und Technologiebewertung/Institute for Future Studies and Technology Assessment, Berlin [in German].

Kuipers, B. (1994). *Qualitative reasoning: Modeling and simulation with incomplete knowledge*. Cambridge: MIT Press.

Luna-Reyes, L. F., & Andersen, D. L. (2003). Collecting and analyzing qualitative data for system dynamics: Methods and models. *System Dynamics Review, 19*(4), 271–296.

Petschel-Held, G., & Lüdeke, M. K. B. (2001). Integration of case studies on global change by means of qualitative differential equations. *Integrated Assessment, 2*(3), 123–138.

Popper, K.(1934). *Die Logik der Forschung*. Berlin: Akademie-Verlag, Mai 2004.

Ragin, C. (1994). Introduction to qualitative comparative analysis. In T. Janoski & A. Hicks (Eds.), *The comparative political economy of the welfare state*. Cambridge: Cambridge University Press.

Raskin, P., Banuri, T., Gallopín, G., Gutman, P., Hammond, A., Kates, R., & Swart, R. (2002). *The great transition: The promise and lure of the times ahead* (A Report of the Global Scenario Group). Boston: Stockholm Environmental Institute.

Rescher, N. (1998). *Predicting the future*. New York: State University of New York Press.

Rotmans, J., & de Vries, H. J. M. (1997). *Perspectives on global change: The TARGETS approach*. Cambridge, UK: Cambridge University Press.

Van Asselt, M. B. A., & Rotmans, J. (2002). Uncertainty in integrated assessment modelling: from positivism to pluralism. *Climatic Change, 54*(1–2), 75–105.

Chapter 5
New Emerging Issues and Wild Cards as Future Shakers and Shapers

Victor van Rij

5.1 Introduction: The Unpredictable Aspects of the Future

Foresight and other forward-looking activities like scenarios, future modelling and planning all play an important role in anticipating future developments. Where appropriate, they are used not only to examine future options but also to shape developments more to our common will. On many occasions, these forward-looking activities fail, however, to anticipate what is referred to as 'high-impact' new emerging issues and wild cards because of their high unpredictability and uncertainty. Therefore, many countries have organized horizon-scanning[1] activities that focus on the identification of high-impact issues and wild cards and their accompanying signals, in order to support more resilient policymaking. In 2007, an attempt was made to align and compare the results of the national horizon scans from three countries (Van Rij 2010); in these scans, the concepts of (new) emerging issues, wild cards and weak signals were used frequently.

In several blue-sky projects that were set out by the EC, the role of these concepts was investigated in more detail. This chapter reports on the findings of the author who was involved in many of these projects and focuses on the role of 'imaginary' wild cards as instruments to shake and shape the future.

[1] 'Horizon scanning' is the *systematic* examination of potential (future) problems, threats, opportunities and likely future developments, including those at the margins of current thinking and planning. Horizon scanning may explore novel and unexpected issues, as well as persistent problems, trends and weak signals.

V. van Rij (✉)
University of Amsterdam & Advisory Council for Science and Technology (AWT-The Hague), Basilicumhof 5, 1115, Duivendrecht, DK, Netherlands
e-mail: vvanrij@live.nl

M. Giaoutzi and B. Sapio (eds.), *Recent Developments in Foresight Methodologies*, Complex Networks and Dynamic Systems 1, DOI 10.1007/978-1-4614-5215-7_5,
© Springer Science+Business Media New York 2013

5.2 (Potential) New Emerging Issues and Wild Cards

The concept of 'emerging issues' is used frequently in horizon scanning and other future-oriented activities and is not well defined in a conceptual way. Analysis of the issues that were scanned in the Sigma scan of the United Kingdom and the Netherlands and OECD/DASTI scan shows that many issue descriptions within these scans are actually fact-based future storylines that envisage either 'promising' developments and events or 'threatening' developments and events that need either 'support and fostering' or 'countervailing and adaptive measures'. Issues described this way are in fact 'potentially' emerging and come close to the concept of 'future narratives', as described and used by Van der Steen in his political discourse analysis.

Van der Steen (2008) sees 'future narratives' as stories 'about what the future, or possible futures, may or will (depending on the narrative) look like' and that connect these possible futures to current issues for political debate (the discourse). They form the counterpart of 'present-day' narratives that stress present-day events, which, according to Van der Steen, 'can already be seen in the world outside, and which are linked by narrators to certain policy solutions'. Different future narratives (potential new emerging issues) have to compete amongst each other and with present-day narratives for the attention of decision makers. A horizon scan should therefore deliver issues that can compete fully with these present-day narratives.

A special kind of 'potential' new emerging issues is the 'wild cards', which not only have a 'high impact' but also a 'very low probability' and a 'very sudden appearance' if they occur.

Examples of new emerging issues that have recently landed in the policy discourse are the development of the Internet, climate change and ageing.

Examples of wild cards are the 'Tsunami-induced Fukushima nuclear plant disaster of 2011', the '9/11 terrorist attack of 2001', 'the fall of the Berlin Wall in 1989' and, to a lesser extent, the 'Iceland volcano eruption in 2010' (see Fig. 5.1). Wild cards, as well as new emerging issues, usually create new challenges for the future that direct the agenda setting for 'Research, Development and Innovation' as well as for socio-economic, environmental, security and safety policies.

5.3 What Are Wild Cards Exactly?

Wild cards have gradually become familiar for many experts who deal with forward-looking activities, such as foresight, future planning and research, ever since we were confronted with challenges, such as Tsjernobyl and Fukushima, the fall of the Berlin Wall, the 9/11 attack, the Arab Spring, and the 'subprime loan', as well as the 'Greek debt' crisis, but also relatively smaller ones, such as the 'New Orleans flooding', the 'Asian tsunami' and, more recently, the 'Iceland volcano'. They tend to interfere with the results of many forward-looking activities that assume gradual development, trends (see text box. 5.1) like energy and/or economic forecasts, and may even interfere with demographic forecasts.

Wild cards are events with a surprising character, a low probability and a very high impact, but we all know they do happen quite regularly.

A wild card can be characterized as an event or series (cascade) of events (with seemingly low probability and high uncertainty):

- That changes the settings of our world completely (high-impact shocks/ disruptions)
- Which we do not foresee or we do not want to foresee, and therefore is seen as surprise

Regarding the surprising aspect, it should be clear that many wild cards may be surprising for many but not for all. In the 9/11 case, nobody knew what was going to happen except for those who planned and executed the assault, and perhaps some of the 'security in-crowd', who for some reason could not or did not want to inform

Box 5.1 Definition of Trends and Hypes

Emerging issues and wild cards have to be distinguished from trends and hypes, which are not issues but may create issues and wild cards if they interact suddenly and change.

Trends

A 'trend' can be defined as a general direction of development over a period of time with a relatively predictable character. Trends usually can be obtained through the extrapolation of historical data to the future by using statistical/ mathematical models. Examples are demographic developments, the global growth of knowledge and global economic development. Trends can have a long-term and strong impact (megatrends), but also a shorter-term character with lower impact. Trends may be a part or the cause of an issue in all sorts of ways. Trends may also refer to developments that are less quantitative but nevertheless visible like design or fashion trends. However, things such as 'ecological farming or living', which may have looked like 'mere fashion trends' in the beginning, can develop into longer-lasting trends.

Hypes

The term 'hype' generally refers to overenthusiasm, excessive publicity around a certain topic, excessive advertising or making exaggerated claims. The US analyst firm Gartner coined the term 'hype cycle' to describe the 'peak of expectations' in regard to (information) technologies. In this phase, 'a frenzy of publicity typically generates over-enthusiasm and unrealistic expectations. There may be some successful applications of a technology, but there are typically more failures'. After this phase of exaggerated expectation comes the 'trough of disillusionment' where people realize that their expectation cannot be met.

Iceland's volcanic cloud 2010 Attack on the WTC 2001

Fig. 5.1 Two examples of wild cards

the people who should have defended against these attacks. Furthermore, we have to point out that in the description of a wild card, we speak about our world, which signifies that the impact of a wild card may be expressed in a subjective character. So not always everybody is affected by the wild card and certainly not in the same way. We may have wild cards with exclusively winners or losers, but many times with both winners and losers.

Last but not least, wild cards can have a very strong impact in many areas worldwide (like the 9/11 event and the financial crisis), and they can have smaller more localized impacts (the Iceland volcano). Many times, when the wild cards happen, the exact impact on the different domains is not clear, because the event may also have a very unclear development when it happens. In case of the Iceland volcano, it was not clear how long it would throw the ash in the sky that disrupted the air traffic. Analysts foresaw more grave consequences if the eruption endured and pointed out that volcanic ash in former centuries even caused famine throughout Russia by diminishing agricultural yields. But fortunately for the world, the worst-case scenario did not happen. A similar story could be told for the BP deep-sea oil leak in the Gulf of Mexico. Here also it took an undetermined period before the leak was under control, while stories arose of complete disaster if the flow from the hole became uncontrollable. The economic and ecological impact of such an event would have been many times larger. The same phenomena can be observed around the recent earthquake and tsunami event in Japan that caused the breakdown of the nuclear reactors at Fukushima Daiichi. The duration, as well as the total impact, was very unclear, while the event was taking place. The impact now stretches from enormous economic damage to health and land-use issues, as well as growing worldwide reluctance to follow the nuclear fission path.

Karlheinz Steinmüller (2004) describes wild cards as the 'earthquakes of the mental landscape', while Nassim Nicholas Taleb (2007) uses the term 'black swans', which he describes as a surprise (to the observer) with a major impact that, after the fact has happened, is usually rationalized by hindsight, as if it had been expected. Specific for black swans is that the event is something that is 'out there' and will happen anyhow and cannot be influenced by will. America was to be discovered as soon as we became aware that the world was a globe (while it was of course known to the people who lived there). The real black swans were found as soon as Europeans set foot on land in Australia. Taleb (2007) stresses the unpredictability of these kinds of stochastic improbable events.

Steinmüller (2007) stresses that a wild card undermines current trends, creates new futures and influences our thinking about the future – and the past – through which it gives rise to new concepts and new perceptions. He also points to the fact that (potential) wild cards guide us to potential early warnings, the weak – or faint – signals that prelude a coming event, and the fact that the imagination of wild cards can be used to stimulate business to examine new less 'conventional' strategic options (out of the box thinking) and to establish more shock proof, resilient strategies (by deeper investigation of what could go wrong).

In this respect, it is important to analyse the phenomenon of the wild card in more detail and to make an initial distinction between wild cards 'that happened' and potential wild cards that may be 'foreseen or imagined'. As well as this, it is important to distinguish between imaginary or even imaginative wild cards that are published and communicated by actors, who want to influence the future discourse, and wild cards that 'are created in a process to seek for resilience or new creative options for decision makers in a given situation'.

5.4 Wild Cards That Happened and That Could Happen

Wild cards, on the one hand, refer to events that have happened and, on the other hand, to phenomena that may happen in the future, and which we can imagine, which we refer to as 'imaginary wild cards'. Imaginary wild cards may be introduced into anticipatory decision-making activity in order to create resilience (Petersen and Steinmüller 2011), but they can be and are also used as instruments to influence the future discourse by many actors in the world. In the SESTI project, articles that describe new imaginary wild cards that clearly were meant to influence decision makers were seen as a primary signal. After the analysis of these imaginary wild cards, indicators were identified which could be monitored to follow the development of the wild card from its imagination to its realization or its disappearance. In the Far Horizon project, as well as in the i-know project, many people were asked to think of potential wild cards which after collection were used to test resilience and to obtain inspiration.

Wild cards that have happened can have natural causes (earthquakes, volcanoes, floods), but they may also be caused by human action or failure (wars and revolutions, financial crisis, terrorist attacks like the 9/11 event, the fall of the Berlin Wall, nuclear or other major impact industrial accidents). The analysis of these wild cards by studying their past narratives is undertaken in Sects. 5.4.1, 5.4.2, and 5.4.3 below.

5.4.1 The Tension Growth Narrative

Many wild cards actually follow (usually with hindsight) a narrative where one or more almost invisible sneaking trends or sequences of less important events lead to a critical tension (or tipping point) which causes a sudden outburst or change of the system. Examples of this are:

- Accumulation of nondegradable toxic substances leading to ecological catastrophes (e.g. the near wild card of the worldwide use of DDT)
- The growing tension between parts of the population and between nations (WW1 – Weimar republic, WW2 – Uganda)
- The subprime loans plus growing perverse incentives in the banking world – causing the piling up and dispersion of 'worthless' subprime debts (which actually at the very beginning could be considered as very high welfare benefits for poor Americans to live the American dream for free), leading to the worldwide financial crisis of 2007
- The growing steel armoury of empires and the perceived scarcity of resources (coal and steel – WW1)
- The growing number of frustrated people leading to revolutions (Arab Spring, fall of the Berlin Wall) or to terrorist power including their political support by frustrated nations – 9/11
- A sequence of discoveries leading to the invention of the electric light

To be able to anticipate these kinds of events, it is important to look at the different types of tensions that may build up in the world and within our societies that may eventually cause a wild card.

5.4.2 The Incident and Accident Narrative

Although many cases seem to follow the rule of growing tensions, there are also many cases that can be described as sudden unexpected events with great impact that we know have occasionally happened, and will still happen somewhere, sometime:

- Volcanoes, tsunamis, cyclones, earthquakes, floods – sometimes wiping out civilizations (Minoan civilization)
- Technical failure, human failure or wrong design in a nuclear or industrial plant (Chernobyl, Fukushima, Bhopal) or deep-sea drilling causing an ecological disaster (Gulf of Mexico)
- Asteroid impact (wiping out complete parts of the Earth's ecosystem)

These kinds of wild cards can be somehow anticipated because we know they can happen (usually from historical evidence) with a very low or unknown probability but with a potential very high impact, for many of them we even know where they can or may happen (like dangerous industrial facilities, volcanoes, earthquakes). Only the exact when, where, and impact is usually completely unknown. It may also be disputed that some of these wild cards actually also follow a path of increasing tension (the risk of deep-sea drilling is taken because of the growing tension between energy supply and demand), while earthquakes and volcanoes are likely to be caused by a build-up of tension within the Earth's crust. Others, however, are more clearly following the path of pure coincidence (like an asteroid impact or technical failure as such), which connects them more to the next category.

5.4.3 The Real Black Swans

A final group of wild cards consists of features that were really unknown before they happened, but were nevertheless destined to happen in the course of the history. These are closest to what Taleb (2007) means by 'black swans'. Examples from the past are as follows:

- The discovery of America (although this may also be explained by growing populations in Europe)
- The first plague epidemics (although this may also be explained by tensions in urban population growth and hygiene plus increase in the mobility of vector populations, e.g. rats on ships)

Also these kinds of wild cards can be anticipated to a certain degree because we can imagine at least some of them, although they may nevertheless happen, we can nevertheless think of what our strategy would be in case of:

- Heavy earthquakes in Paris or London following the formation of a new earthquake fault line in Europe
- Alien life forms show up

- A life elixir is discovered
- The sun heats up (sudden change of the solar cycle), shift of the Earth's axis

An interesting aspect is that the 'black swan' (Taleb 2007) *refers to (unknown) events that inevitably (may) come to us from the physical world but are unknown to us. So, the ones that come upon us are in some sense inevitable (written in the stars).* As said before, black swans come close to the accident narratives, but may be distinguished from them by the way we can still influence their occurrence or their impact. The ability to do this is growing with our growing understanding of the universe.

5.5 Nature and Human Causes

If we think about anticipation or about the phenomena of wild cards, it is very important to see that they can be caused by nature (our environment) or by humans (sometimes through the disturbance of nature, as is assumed in the climate change issue and connected wild cards, such as those too hot to handle (Lynas 2007)).

This distinction is important because wild cards caused by nature usually call for adaptation or defence measures, while human-caused wild cards may be prevented if foreseen. Moreover, the distinction becomes even more important if we realize that human-caused wild cards are the product of human inventions, our (collective) minds, behaviour and action. For the search of potential future wild cards, this aspect is extremely important because it leads to the conclusion that especially the 'human-caused' wild cards that people imagine and/or foresee become more or less part of the foresight and governing process as soon as they are formulated and communicated. Their successful communication may already influence the policy discourse, while realizing them will certainly influence our line of thought and the course of history. A terrorist action such as 9/11 should be seen in this light: the terrorists, but also the people who let them do it, were acting consciously with a future narrative in mind that was a surprise for the rest of the world.

This implies that potential wild cards, communicated via the media, are in fact instruments that are deliberately or unconsciously brought out to influence the discourse on the future or to even create futures but also that foresight processes that stimulate the imagination of new wild cards are having the same function.

Although many examples of imaginative wild cards seem to have a dominantly negative impact, this does not always have to be the case, since we can also see many positive wild cards: climate change in an unexpected way can bring us, for instance, better conditions: volcanoes can provide fertile soil. The wild card of the Internet usually is seen as a blessing.

5.6 Imaginative and Imaginary Wild Cards as 'Instruments' to Shape and Shake the Future

Thinking about wild cards expands our thinking on what may go wrong but also on what may be a very useful development to achieve (policy) objectives. This makes it possible to take precautionary measures (to prevent, mitigate or adapt (to) the impact) and gives us hints on 'new ways' to reach objectives. Thinking and inventing wild cards can, therefore, be used to develop more resilient policies as well as to create more options to achieve objectives.

When we look through a great number of future wild cards that have been identified and produced in horizon scans and in blue-sky projects, such as SESTI, Far Horizon and i-know, then we can see that almost all of these cards are actually described as future narratives telling a story that leads to something people wish for (wish cards) or something they fear (fear cards), often indicating developments or events that may lead to the desired or the feared situation. Usually these narratives are complemented by some or a great deal of factual material that shows that the wild card could turn up and is plausible. Therefore, depending on their (un)desirability and plausibility, they can stimulate self-fulfilling or self-denying actions amongst the audience. Since each potential wild card is created by humans, we have to be aware that the description is in fact a message (or tool) of communication to influence the discourse on the future coming from an individual or group of individuals who may have a specific interest or objectives with their creation. This is especially the case with the deliberately published ones. Published wild cards can be seen as future-oriented lobby or propaganda mechanisms (to obtain support) or marketing cards (to sell an idea, product, process or lifestyle) which can only be assessed properly if we take the context of the creator and his/her specific interests or ideals into account. To assess these imaginary and sometimes imaginative wild cards, we should not only look at the factual validity but also at the (potential) emotional and interest-driven response that the communication may evoke (or evokes) amongst the target group of the communication. Another important aspect is the interests and motives of the initiator and/or publisher of the imaginative wild card.

To explain this idea, we give some examples.

Many years before World War II, Hitler started to publicly communicate a potential wild card that he said could be foreseen. The storyline of this wild card was that there was a growing 'International Jewish Conspiracy', which, according to Hitler, had the objective to undermine Europe, and which according to his forward look, would cause a world war. This would, in turn, be answered by the extermination of Jews in Europe. The communication of this wild card was not based on facts (although the fierce protests of the Zionist communities in the United States against his anti-Semitism were useful to build the construction of the wild card), but it still had an enormous effect because of its emotional charge which can be described as a mixture of fear and hate.

It fuelled the anti-Semitism that existed in large parts of the European populations and was used to activate this part of the population to create the self-fulfilling prophecy of a worldwide war predicted by Hitler and the extermination of the majority of the Jewish population of Europe during that war. The example shows that without any factual basis, individuals and groups of individuals can use imaginative 'wild-card-like narratives' to persuade people to support their interests and objectives.

In 1935, the wild card prophecies of Hitler were illustrated in the Nazi propaganda.

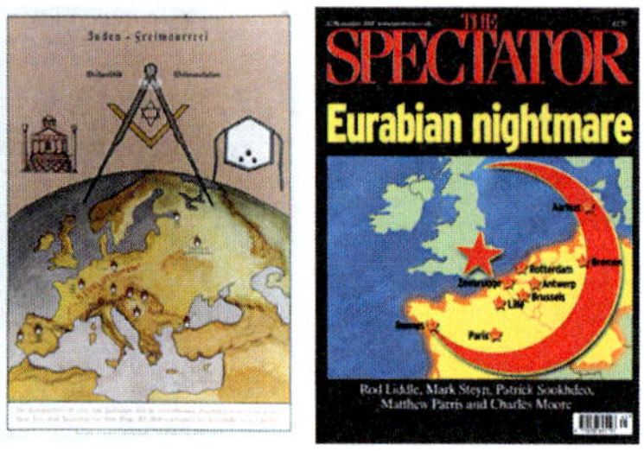

Propaganda card (1935)

The Nazi's presented the prevision of a Jewish/freemasonry conspiracy causing uproar throughout Europe to obtain Jewish domination of Europe

Source: no 64 in the series 'Erblehre ressenkunde', published by the Verlag fur Nationale literatur, Stuttgart, 1935.

Propaganda card (2005)

Right-wing politicians present the prevision of Islamic domination of Europe, initiated by riots and terrorism; an article in *The Spectator* (November, 2005) visualizes the prevision

One could claim this would be possible in the age of public ignorance, but a more recent example is given by politicians in many European countries who regularly communicate the wild card about a Europe that in the coming decades will be conquered by Islam and that European countries should defend themselves against this to avoid inevitable Muslim domination. Although this storyline also has no strong factual basis, in some countries, it still attracts a large number of voters, once again because of its emotional charge which provokes fear and uses the growing unrest on migration issues in Europe.

Here there is also a theoretical possibility that the call of this wild card may have 'self-fulfilling' reverse aspects (like the recent killings of people of Turkish origin by neo-Nazis in Germany and the shooting of Norwegian young Socialists in 2011 by an ultra-right-wing person, who even quotes politicians who are playing this wild card).

A final example is Al Gore's communication of his wild card known as 'the inconvenient truth' that was even communicated by means of a movie that visualized the fears associated with the human-caused global warming. Here we see a mix of facts with emotion provoking contents which probably influenced the discourse on climate change strongly against any remaining climate scepticism.

It is not only politicians who are using the hypothetical wild cards to create or influence the future discourse. Also scientists may sketch wild-card-like narratives on future developments, which may be regarded as lobby mechanisms to obtain funding. In very recent times, we cannot escape from the thought that speculators are deliberately throwing wild cards on the financial crisis for their own profit.

A search on the Internet and YouTube in 2009–2010 delivered many websites with elaborate descriptions of wild cards bearing emotional titles, such as 'economic meltdowns', 'dollar crisis', 'super inflation' and 'Financial Armageddon', many times advising people to buy gold through advertisements on the same website.

Some of these hypothetical wild cards can be directly traced to financial advisors who stir up emotions with a small factual basis to influence buying behaviour, not aware that they may cause serious self-fulfilling prophecies by doing so. It may be clear that, after a certain number of these kinds of messages, a critical point may be reached and an avalanche may be caused that may create the final worldwide 'bank run'.

5.7 Why to Look for Wild Cards

The fact that wild cards are dealing with rare actually happening events coming from a multitude of possible events makes it understandable that some people conclude that it is useless to search for future wild cards since it is like looking for a needle in a haystack. Often this conclusion is followed by the idea that if the wild card occurs nothing can be done about it, so why to bother or, worse, spend (taxpayers) money on it.

This makes it hard to explain why we should look for future wild cards (and if we look for future wild cards, which wild cards should we look for and how). Still it is strange that in our private lives, we are quite regularly looking for the wild cards and their early warnings. A flat tyre is something that happens only rarely, but still we have our tyres checked regularly for signs of wear; health checks and health insurance are also quite common for people who are healthy. The reason for this is that we do not like surprises, and somehow would like to protect or insure ourselves against negative surprises.

It is proven that this precautionary behaviour of mankind can be rewarding, and it should be worthwhile doing something to implement this way of thinking in government and to use the search for wild cards to:

- Make policies more resilient to the occurrence and the effects of wild card events (more adaptive to what could change suddenly)
- Enable the monitoring of early warning signals of wild cards to timely adapt to or mitigate the impact (preventing damage to what we value)
- Support safety measure investments (such as monitoring systems, higher dykes, affluence measures, earthquake-proof building mechanisms, escape routes)
- Counteract undesirable lobby cards and human-caused wild cards that are still in the phase of construction (either imaginative or planned for real, as with terrorism)

However, one problem seems to be that wild cards, by their nature, may be unforeseeable. As Taleb points out, humans have the tendency to rationalize the wild card event or fact by hindsight, as if it had been expected, while this was obviously not the case (otherwise it would not have been a wild card).

Although this may be true, it does not mean we should not watch out for the future wild cards we can imagine and that we consider to be, plausible enough, to investigate further. The fact that major volcanic eruptions occur, many times in a century with

Box 5.2 Can Steps Be Taken to Anticipate Wild Cards?

With hindsight, we could give the following answers:

Katrina or more precisely the New Orleans flooding: foreseeable? Yes, hurricanes are very regular phenomena in the region. Could we have done anything? Yes, better water flooding protection (as in the Netherlands) and no construction of houses, etc., on unsafe areas.

9/11: foreseeable? Yes, al-Qaida had already hit large US targets in Africa; the airplane suicide scenario was known. Could we have done anything? Yes, better intelligence on 'real' dangerous groups, better safety measures in planes rather than useless ones in airports (leaves other targets such as trains, tunnels, and metros unprotected).

Financial crisis: foreseeable? Yes, as well as the next ones (Europe state debts, US student loans), subprime building up, and reselling of loans though the financial system was observed as very risky; could we have done anything? Yes, direct support of subprime lenders by the US government at the start could have prevented the cascade that followed. More shields between risky loan activities and regular banking could have been installed in advance to contain risk.

The Iceland volcano: foreseeable? Yes, volcanic activity is regular in Iceland, and eruptions with fumes that could harm Europe were already known from history. In 1783, the same volcano caused the starvation of thousands of people, together with lung diseases in England because of ash clouds; the effect of volcanic ash clouds on aeroengines was known from near accidents elsewhere in the world, but the present incident could have been even worse, because history also tells us stories of lost yields and famine caused by the short-term climate influence of volcanic ash in the upper layers of our atmosphere. Could we have done anything? Yes, having a fume monitoring system in place, and alternative means of travel ready (buses, extra trains); if the fume clouds persist, we also should think about maintaining additional food storage and measures for people with lung diseases.

Oil spill in the Gulf of Mexico: foreseeable? Yes, oil drilling has always had the risk of accidents (many accidents on land which were difficult to fix), and it is in fact logical that the risks are much higher with deep-sea drilling and the ecological risks are much higher. Could we have done anything? Yes, we could abandon the idea of searching for oil in places where exploitation is too risky (especially if we do not have clear cut scenario's on how to close a deep-sea leak) and focus more on the safe alternatives.

grave consequences, means that they are serious wild cards to take into account, which can even be searched for systematically. The fact that tsunamis and earthquakes happen quite regularly over the centuries in certain regions is known as well. Also, wars and genocides happen more frequently than we would like to admit to ourselves.

With this kind of events, the questions are as follows: Could we not have foreseen the events? (and more important) Could we have done anything about it? With hindsight (see textbox 5.1) we can see that in many cases, damage could have been diminished, and life could have been saved by taking precautionary measures and

Box 5.3 Foresight Solutions for Wild Card Effects

With foresight the answers may be:

A worldwide food crisis, some of the largest agriculture production areas are contaminated or spoiled for one or more years (by large volcanic clouds, nuclear accidents, drought and so on). Through a worldwide agreement, the Earth's ecosystem could be re-greened to 'overproduce' food with the aim to create a substantial reserve if no calamities happen. The stored food that is not used over a period in time could be used as biomass for energy production. In this way a continuous buffer is available to cope with food production catastrophes.

Earthquake 8.0 Istanbul: Earthquakes in this particular region are regular. Can we do anything about it? Rebuilding earthquake-proof, prepare evacuation scenarios (taking into account disrupted infrastructures).

Eruption of Vesuvius (and other volcanoes in densely populated areas): Vesuvius is irregularly active and highly dangerous. Can we do anything about it? Evacuation measures (prepare additional escape infrastructures) and shrinking the built-up area in the longer term (people should not live in this area).

EC financial crisis: The growth of countries' national debt was known. Can/could we do anything about it? As in the subprime crisis, direct support for the debt countries stops the initiation of cascade effects; anti-speculation measures to protect the euro should be in place (monitoring money flows, who buys and sells what); debts should be recovered by careful economic-financial schedules that are controlled by the supporting countries (EC), fixation on budget cuts should be carefully weighted against increase of the usually low tax rates and tax discipline in countries such as Greece. Weakening the state could even worsen the control and regulation needed to realize the state income and paying of debts.

US student loan and credit card crisis: Debt growth is visible; selling and reselling of credit card debt bundles should be forbidden; absolute shields should be constructed between this kind of banking business and normal banking. Present credit card debts should be resolved (diminished) in a 20-year plan and limited in such a way that payment is absolutely guaranteed by secure assets of the credit card holders. The whole concept of student loans creates a long-term risk because of the not proven future assumption of economic profit for the individuals involved.

by the awareness raising of authorities and people concerning the things to do if the wild card plays out (especially with wild cards caused by natural events).

The problems with this approach lie in the fact that people prefer not to look at possible disturbing events in the future and that large investments seem only to be feasible if things really went wrong, like the billion dollar investments made in the Netherlands on the Delta flood prevention programme that was initiated after the 1953 flooding in the Netherlands, which entailed not only the loss of more than 3,000 lives but also enormous economic damage. Therefore, it is important to realize that through the systematic examination of the plausible wild cards not only lives but also billions of euros could be saved.

Another aspect to consider concerning imaginative and human-caused wild cards is that they are in fact constructs of people minds, with good or bad intentions, stimulating good or bad emotions, and influencing decisions.

The monitoring and analysing of these cards in advance can therefore be done to make them more transparent, and to distinguish their emotional, propaganda and factual basis, and possibly to countervail negative cards and foster positive ones.

5.8 How and Where to Look for Wild Cards

As said earlier, future wild cards can be obtained in different ways. They can be either developed or obtained by deliberately asking for them, and they can be searched for in the ongoing communication that takes place through different media. These approaches will partly lead to a different result. Where the 'searched' wild cards actually represent an image of the attempts of actors to push their wild card narratives on the agenda to influence the future, the wild cards we ask for may be either a new creation on the spot even with suggested solutions (an example of this is given in text box 5.3) or a selection of what the respondents memorize from the wild cards they have somehow heard about from communication by the media (which will give a partial reflection of a search).

The development of wild cards is usually done in special workshops where stakeholders or experts are asked to imagine and create wild cards that disrupt either their present activities or the entire world. The purpose of this activity is to create more resilient strategies that may combat the effects of imagined wild cards. Usually this is done in brainstorming sessions or through instruments on the web (as is done in the i-know and Far Horizon projects), but it may also be the outcome of lobbyist meetings or interest groups to select favourable wild cards in order to impose futures on others by using wild cards as instruments, as we showed before. A terrorist group such as al-Qaida probably uses this brainstorm technique and came up with the 9/11 event. Counterterrorism foresight should, however, do the same to imagine the wild cards that terrorists may cause.

The search for wild cards is usually done by professional scanners who scan the media for wild cards that are published with the intention to influence the future discourse. Sources may be:

- The Internet (blogs, experts, special future-oriented websites) and other (public) media
- Literature (especially science but also future fiction)
- Conferences of experts and/or futurists

Box 5.4 Origins of Wild Cards

Natural Causes Environment

- Earth, land (volcanoes, landslides, gas eruption, mud volcanoes, earthquakes)
- Air (climate change, dust, tornados, storms, etc.)
- Water (drought, floods, pollution, natural causes)
- Biosphere (epidemics, plagues, zoonoses, starvation, infertility, etc.)
- Outer space (asteroids, extraterrestrial life, no or sudden solar activity)

Human Causes

- Society (value shifts, movements, hypes, social trends, demography)
- Science and technology (breakthroughs, new technologies, etc.)
- Economy (crisis, prosperous developments, etc.)
- Politics/public services (everything that can go wrong and right)

Rich sources are also established by specialist sites, such as i-know, the Sigma scan, the Netherlands Horizon scan 2007, the OECD-DASTI scan and many others from military and security agencies.

As mentioned earlier, wild cards find their origin either in natural causes (the physical and biological environment) and are then communicated by scientists or in human causes which can be split up into the other STEEP domains: society (including values), science and technology, economics and, finally, politics and government.

The search for wild cards usually follows these domains with some further refinement.

5.9 Assessment

While a search for wild cards and new emerging issues may deliver an overwhelming amount of possibilities, it is clear that certainly in the case of a search, assessment should take place to deliver the most plausible and impact-rich wild cards and new emerging issues, which should be forwarded to the policymakers. This makes it necessary to think about the way to assess them on major aspects as well as to think about indicators (signals) that may be used to follow their emergence or disappearance. As discussed before, we have to be aware that it is especially those wild cards and issues that deal with human causes, which should be given special attention because they are the product of human ideas, decisions and communication in combination with (perceived) facts which makes it likely that many imaginary cards and issue descriptions may be actually seen not only as tools for initiating the thinking, debating and shaping of the future but also as enablers of self-fulfilling and

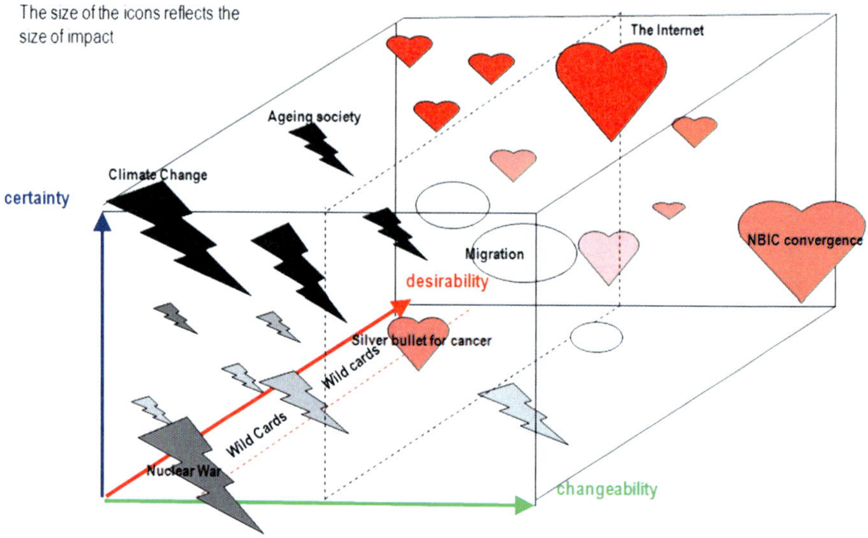

Fig. 5.2 Assessment of emerging issues and imaginary wild cards (Explanation: shaded objects have grown to certainty (*upper* plane of the cube); thunder signifies undesirable issues, hearts desirable (plane at the back) and ovals neutral. Issues changeable by policies can be found on the *right* plane, and wild cards are almost not changeable in the *left* plane)

self-denying prophecies that are initiated either by terrorists and fanatics (e.g. Nazis) or by idealistic lobby groups (e.g. the Club of Rome in the 1970s), merchandisers (selling sun-protection cream), speculators (stimulating the prolongation of the financial crisis) or policymakers.

A first important aspect of these wild cards and emerging issues is, therefore, who initiated them and for what reasons. Secondly, it should be clear that their strength and potential impact lie not necessarily in the facts they are presenting but may also be found in the evocation of interests and emotions (such as fear and hope or inspiration). The strength may depend on the shape and contents of the particular wild card or issue, the plausibility of the narrative, its foreseen impact (especially its emotional charge) and the personality of the conveyor of the communication but also on the 'mood of the audience' and the historical 'context' in which the wild card or issue was launched.

Within the SESTI project, this led to the following list of questions that were used to assess the strength of emerging issues and imaginary wild cards that were identified:

- Who initiates and communicates or tries to block (for what reasons)?
- What is the size of the impact and on what values (with respect to lives, money, emotions, ethical standards, etc.)?
- What is the changeability of the impact? (What can we do about it?)

- What is the plausibility/certainty/probability and on what is it based (monitoring/initiating weak signals – can we see if it is really going to happen and how?)?
- What is the desirability, and for whom (fear and wish cards – lobby cards again)? Whose and what values are at stake?
- What is the time frame in which it is expected to occur (evolution of tensions)?
- What are the interactions of the wild card/issue with other future issues (direct–indirect effects, enhancing and buffering effects, double-sided coins, etc.)?
- What may be the early indicators (early warning signals)?

Assessment of emerging issues and imaginary wild cards with respect to these questions can be visualized in a three-dimensional graph (see Fig. 5.2 below), while their development in time could be visualized by sequencing the graph over time in a video (Van Rij 2010). In this video, imaginary wild cards may stay for a long time or, even forever, near the (un)desirability axis (left under in the uncertainty plane). But they may suddenly jump into certainty if they are becoming reality.

5.10 Strength of a Wild Card/Issue: The Final Test Discourse/Decision Making

Imaginative wild cards and new emerging issues usually have to struggle for attention in the discourse of decision makers against an overwhelming number of present-day narratives (Van der Steen 2008). The intriguing question is why some imaginative wild cards and new emerging issues have succeeded in entering the discourse, while other ones with grave consequences were kept out till the moment they materialized, with grave consequences (e.g. the subprime loan crisis of 2008 that was identified in earlier scans in the United Kingdom and in the Netherlands).

It is clear that the strength of imaginative wild cards is not solely bound to their 'factual' or 'evidence' basis but that other factors may be also very decisive for their effectiveness. Although some of these factors come close to what we already know about success factors for foresight projects in general, we conclude on the basis of the blue-sky projects (like SESTI) that many other factors may be as important. For foresight projects, many authors stress the importance of the help of what is called a foresight 'champion' (Loveridge 2009; Georghiou et al. 2008), mainly to bring the message to the policymakers, as well to ensure the involvement of decision makers in the process. Also, there seems to be a golden rule that only foresight that is asked for may be effective, which was stressed by the organizers of the national horizon scans of the United Kingdom and Denmark.

These factors, although useful, are not enough to bring in all relevant 'emerging issues' and 'imaginary wild cards' into the policy discourse. This is especially because not all 'plausible, impact-rich narratives' that may be on the brink of realization are communicated by champions, while some of them also may run into barriers that are connected to vested interests and embedded emotional aspects of the decision makers, which may lead to denial (Markley 2011) and even hostile reactions. On top of this, it is observed that the emotional aspects may lead to completely contradictory reactions not only between different groups but also in the same groups after some time (from belief to unbelief and from fear to euphoria and vice versa).

Nevertheless, we can provide a list of factors that, according to the blue-sky projects, have to be kept in mind when assessing the strength of a wild card:

- *Plausible storyline* (thus with logical connections of the story to the impact and to perceived facts, observations and evidence), including some lines for action (to diminish negative and increase positive impacts). In the case of emerging issues, the perceived plausibility may be supported by direct evidence (signals) that the issue is really emerging which leads to more or less *certainty*.
- *High impact* on important interests and values (of the decision makers and their clients).
- Overall *desirability* (is it something which needs to be avoided or something we desperately want?).
- *Changeability* (can something be done to prevent, to avoid or to take opportunities?).

Although these criteria seem to be important for the assessment of the strength of imaginary wild cards and emerging issues, they are not enough to guarantee that the issues will be taken up in the policy discourse or moreover will be able to change the mindset of the decision makers towards the issue. The reason for this is that we are dealing with a communicative process, where the cognition, psychology and interests of humans play a major role in the successful or persuasive transmission of messages. This applies to the senders as well as to the intermediaries and those who need to receive and react to the message. Because the senders want to persuade the receivers to take their message into account, and many times to change their attitude to the issue or wild cards that are communicated in the message, their success will depend on the rules for persuasive communication, as described by Perloff (2010) in his book *The Dynamics of Persuasion*.

For the senders and/or intermediaries, this means they need not only authority but also credibility and preferably a reasonable dose of social attractiveness (charisma), which seems to confirm the usefulness of foresight champions, while on the receiving side, we may have to deal with barriers and facilitating factors that should be taken into account by the senders, for example:

- The cognition, interests and values of the receiver(s), some connection to the present experience
- Fear evoked by the narrative which may lead to action preparedness but also inertia, denial or hostility
- Hope which may also lead to action preparedness but also to unbelief
- The available time of the receivers

Connected to this, it is very important to think about the shape of the message (that contains the 'wild card' or 'emerging-issue' future narrative). Here attention is needed for language (taking into account cognitive levels, cultural and emotional

Box 5.5 The Discourse on the Future

Discourse refers to the (continuous) communicative debate that takes place in the political arena or in a decision-making process. This debate is unique for human beings with their ability for speech and writing (and nowadays audio-visual presentations and computer simulations). The outcome of a (political) discourse is partly dependent on the contents and shapes of the communicative expressions of the participants, their cognition and values in a wide sense, as well as the power relationships between the participants. Issue descriptions, together with the early warning signals around them, are communicative expressions that finally will have to find their place in this (policy) discourse to become relevant.

Early warning signals, (potential) wild cards, trends, hypes, and (potential) emerging issues all have the potential to influence the discourse on the future in a different way. 'Hypes', for instance, have the tendency to influence the discourse strongly but for a short time, while 'strong trends', 'impact-rich issues' and especially 'wild cards that have happened' are dominating the discourse with more persistence. But 'imaginative wild cards' and '(potential) emerging issues' including their 'early warning signals' have to fight for attention in the discourse, except for situations in which they fit or are strongly connected into the present-day discourse. It is clear that this fight can only be won through a future narrative that is strong enough to draw the attentions of the participants and agenda setters of the discourse. This fight is not only between future narratives but also between future narratives and their present-day competing opponents, which may include huge impact wild cards like the Japan earthquake. The occurrence of these wild cards may, however, mitigate and even enhance the strength of the future narratives.

aspects), the use of symbols, the size and timing of the message, visual and/or auditory support.

Although we have some clues concerning what aspects are influencing the success of 'emerging issues' and 'wild cards' to reach the policy agenda effectively, there are still a great many puzzles that need to be sorted out.

The future storylines of wild cards and emerging issues can very easily go into reverse by small changes in either the factual or the emotional aspects. The world would have changed in a completely different way if there had been evidence that the US administration had known about 9/11 in advance and deliberately let it go ahead. If such evidence would turn up later, it would be a new wild card.

The positive attitude towards widespread broadband telecommunication using high-frequency transmitters may turn around if the majority of the people start to think that the frequencies may cause serious health problems, with but also without solid evidence.

The narratives may therefore contain tipping-point elements but also kind of flip-flop elements, which should be taken into account when monitored.

Further investigation with the help of disciplines, such as policy sciences, discourse analysis, historical science and persuasive communication science (psychology and sociology), could be helpful for the analysis and monitoring of emerging issues, as well as of imaginary wild cards that are communicated in the media.

5.11 Early Warning Signals

For decision makers, it is important to know whether new emerging issues and (imaginative) wild cards become real. Horizon scanning focuses therefore not only on what may be imaginative plausible wild cards and new emerging issues but also on the signals that indicate their realization.

Many times these signals are difficult to spot. Therefore, they are usually referred to as 'faint' or 'weak' signals.

In the SESTI project, 'descriptions', in any of the media, of new plausible future narratives which claimed high impact by foreseen 'sudden events', as well as by 'foreseen or ongoing gradual developments', were considered as primary signals for wild cards or for emerging issues, while articles or events that confirmed or denied the storyline were considered as secondary signals.

For physical-caused issues and wild cards, primary signals are usually societal and communicated by scientists, who may foresee growing risks (based on scientific findings), while secondary signals are usually physical and may be measured using scientific methods. For societal-caused issues and wild cards, both the primary and the secondary signals are societal.

5.12 Physical Signals

Many disruptive events in the physical world announce themselves with subtle preceding events which may serve as weak or faint signals that may be used as early warning signals. An engine may produce a strange sound before it collapses. Diseases announce themselves with symptoms. The recognition and interpretation of these physical signals find their basis in the '(intuitive) experience with' or the 'historical analysis of' disruptive events. The historical analysis tries to identify which physical abnormalities preceded the past event and applies correlative knowledge and explanation to diagnose future events of the same type. For many of the nature-caused wild cards, scientists in many disciplines are searching for physical signals that may be used as indicators for volcanic eruptions, storms, tsunamis and earthquakes.

A recent example of the search for early warning signals for earthquakes is given by Freund Friedemann, who uses the concept of electrostatic charging of fault lines in the Earth's crust to obtain a set of signals that are indicators for the growing risk of earthquakes (Friedemann 2011). Since physical early warning signals are usually recognized by 'experience with' or 'analysis of' past events of a certain type,[2] it is clear that most physical signals are related to 'known' events or events that are 'analogous' to known events.

5.13 Societal Signals

Societal signals are usually quite different from the physical signals because they are produced by humans. In many cases, they are meant to be communicated, translated and interpreted even with the purpose of influencing the future discourse. These signals may contain a rational factual basis which may contain descriptions of physical signals, coming from scientific communities, but, at the same time, they may also be signifying strong interests or emotions. The signal given by the former vice-president of the United States Al Gore with regard to what he called 'global warming' is a clear example of this. In moments of tension or crisis, even purely emotional signals without factual basis or interests may occur, which may have a high impact for people concerned.[3] The interests and emotional charges of signals are important aspects of their prospective significance and sense-making. The societal signals may be seen as intended or not intended precursors of self-denying or self-fulfilling prophecies of (collective) behaviour including wild cards. This leads to the conclusion that a first deliberately published (or announced) description of a plausible wild card or new high-impact emerging issue is in fact a primary signal for its realization, while subsequent communications, events and actions that refer to the contents of this first signal can be seen as secondary signals that either deny or confirm its further development. The more unintended or even unconscious the social signals are, the more they resemble physical signals.

[2] Interpreted within the context of a vast body of natural scientific knowledge

[3] During the silent memorial service for the victims of World War II on the Dam Square in Amsterdam in 2010, crowds of people were gathered for the traditional 2 min of total silence. During this period, some person started to shout some obscenities, which drew the attention of some security officers who wanted to remove the person. Because of the local commotion, a fence collapsed with a loud bang, and a lady in the audience panicked and screamed loudly out of sheer terror. The majority of the people present at the memorial service reacted with a stampede, causing many casualties. The combined signals of the loud bang of the fence and the scream of fear provide an example of a societal signal preluding, and even causing, an event with potential high impact, which in itself was completely emotional.

5.14 Concluding Remarks

The concept of wild cards is very important to understand why forecasts based on linear and more complex modelling have a very limited predictive value. Wild cards are normal phenomena that may be rare within each domain, but which happen quite frequently if we take into account the full spectrum of (STEEP) domains. They change the rules of the game and have a disruptive effect on the world but also on the outcomes of models that are based on more or less business-as-usual scenarios and scenarios that assume gradual changes in the main drivers within the domains. Horizon scanning provides a method to identify plausible future wild cards that may happen. The identification and description of these plausible wild cards may lead to the identification of early warning signals that may be monitored to anticipate an event but may also lead to resilient policies that focus on prevention, mitigation and/or adaptation. With wild cards that come from the physical world like flooding, epidemics, earthquakes and so on, these signals can be identified by systemizing experience with wild cards that happened in the past (searching precursors) and by natural science research on the phenomena that should lead to a greater understanding on the primary causes of these events.

For wild cards that are caused by humans, the situation is more complex since the precursors lay within the complexity of human communication. Here it is important to identify human activities and future-oriented communication (on these activities) that may initiate wild cards by spreading their narratives. Identification of signals may be based on the analysis of major human-caused wild cards in the past and on social scientific research that should lead to a better appreciation of the communicative processes that lead to the realization of imagined and imaginative wild cards. The 'strength' of these imagined and imaginative wild cards seems to be dependent of many factors that are also recognized in the field of persuasive psychology as: the 'personality of the initiators, enhancers and opponents'; 'the plausibility of the storyline (future narrative), including its evidence base and its logical structure'; 'the emotional charge, including its shape which may be enhanced with visual and auditive signals'; 'the interests and powers involved in the storyline'; and 'the mental situation of the audience that is supposed to activate the wild card'.

While understanding and monitoring potential wild cards is important to create resilient policies, they are also important as a tool to create and shape policies at will. Collecting, and thinking on, wild cards does not only open our mind to what possibly may go wrong but also to new opportunities and ways of thinking which may be useful to create more options for our future actions. Horizon scanning and foresight identifying and using wild cards can, therefore, be seen as tools to shape the future more to our will. At the same time, imagined wild cards may be disabused by individuals for their own benefit (influencing the purchasing behaviour of individuals and organizations) or for spreading their destructive ideology (influencing people to criminal and destructive behaviour). Hence, their monitoring has very high political and economic relevance.

References

Friedemann, F. (2011). Pre-earthquake signals: Underlying physical processes. *Journal of Asian Earth Sciences, 41*(4–5), 383–400.

Georghiou, L., Cassingena, H. J., Miles, I., & Popper, R. (2008). *The handbook of technology foresight: Concepts and practice*. Cheltenham/Northampton: Edward Elgar.

Loveridge, D. (2009). *Foresight: The art and science of anticipating the future*. New York: Routledge.

Lynas, M. (2007). *Six degrees: Our future on a hotter planet*. London: Fourth Estate.

Markley, O. (2011). Staying resilient in a wild-card world. *Journal of the Institute of Noetic Science, 6*, 1–7. http://noetic.org/noetic/issue-six-january/staying-resilient-in-a-wild-card-world/….

Perloff, M. R. (2010). *The dynamics of persuasion: Communication and attitudes in the 21st century*. San Diego: Taylor & Francis.

Petersen, L. J., & Steinmüller, K. (2011). Wild cards, state of the future and futures research methodology version 3.0. In J. C. Glenn, Th. J. Gordon, & E. Florescu (Eds.), *The millennium project (2011)*, Washington. http://millennium-project.org/millennium/2011SOF-FRM.html.

Steinmüller, K. (2004). *The future as wild card: A short introduction to a new concept*. Essen/Berlin: Z_punkt GmbH, Büro für Zukunftsgestaltung.

Steinmüller, K. (2007). Thinking out of the box – Weak signals and wild cards for European regions. *Futura, 2*, 22–29.

Taleb, N. N. (2007). *The black swan: The impact of the highly improbable*. New York: Random House.

Van der Steen, M. (2008). Ageing or silvering? Political debate about ageing in the Netherlands. *Science and Public Policy, 35*(8), 575–583.

Van Rij, V. (2010). Horizon scanning: Monitoring plausible and desirable futures. In Roel in 't Veld (Ed.), *Knowledge democracy* (pp. 227–240). Berlin, Heidelberg: Springer.

Websites

http://ec.europa.eu/research/social-sciences/forward-looking_en.html
http://sesti.info/
http://farhorizon.portals.mbs.ac.uk/
http://wiwe.iknowfutures.eu/iknow-description/
http://www.innovation-futures.org/
http://www.sigmascan.org/Live/
http://www.horizonscan.nl/uploads/File/COS_binnenwerk%20engels_06(1).pdf
http://www.oecd.org/dataoecd/22/56/39991237.pdf
http://app.hsc.gov.sg/public/www/home.aspx

Part III
System Content Issues

Chapter 6
Forms of Reasoning in Pattern Management and in Strategic Intelligence

Tuomo Kuosa

6.1 Introduction

The contemporary world is full of information. It has been said that in the seventeenth century, an average person acquired the same amount of information in their whole lifetime about their world as we get from a single newspaper every day (Scholte 1996). The amount of information flowing constantly around us is huge, but only a small fraction of it is useful or valid for us as such. Not so long ago, information and knowledge were scarce and therefore very valuable. Nowadays, most information is free and easy to access, but a rapid understanding of it is rare (Weick 2001).

Sense-making requires more than just reading empirical data. It calls for valid expertise, good methods, time and resources (Kauffman 2000; Foreman-Wernet 2003). Strategic intelligence solutions involve services and consulting which help enterprises and consumers to always obtain the most valid and up-to-date information and strategic understanding of the issues in which they are interested. Why is strategic intelligence emerging? The world is not only full of loose information but it is also more complex, interdependent, hectic, nonlinear, coevolutive and less stable (Casti 2000; Kauffman 2000). The structures and processes of social systems involve increasingly large networks like the Internet (Kauffman 2000). There usually exists a whole network around a certain issue, which is called the network's macro-level. On the meta-level, the whole network usually further self-organises into local clusters. Inside these meta-level clusters are clusters of micro-level agents that are, for practical reasons, more strongly linked with their 'neighbouring' nodes than the nodes in distant locations (Cilliers 1998). These micro-level agents, for instance, individuals, are often called *complex adaptive systems* (CAS), since they are able to share knowledge, change their behaviour or learn owing to their local interactions

T. Kuosa (✉)
Alternative Futures, YATTA, Eteläranta 6, Helsinki 01300, Finland
e-mail: tuomo.kuosa@gmail.com

M. Giaoutzi and B. Sapio (eds.), *Recent Developments in Foresight Methodologies*, Complex Networks and Dynamic Systems 1, DOI 10.1007/978-1-4614-5215-7_6,
© Springer Science+Business Media New York 2013

(Kauffman 1995). The complexity or unpredictability around the network clusters or individual CAS is revealed by their underlying relationships with other network systems, synchronous self-organisation processes, relevant feedback loops, coevolutions, etc. Eve Mitleton-Kelly (2003) uses the concept of *complex evolving systems* (CES) to describe actions and learning in this whole rugged landscape.

Because the members of each network cluster share more knowledge in their local interaction, not all the clusters of the whole network have the same information. The dissonance of information increases as the whole network grows. At the same time, however, its ability to preserve information is increasing, thanks to the clusters, CAS, links and delays. This sets a challenge for any data management, sense-making or strategy work in the world of networks. And the challenge becomes greater if we add in Malcom Gladwell's point of view here. He claims that the spread of ideas, behaviour and the like between CAS, clusters and the whole network can be compared to the contagiousness of viruses in a population, which makes any linear 'ivory tower' predictions very difficult (Gladwell 2000; Barabási 2003).

What does this mean for strategy work, foresight or the management of an organisation? We need to accept that no one can steer, determine or even predict the development beforehand, and it is very difficult to obtain all the relevant information on time (Cilliers 1998). Furthermore, in this kind of environment, an actor can no longer rely on a single strategy and single method (Nicolis and Prigogine 1989). Thus, appropriation of the change and proactive strategies require ever faster, broader and more in-depth understanding of general transformations, and this cannot be accomplished without proper methods of observing, reasoning, understanding and influencing the complex processes.

Strategic intelligence can be understood as a form of meta-knowledge, which reveals large, complex or complicated issues or transformations in a more understandable form. Strategic intelligence reveals something that is unseen from plain information alone. Metaphorically, it can prove that individual trees together make up a large forest that is more than its sum. The approach in strategic intelligence can be seen to have some similarities with general inductive reasoning, which refers to inducing the universal from the particular or, more practically, recurring phenomenal patterns from limited observations. In strategic intelligence, it is possible to use narrative, semiotics, fractal or statistical mathematics, art and other visualisations, metaphors, analogies, etc., because these all have some abilities to express complicated or complex issues in a simplified way.

This chapter focuses on one form of strategic intelligence, called 'pattern management'. It is an approach which may be more based on empirical data and formal structures than other forms of strategic intelligence, but at the same time, it may be seen as a heuristic and creative approach. The domain of 'pattern management' is divided into three categories which reveal different sides of its existence. The first category is *empirical calculation*, which is common especially in enterprise consulting. The second one, *theory proving with observations*, is especially common in the natural sciences. The third one, *real combining*, can be considered common especially in qualitative research and in narrative. The categories vary according to their approaches to reasoning, methods used and especially the understanding of the 'truth' or the type of pattern that is looked for.

6.1.1 Sense-Making in Pattern Management

Pattern management (PM) is a fairly new concept. One of the first developments was Kamran Parsaye's (1999) article, where he drew a line between data management and pattern management. According to Parsaye, when recent data is put into an operational system and merged with historical data gathered over time, we have data management. When all this data, analysed over time, is being merged with historical *patterns*, we have 'pattern management'. Thus, PM is not knowledge management, data mining or construction of knowledge-based systems. PM deals with patterns after they have been discovered by data mining. Parsaye (1999) gives a simple analogy, 'consider data as grapes and patterns of knowledge as wine. Data mining is then the wine-making process, (…) and the data mining tools are like wine-making equipment'.

Parsaye's definition of PM is accurate from the point of view of managing knowledge, but it is possible to have a more versatile approach here as well. David Snowden (2002) has discussed the management of patterns as a more anticipatory and proactive process. From Snowden's point of view, patterns may even be seen as something more tangible than knowledge, understanding and beliefs alone.

> We need to identify the early signs of pattern formatting and disrupt those we find undesirable while stabilizing those we want. If we are really clever then we seed the space to encourage the formation of patterns that we can control. These patterns are, to use the language of complex adaptive systems theory, emergent properties of the interactions of various agents. By increasing information flow, variety and connectiveness either singly or in combination, we can break down existing patterns and create the conditions under which new patterns will emerge, al-though the nature of emergence is not predictable. (ibid, 107)

Snowden continues: 'Most humans make decisions on the basis of past or perceived future patterns, not through rational choices between alternatives. An understanding of patterns, is therefore, key to managing behaviour within organizations and in relationship to markets and environmental factors' (ibid). Therefore, patterns are not only knowledge, an understanding and beliefs of development but also something more tangible, such as proactivity with emerging paths and trends in a complex environment (see Aaltonen 2007; Kuosa 2007).

Other even more versatile and tangible descriptions for patterns managed in the process can be given. In Csikszentmihalyi (1996), I have linked PM to the rugged landscape between the complex adaptive systems (see Cilliers 1998; Kauffman 2000) and to managing knowledge of physical objects and more tangible transformation processes. In this sense, pattern can be understood as a phenomenon (Gladwell 2000) or even an object, which may not be visible or tangible as such (see Csikszentmihalyi 1996), and it can also refer to an existing, changing or emerging path of transformation. Here, management transforms finding the patterns into a process. It contains all the actions of observing, reasoning and understanding the issue at hand.

As an example of managing a phenomenon, a pattern can, for example, refer to findings in consumer behaviour. Those who buy diapers for babies will probably need to buy baby food, milk and towels as well and vice versa. The phenomenon of probable consumer types can also be rationally categorised according to the consumers' age, sex, income, education, values, etc. In addition, the consumer types can also be drawn automatically from empirical data of customer purchases, given that many customers use loyalty cards. This kind of knowledge can be used efficiently in marketing and product placement.

6.2 The Main Categories of Pattern Management

PM is, above all, a common logic of observing, reasoning and understanding our surrounding world. The theory of PM is not a closed and sophisticated collection of methods and procedures or a strict system description. It utilizes various forms of inductive, hypothetic-deductive, abductive or analogy- or case-based reasoning used within various fields of everyday life and science. Reasoning is an old field of philosophy with many well-established theories alongside its controversial issues. Rather than try to solve or further attend to these discussions, I attempt here to show how versatile, but at the same time unifying, PM can be. For classifying the different approaches related to PM, I have established the following main categories for pattern management (see Fig. 6.1).

Firstly, we can divide PM into two general categories. The first one is *empirical calculation* (EC), which refers to the quantitative search for increases or decreases with a large amount of data. The second one is *synthesising empirical and rational data* (SER). This can be further divided into two special types, which are *theory proving with observations* (TPO) and *real combining* (RC).

6.2.1 Empirical Calculation

By empirical calculation (EC), I mean the quantitative search for increases or decreases in the frequencies of certain issues with a large amount of data. When the work is started according to EC, there does not have to be time series or any hypotheses of the possible findings in advance, but the research theme, database and observing method are usually very well known. In other words, EC does not refer directly to time series analysis or statistical extrapolation. The logic of EC is more open and explorative and less fixed to historical findings. Nowadays EC, or data mining in its narrow definition, is mostly done by computing, but it can be done by using human observations alone.

To give a few examples, IBM and Google are companies which use EC on a large scale in their enterprise consulting work. For instance, IBM has developed many different kinds of multiphase data mining software tools for drawing rising peaks of development from large databases. IBM uses several methods, such as Public Image Monitoring, OmniFind, and Web Fountain (IBM 2006), for pinpointing the rise or waning of interest in discussion topics from the Internet or for drawing the most interesting Internet sites from up-to-date download statistics. In addition, Google uses its own database, which is collected from Google's own search service, in order

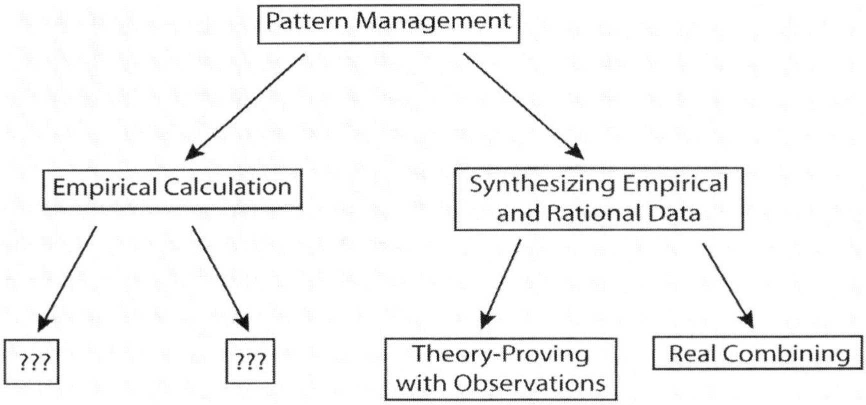

Fig. 6.1 Main categories of pattern management

to make sense of the changes in the topics in which people are interested nationally or internationally. According to the founders of Google, Sergey Brin and Larry Page, Google's next grand goal is the reorganisation of world knowledge into one search engine. If this attempt succeeds, there may be a new renaissance of EC ahead.

Alongside with enterprise consulting, EC or data mining has been used in technology assessment. There, EC can be done by searching for developing technology topics, for example, from refereed journals, patent applications, media discussion topics and internet downloads, in order to try to estimate when the time is right to expect a breakthrough in something or to start one's own R&D project.

EC is very much an umbrella concept. It defines general approaches and the logic behind them, but does not mark exact boundaries between life and science. Neither is it just confined to a set of methods – such as data mining, some other approaches of automatic calculation, and specific forms of quantitative research – used either as an insider or outsider to the system. Hence, we might say that we are talking about different kinds of wine-making equipment here, as Parsaye suggested. Nevertheless, it might be possible to try to divide EC into further subcategories.

6.2.2 Synthesising Empirical and Rational Data

The roots of the synthesising empirical and rational data (SER) approach can be found in Immanuel Kant's (1724–1804) philosophy, in which he wanted to combine rational and empirical reasoning. According to Kant, loose empirical knowledge is unhelpful, unless we have the capability to reach conclusions and to discover the phenomena behind the findings. Thus, it is the representation that makes the object possible rather than the object that makes the representation possible (Kant 1783 [1977]). Kant's approach introduced the human mind as an active originator of experience rather than just a passive recipient of perception. When we see a box as three-dimensional, the shape of the box may not be part of the box's nature. There need to be not only empirical observations from the surrounding world but also synthesising by an intelligent agent who can put the observed pieces together in order to make findings and reasoning. Here, Kant and Hegel (1770–1831) were impressed by the astronomy of their own time and the Copernican revolution and by the fact that the locations, formation, size and weight of planets could be drawn from the data of indirect observations from the surrounding space by reasoning and by synthesising theories (Redding 2002).

Today, astronomy is more advanced compared with the time of Kant and Hegel, but very similar principles are still steering the rational and inductive processes of reaching for the phenomena behind the loose observations. How is the existence of black holes, wormholes, dark matter or planets in distant solar systems deduced when no one can reach the substance under investigation or even obtain direct observations of the subject with telescopes? The answer is that the astronomers observe and collect data from the surrounding local space. Related to the research matter, they observe the changes in radiation, bending of light, compare gravitation fields, shadows and light spectrums, reflections of infrared light, etc. (Valtaoja 2002; Hubblesite, Newscenter 2006). Information about the kinds of findings that are needed to prove that a phenomenon exists is embedded in theories of astronomy. If the findings are not explained fully by the theories, then the theories have to be changed. The scientific work of astronomy is one example of theory proving with observations, the first form of SER to be discussed.

6.2.3 Theory Proving with Observations

The approach of theory proving with observations (TPO) resembles the logical reasoning method of abduction more than the other forms of pattern management. This reasoning method is more complex in its structure and can involve both inductive and deductive arguments. The main characteristic of abduction is an attempt to favour one conclusion above others by either attempting to falsify alternative explanations or by showing the likelihood of the favoured conclusion with a set of more or less disputable assumptions. As reasoning in TPO resembles abduction, it should also be noticed here that Karl Popper's (1902–1994) approach of critical rationalism – the principle of theory's falsifiability – should be strongly embedded in TPO as well. According to falsifiability, a theory can be considered reliable only if there is an opportunity given for falsifying the theory by a contrary case. Nowadays, the principle of falsifiability is strongly embedded in mainstream scientific method.

Astronomy has already been suggested as a form of TPO. Most recently (Hubblesite, Newscenter 2006), the astronomers of the Harvard-Smithsonian Institute proved the existence of dark matter by locating its 'fingerprint' from a location they called 1E0657-556.

Crime scene investigation (CSI) is another possible example here. Crime scene investigators try to figure out what really took place at the moment of a crime such as murder. At first, CSI tries to collect all the valid data that is possibly related to the issue. They try to identify where the blood stains were found; what kind of splashes or hit angles, finger- or footprints, scratches and marks were found; who has the motive; and who has the alibi. The collected information is then embedded into criminal psychological theories (Hare 1999). When all this information is put together, there will usually be several alternative scenarios for the crime. The final phase of the investigation process is a puzzle, in which pieces of information must be put together in order to favour one conclusion above others by either attempting to falsify alternative explanations for the chain of events or showing the likelihood of the favoured conclusion with a set of more or less disputable assumptions. Furthermore, the favoured conclusion in CSI can be falsified by contrary observations.

Codebreaking can be seen as one form of TPO type of pattern management. In cryptography, there is usually a mathematical model, a cryptographic key, used when a secret message is hidden in a message.

There are many forms of cryptography and codebreaking in the world. Karl Weick (2001) described a codebreaker's work in the following way: 'The object of a codebreaker is to duplicate the exact pattern of colored pegs inserted into holes that has been created by the codemaker but is concealed from the codebreaker by a shield. The codebreaker ventures hypotheses as to what the pattern might be and, on the basis of information supplied by the codemaker, refines the hypothesis until the codebreaker's hypothesis exactly matches the codemaker's original pattern.'

6.2.4 Real Combining

Real combining (RC) is another form of SER. The main difference between TPO and RC is in reasoning. TPO is based very much on abduction and falsifiability (or hypothetic deduction). RC relies mainly on the use of analogies, metaphors and other approaches for finding interconnectedness, similarities and possibilities to combine qualitative data into meta-knowledge, with a common storyline and understanding (Table 6.1).

Reasoning in analogical thinking goes, for example, from one particular to another or from a theory in one field to a theory in another field. Analogy refers to picking or pointing out one similarity between two things that are otherwise different. Metaphor itself is a rhetorical concept, which comprises the subset of analogy, and it is related to comparison between thoughts. In some cases, RC may also resemble inductive reasoning, when the attempt is to find theories which explain various particular things and interrelationships.

The form of reasoning and refining understanding, which I here call real combining, is common in narrative and some forms of literature, as well as in many academic fields, especially in qualitative research. Here, I provide two different

Table 6.1 Categories of pattern management

Categories of pattern management	EC: empirical calculation	TPO: theory proving with observations	RC: real combining
Research material	Quantitative and statistic raw data. Quantitative search for increases or decreases in the frequencies of certain issues with a large amount of data	Any data or perceptions can be equally used as evidence. The objectives determine the required or relevant data sources	Qualitative, perceptions, linkage, literature, talk, tacit knowledge, weak signals, intuition, wisdom, interpretations
Reasoning	When the work is started according to EC, there does not have to be time series or any hypotheses of the possible findings in advance, but the research theme, database and the observing method are usually very well known. Mostly inductive, statistical and computer-based	Mainly abductive. The main characteristic of abduction is an attempt to favour one conclusion above others either by attempting to falsify alternative explanations or by showing the likelihood of the favoured conclusion with a set of more or less disputable assumptions	Analogies, case-based reasoning, creating linkages and metaphors, which themselves are rhetorical concepts, which comprises the subset of analogy, and it is related to comparison between thoughts. Inductive reasoning, when the aim is to find theories which explain various particular things and interrelationships
Type of the pattern/truth/ phenomenon	Changing existing (emerging)	Permanently existing (emerging)	Invented subjective (emerging)

examples of reasoning according to RC. The first one is Amazon.com, which uses automatic RC. When one starts selecting books for a shopping cart in Amazon, the programme starts suggesting new books – even from new themes – which have often been purchased or viewed by other customers who bought the same books one has already selected for his/her shopping cart. Therefore, the software used by Amazon.com makes comparisons and finds relations between various themes automatically by utilising some form of meta-knowledge, that is, subjective meta-information (Johnson 2001).

Another example of RC could be *The Kalevala* (Lönnrot 1835), the national epic of Finland. Elias Lönnrot spent years of his life walking around Karelia, talking with people and gathering oral stories in his notebooks. In the end, he was able to identify the common denominators of the stories and give them a literary and smoothly running storyline, creating one of the mightiest epics in the world, which in contrast to many other epics, for example, *The Iliad* and *The Odyssey* by Homer or *The Lord of the Rings* by J. R. R Tolkien, is more heavily based on the oral tradition of the people than on the creative work of the author (Aaltonen 2007).

6.3 Existing, Changing and Invented Patterns

There are at least three kinds of patterns: existing, changing and emerging. Any of these can be managed with the types of pattern management we have identified. However, some types of PM are more suitable for managing certain types of patterns than others. When EC is used, the pattern or 'truth' is understood as something that is changing and can be analysed with quantitative approaches. Therefore, EC can be used for locating existing patterns or for analysing changing patterns, such as how the consumer types drawn from actual shopping change over time.

When the TPO approach is used, there is usually a belief that one 'permanent' objective 'truth' can be located or that there is at least one 'permanent' 'truth' that is objectively less disputable than the others. Therefore, TPO is also suitable for reasoning out existing patterns – something that can be seen as objective or tangible: a finding, a pattern, a path, an object or a phenomenon. As RC is a more subjective and qualitative form of pattern management, the patterns drawn according to it may be different or more subjective. Should we call the patterns or the 'truth' that is looked for in RC 'invented'?

The dots in the pictures of Fig. 6.2 represent (loose) observations or raw data. If the dots are very close to one another, they are believed to have some common

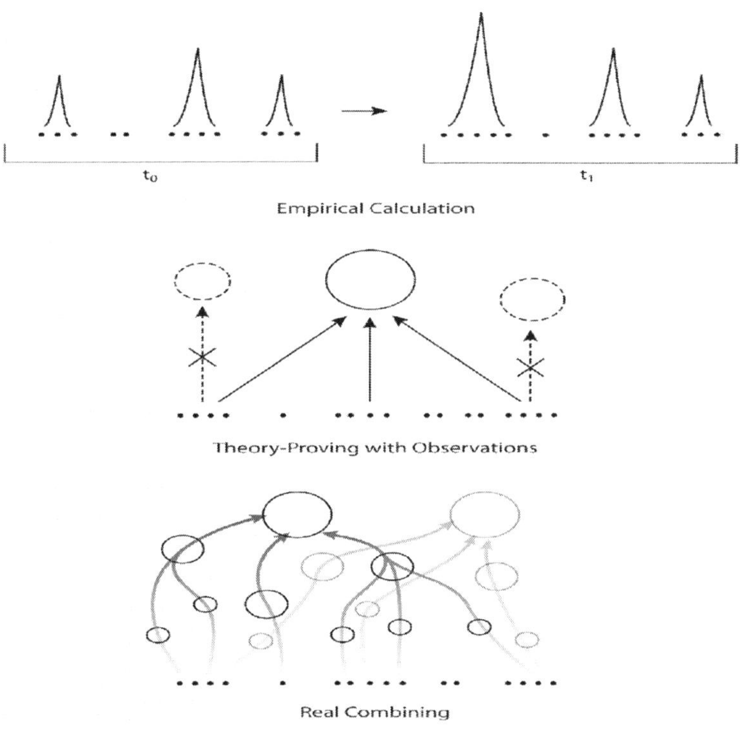

Fig. 6.2 Three general types of reasoning in pattern management

denominators. If they are separated, they are believed to have less in common. In EC, the method of the management of observations into patterns is mostly quantitative. In TPO, the observations are used either for falsifying alternative explanations or for showing the likelihood of the favoured conclusion by giving a set of more or less disputable assumptions.

In RC, the method for drawing patterns from observations is mostly qualitative and structural. The observations are used as building blocks in order to obtain a common storyline or understanding of the issue.

6.4 Sense-Making of Emerging Patterns

In addition to the existing and changing patterns, there remains one more form of patterns: the patterns which are potentially emerging. The processes of managing emerging patterns take us close to the fields and concepts of anticipation, proactivity, prospective thinking, appropriation, foresight and futures research. However, it has been difficult to find any formalised descriptions or methods for such management of emerging patterns from these fields. I have not found any descriptions of such an approach from the fields in the most well-known or recent sources, such as Godet et al. 2000; Armstrong 2001; Kamppinen et al. 2002; Glenn and Gordon 2003; Bell 2004; and EFMN 2005. Usually, the need for the process seems to be understood, but the methods of management are lacking, or they do not fit such a process (EFMN 2005).

'Emerging pattern' here refers to something that is only a potential seed of transformation at the moment but which is shown to have good opportunities to start growing in the future. A simple physical example of an emerging 'object type of pattern' could be an embryo, which, according to all valid and accessible knowledge, is believed to have a good chance of growing into adulthood.

An example of an emerging 'phenomenon type of pattern' could be virtual consuming. It is a minor field of consuming at the moment, but it is possible to locate many reasons, driving forces and supporting factors why it is conceivable that it will expand and partly change the world of consuming in the future.

Time and place dependence is weakening in, for example, consuming, work and communication; the role of expertise has been growing in society; the values of young people are already different from those of their elders[1]; there is the continuing development of ICT[2]; software and games seem to be becoming more realistic and interesting. This approach resembles both RC and TPO.

[1] What happens when today's teenagers are in their 40s and really start to make national politics?

[2] Faster and more easy to use computers will be seen.

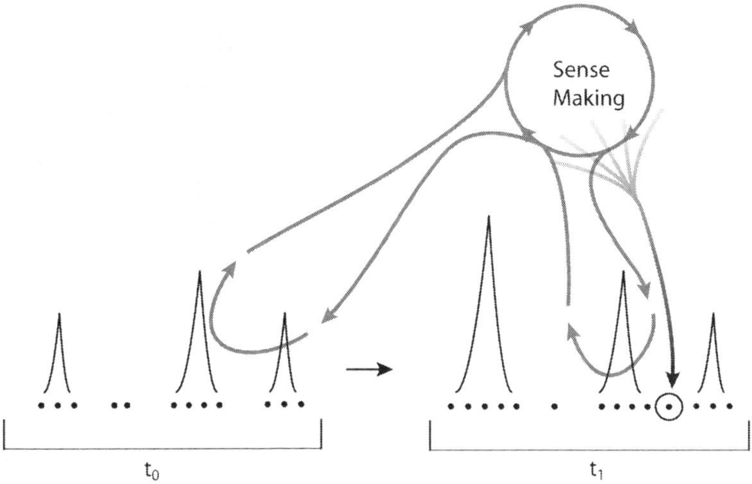

Fig. 6.3 Managing an emerging pattern

6.4.1 Making Sense of Emerging Patterns

In Fig. 6.3, the process of managing an emerging 'phenomenon type of patterns' is shown from another point view. Here, the PM process starts with EC both in time 0 and in time 1 and is continued with the sense-making process, which may here resemble RC more than TPO. In the figure, the located patterns are not the same in t_0 and in t_1. Some of the patterns are weakened and some are strengthened, something has emerged, and something has declined. When there is finally more understanding of changes in patterns in time, plus more understanding of the drivers of change, there is a fruitful stage for sense-making emerging 'phenomenon type of patterns'. It may, therefore, be possible to locate something which is unformulated or weak at the moment but which nevertheless has very strong support, demand or capacity to be developed.

The best examples of approaches and methods for managing emerging patterns come from the fields of risk assessment and horizon scanning[3], fashion and consumer behaviour intelligence and technological forecasting. From the domain of futures research and foresight, the best example of such work comes from John Naisbitt´s megatrend management (Naisbitt 1985, 1997, 2004; Naisbitt and Aburdene 1991).

---

[3] A large international top-level risk assessment and horizon scanning symposium (RAHS) was held in Singapore on 19–20 March 2007. The symposium set a clear picture of the state of the art within the utilisation of recent complexity research and modelling for contemporary international risk assessment and foresight.

Naisbitt has a company which goes through and analyses broad selections of world newspapers. The aim of his process has been to find knowledge which tells us about the rising peaks behind raw data. Naisbitt and his colleagues set these peaks into a framework of platforms that claim to provide the knowledge of megatrends or other great changes (Aaltonen 2007). Within his approach, changes are constructed from the bottom up, from the grass-roots level, by clustering – as in a puzzle. A new phenomenon or idea that does not manage to gain support in the ongoing development process dies away – just like useless pieces are not put into a puzzle. Missing pieces are, however, looked for very hard.

Another example of PM of emerging patterns is found from trend analyses made in fashion houses or the clothing industry in general. Here, we can utilise Naomi Klein's (2001) description of the work of trend analysts or cool hunters in fashion houses like Nike and Tommy Hilfiger. According to Klein, such fashion houses have hired signal detectors who observe and interview especially young avantgardist individuals from marginal groups. They also observe music videos of MTV and hip-hop magazines such as Vibe. By 'young avantgardist individuals from marginal groups', Klein is referring to, for instance, the poor young men from the black ghettos of the big cities – strong figures, who hang around basketball courts. They are influential opinion shapers in their communities. When these people start representing something, using certain colours, styles, patterns, shapes and designs in their community first, their style is believed to be gradually adopted by the entire community, as people are group animals. Later on, the fashion of the ghetto will have an effect on the fashion of the whole country and even international clothing markets (see also Gladwell 2000). What is fashionable among avantgardist groups in the spring might be fashion on the national or international level in the following fall. This synthesising rational and inductive process made by the trend detectors, of course, requires very diverse observation work. The company could not trust the observation of just one 'ghetto' or one observing method (Klein 2001). There has to be a considerable amount of information collected from different sources, which needs to be embedded in the available theories of fashion and group behaviour.

Such trend detectors are used not only within the fashion business: Nokia (Merriden 2001) uses anthropologists for observing people and their lifestyles in, for example, parks, streets and shopping malls. The observers are supposed to identify early information about psychological changes in human behaviour, individual value systems and key drivers of customers (what excites and motivates people, and what are the ways people want to communicate and establish groups?). By synthesising this information at an early enough stage, there is a better chance for a mobile phone company to be prepared for emerging or immerging (declining) consumer needs (Merriden 2001).

Many intelligence agencies, such as the Pentagon and especially the Central Intelligence Agency (CIA 2000), have developed sophisticated systems for data gathering, analysis and risk evaluation. All the forms of patterns (existing, reasoned and emerging) seem to be used simultaneously, alongside all the forms of PM reasoning (EC, TPO and RC). To give one example of these approaches, the

CIA tries to identify possible central nodes or figures in terrorist networks by searching subjects of sent e-mails or Internet downloads and connecting this information to certain people. It also uses anthropologists for observing and interviewing local people in possible crisis areas, such as Iran. The stories that people tell there are especially important in the approach. In this way, the local silent knowledge (weak signals and emerging issues) at the grass-roots level is gathered in order to understand the early changes in public opinion. Certain paths in common storylines are believed to indicate a certain growing social phenomenon in the social context (CIA 2006).

The CIA observes global statistics as well. It has a special interest, for instance, in the demand and supply chains of certain chemicals or equipment which can be considered necessary for preparing terrorist action. It has been said that, within this kind of statistical and multisource information collection and synthesising, the CIA has been able to expose a large-scale cocaine poisoning process which took place in Columbia. The poisoned cocaine was meant to be shipped to the North American markets. The work of the CIA could be given here as an example of a multi-approach process, where all the pattern management's forms of reasoning have been used simultaneously in order to ensure the reliability of the findings.

6.5 Conclusion

Reasoning is a mental process, which informs our imagination, perceptions, thoughts and feelings and links our everyday experience with universal meanings. Thus, reasoning is a vital part of the process of sense-making, understanding and internalising. In philosophy, there are many structured forms of reasoning under its main forms: inductive, deductive and abductive reasoning. In addition, there can be found some special approaches of reasoning, such as analogies and their prominent everyday forms, like case-based reasoning.

In this chapter, they have been discussed, and some theoretical forms of reasoning in philosophy have been merged with the findings of reasoning in some real-life cases, as well as with some practical methods or common-sense approaches. In the process, some methods and approaches have appeared to have more common denominators with some forms of theoretical reasoning than with others. An especially meaningful finding has been the deviation of the 'aims or objects' in different approaches and processes. What kind is the 'truth' or the pattern being rationally searched for?

Successful involvement in the present networked world, which is more hectic, interconnected, coevolutive, unstable and full of loose and rapidly changing information, is difficult. It is especially difficult if we want to predict anything, or if we are strategic actors, or we want to manage an organisation proactively in this complex, evolving, rugged-landscape system. Strategic intelligence, and especially its most structured but open form, pattern management, is a multi-approach attempt to answer, or help to answer, this challenge.

References

Aaltonen, M. (Ed.). (2007). *The third lens. Multi-ontology, sense-making and strategic decision-making*. London: Ashgate.

Armstrong, J. S. (Ed.). (2001). *Principles of forecasting: A handbook for researchers and practitioners*. New York: Springer.

Barabási, A.-L. (2003). *Linked: How everything is connected to everything else and what it means for business, science, and everything*. New York: Plume.

Bell, W. (2004). *Foundations of futures studies: Human science for a new era* (Values, objectivity, and good society, Vol. 2). New Brunswick: Transaction.

Casti, L. J. (2000). *Five more golden rules: Knots, codes, chaos, and other great theories of 20th-century mathematics*. New York: Wiley.

CIA. (2000). *International crime threat assessment*. Washington, DC: Office of Public Affairs CIA.

CIA Factbook on Intelligence. (2006). The George Bush Center for Intelligence, Office of Public Affairs CIA, Washington, DC. http://www.cia.gov/cia/publications/facttell/index.html

Cilliers, P. (1998). *Complexity and postmodernism: Understanding complex systems*. New York: Routledge.

Csikszentmihalyi, M. (1996). *Creativity: Flow and the psychology of discovery and investigations*. New York: Harper Perennial.

EFMN. (2005). *Foresight in Europe and other regions of the world*. The EFMN (European Foresight Monitoring Network) annual report 2004–2005.

Foreman-Wernet, L. (2003). Rethinking communication. In B. Dervin, L. Foreman-Wernet, & E. Lauterbach (Eds.), *Sense-making methodology reader*. New Jersey: Hampton Press.

Gladwell, M. (2000). *The tipping point: How little things can make a big difference*. New York: Little Brown.

Glenn, J. C., & Gordon, T. (2003). *Futures research methodologies – V2.0.AC/UNU*. Millennium project, Washington, DC.

Godet, M., Monti, R., Meunier, F., & Robelat, F. (2000). *Scenarios and Strategies : A toolbox forsScenario planning*, Cahiers du LIPS, LIPS working papers, Paris.

Hare, D. R. (1999). *Without conscience: The disturbing world of the psychopaths among us*. New York: Guilford.

Hubblesite, Newscenter. (2006). NASA Finds Direct Proof of Dark Matter, 21 Aug 2006. http://hubblesite.org/newscenter/archive/releases/exotic/2006/39/fastfacts/.

IBM. (2006). Public image monitoring, *OmniFind*. http://www-306.ibm.com/software/data/integration/db2ii/editions_womnifind.html.

Johnson, S. (2001). *Emergence: The connected lives of ants, brains, cities, and software* (pp. 122–129). New York: Scribner.

Kamppinen, M., Kuusi, O., & Söderlund, S. (Eds.). (2002). *Futures studies: Principles and practices (in Finnish)*. Helsinki: SKS.

Kant, I. (trans: Paul, C.) (1783) [1977]. *Prolegomena to any future metaphysics that will be able to come forward as science*. Indianapolis: Hackett.

Kauffman, S. (1995). *At home in the universe: The search for the laws of self-organization and complexity*. New York: Oxford University Press.

Kauffman, S. (2000). *Investigations*. New York: Oxford University Press.

Klein, N. (2001). *No logo*. London: Flamingo.

Kuosa, T. (2007). A few extensions to path-dependence and emergence in complex social systems. *Emergence: Complexity and Organisations (E:CO)*, *9*(6), 3–16.

Lönnrot, E. (1835). *The Kalevala*. Finland

Merriden, T. (2001). *Business the Nokia way: Secrets of the world's fastest moving company*. Oxford: Capstone.

Mitleton-Kelly, E. (2003). Ten principles of complexity and enabling infrastructure. In E. Mitleton-Kelly (Ed.), *Complex systems and evolutionary perspectives on organisations: The application of complexity theory to organisation*. Oxford: Pergamon.

Naisbitt, J. (1985). *Ten new directions transforming our lives*. New York: Warner Books.

Naisbitt, J. (1997). *Megatrends Asia: The eight Asian Megatrends that are changing the world*. London: Nicholas Brealey.

Naisbitt, J. (2004). *Preview of the future – Where is the world headed*? Naisbitt's seminar in Kalastajatorppa on 15 Dec 2004, Helsinki, Kontakti.net. (unpublished report written by Tuomo Kuosa).

Naisbitt, J., & Aburdene, P. (1991). *Megatrends 2000: Ten new directions for the 1990s*. New York: Morrow.

Nicolis, G., & Prigogine, I. (1989). *Exploring complexity: An introduction*. New York: Freeman.

Parsaye, K. (1999, January). From data management to pattern management. *DM Review Magazine*.

Redding, P. (2002). Georg Wilhelm Friedrich Hegel. In E. N. Zalta (Ed.), *The stanford encyclopedia of philosophy (summer 2002 edition)*. http://plato.stanford.edu/archives/sum2002/entries/hegel/.

Scholte, J. A. (1996). Beyond the buzzword: Towards a critical theory of globalization. In E. Koffman & G. Youngs (Eds.), *Globalization: Theory and practice*. London: Pinter.

Snowden, D. (2002). Complex acts of knowing: Paradox and descriptive self-awareness. *Journal of Knowledge Management, 6*(2), 100–111.

Valtaoja, E. (2002). *At home in the universe (in Finnish)*. Helsinki: Ursa.

Weick, E. K. (2001). *Making sense of the organization*. Oxford: Blackwell.

Chapter 7
Micro-Meso-Macro: From the Heritage of the Oracle to Foresight

Péter Alács

7.1 The Methodological Construction of Foresight

While 'foresight' has become a vogue word for some successful participatory, future-oriented activities, the need for abstract definitions or theoretical underpinnings has arisen to improve efficiency. Defining foresight would be certainly easier if foresight could be included in a special category of practice or if foresight fell under some of the major categories of academic activities. Indeed, some of the foresighters regard foresight as a practice, while others consider it to be more of a science.

Practice-oriented foresight approaches (Horton 1999) focus mainly on strategy building and usually confine their scope to a specific field of practice like technology or regional development. Their objective is to aggregate stakeholders' opinions and experts' knowledge about the specific field to support executive decisions. Scientific-oriented foresight approaches (Slaughter 2003) are derived from future studies and social sciences. Here, foresight is distinguished from other categories of future studies by the application of participatory methods and the shifting of the focus from an objective notion of the future to subjective interpretations of the future itself ('future in the present').

In spite of the many similarities, these two kinds of approach differ fundamentally. The centre of the differences is the emphasis on the achievement of a consensus or the exploration of the information about the future. Practice-oriented approaches need a kind of consensus among stakeholders to ground strategic plans and find pure autotelic exploration. Scientific approaches, on the other hand, refuse to seek consensus because of the presumed distortion of the opinions of the process, and so the loss of objectivity.

P. Alács (✉)
Jövőkutatás Tanszék, Fővám tér 8, Budapest, H-1093, Hungary
e-mail: petersandor.alacs@gmail.com

M. Giaoutzi and B. Sapio (eds.), *Recent Developments in Foresight Methodologies*, Complex Networks and Dynamic Systems 1, DOI 10.1007/978-1-4614-5215-7_7,
© Springer Science+Business Media New York 2013

The main differences between the two kinds of approaches to foresight explain the limits of their application. Consensus distorts information, and simple information gathering fails to meet the needs of decision support. In this chapter, we develop a more comprehensive notion of foresight that involves both kinds of approach. Realizing that foresight cannot be definitely interpreted as a kind of practice and cannot be efficiently applied as a scientific category, we decline an approach to the notion of foresight from the application point of view. Foresight, by its special nature, should have a special place in our way of thinking; therefore, foresight is interpreted as an intellectual activity. This differs from the practice-oriented approach, in that it has no definite output, hence no measures of efficiency. And this also differs from a scientific-oriented approach because the notion of foresight, in this context, will not have objective statements.

We consider foresight as an activity consisting of several methods that are scientifically based and carefully applied for certain purposes. Thus, foresight is more a collection of methods and know-how than a practical or scientific solution to a given problem. Its efficiency is determined by its applicability and by how the methods are used. This approach is called the 'methodological construction of foresight'. As this construction is based on the special use of scientific methods, firstly, the special nature of these methods is considered, which may lead to a three-level categorization of the theoretical advances in science. This three-level categorization can be used to construct foresight through approaching, the issues of complexity, uncertainty, and time, at the above levels.

7.2 The Ladder of Theoretical Advance

The ladder of theoretical advance is a simplified representation of the scientific method (Wilson 1952). The construction of the ladder is inspired by a three-level approach from evolutionary economics (see Dopfer et al. 2006). In evolutionary economics, the notion of innovation is very similar to the way foresight could be perceived. The similarities stem from the existence of nonequilibrium conditions in the environment that enable innovators to emerge. The application of this idea to foresight can be seen by presenting the ladder of theoretical advance as follows:

- Work with aggregates, that is, discover a phenomenon
- Look behind aggregates, that is, identify the structures behind the phenomenon under study
- Explore the dynamics, that is, understand how the system works

In the first stage, the phenomenon is discovered and then described. Here a common agreement about the definition of the phenomenon is necessary. In the natural sciences, agreement is usually a controlled experiment that could be repeated and objectively measured and therefore can be generally accepted. In the social sciences, there is no such clear way to define a phenomenon, but this does not imply that an agreement could not be achieved. This level also includes a basic study of the phenomenon to the extent of its identification and its distinction from other phenomena. We should realize that this level bridges the subjective state of mind and the collective knowledge about the phenomenon. This means that on this level we meet objective aggregates and also subjective narratives in methodology.

In the second stage, the phenomenon is specified in detail. This implies an exhaustive study of the phenomenon and its relationships to other phenomena. If the studied phenomenon is too complex to be understood at the first glance, different dimensions, projections, or confinements are studied in parallel that, by the nature of the phenomenon or because of the limits of our mental capacity, may not necessarily merge into a single level.

In the final stage, rules, and the relationships between specified elements and time, are revealed, and the dynamics of the system are defined. This stage represents the extent to which the different segments of parallel studies could be joined. This level is not necessarily unique with respect to a given phenomenon, but under some well-chosen constraints, we may regard the level unique as it maximizes understanding.

The three stages of this ladder are also referred to as the roughness or the complexity of its variables at the macro-, micro-, and meso-levels. In the first stage, aggregates are considered macro-variables because these represent information compression or integration with respect to a special target or to a certain constraint. This is needed to cleanse information in order to find the relationship between the variables and subjective notions, that is, the ideas about the phenomenon. In the second stage, detailed studies are carried out, where information is available in its most elementary form. Between the macro- and micro-levels, the final understanding of the phenomenon occurs on the meso-level.

We should notice that between the micro- and macro-level, several inter-levels could be defined. As we move from the micro- to the macro-level through aggregation, we may identify increasingly fewer variables and increasingly fewer notions. This certainly helps us to identify or grab the phenomenon, but does not necessarily help us to understand it: through aggregation, important dynamic relationships could be hidden and lost. The dynamic completeness is revealed on the macro-level, but the usual segregation of this level prevents a comprehensive understanding. Thus, the meso-level is defined where complexity is limited, but the dynamic diversity is still preserved. We see that the objective for science is the meso-level because this level incorporates most of the understanding of the phenomenon (Alács 2006).

7.3 The Three Levels of Foresight Activity

Foresight is, however, not science. We consider foresight not as an activity to predict the future (which would be certainly a scientific activity) but the activity to find today the right decisions that may lead to a better future. Foresight aims to provide 'useful' conclusions for the future to any stakeholder or decision maker. A process for the foresight activity, therefore, should consist of the following three steps:

- *Meso-level* – assessing the available knowledge. Identifying the seeds of change
- *Micro-level* – thoroughly studying the knowledge and its distribution among the stakeholders. Determining the key dimensions of desirable futures
- *Macro-level* – constructing a general view that helps to deduce the rules for the specific situations

In the first stage, the knowledge should be identified as the basis for the foresight activity. This should also include the state-of-the-art scientific achievements, that is, meso-level knowledge. Also experts' opinion is considered here as meso-level knowledge. Unlike in science, in the case of the foresight activity, the meso-level consists of several types of information.

In the second stage, this knowledge should be studied by involving the stakeholders. However, the great help provided by including the scientific results of the meso-level does not make the micro-level of foresight superfluous. Here the elements of possible futures are present, and by using these elements, the several dimensions and factors of desirable futures must be considered, that is, choosing the way of describing the desirable future.

In the final stage, these logical operations are constructed in one aggregated view. These operations are in accordance with the chosen direction of describing the future in the second stage. In science, the macro-level bridges our subjective ideas with the common agreement about the studied phenomenon. In foresight, the direction is just the opposite. The macro-level of foresight represents the individual and common efforts of integrating micro-information according to the chosen dimensions and creating principles to ease decision making.

Foresight is not necessarily an organized activity; it is deeply involved with human nature. This kind of activity we leave to psychology. Excluding beliefs from foresight activity does not necessarily mean that we have prevented the application of the activity from expanding the influence of certain interests. This may be achieved by:

- Preventing the overlap with scientific results in the first stage as an input of the activity
- Widely involving stakeholders, in the second stage, to determine the key factors for the description of the desirable future (Note that stakeholders should agree by consensus only about the key factors and not about the desirable future!)
- Developing the way in which these key factors can be individually integrated into the world view of every single stakeholder in the third stage, in order to ease decision making in every possible future situation

7.3.1 Foresight and Information Theory

In terms of information theory, the three-level approach can be interpreted as the creation of the syntax, context, and semiotics of the foresight-related information. The syntax of the foresight information is defined on the meso-level. This means that dynamics, trends, prediction, forecasts, or seeds of change are not discovered by foresight but only realized and coded for further process. The meso-level does not stand for purely scientific but for purely dynamic purposes. This implies the coding of the dynamics experienced by the stakeholders rather than the scientifically understood part of dynamics, although the wider use of the latter would be favourable. The syntax of foresight is the gathered meso-level structures that define the 'grammar' for the later levels.

On the micro-level, stakeholders learn about the regime of possible futures. During this encounter, they should decide about the key factors of their desirable future, and they should come to a consensus about these factors. The consensus enables them to create a common context of the foresight information. The context will be defined through one or more emerging meso-level structures based on the chosen micro-level factors of the desired future.

Macro-level information focuses on the reasons for change instead of change in its dynamic sense. As a single stakeholder does not necessarily realize all the possible reasons for change and what will emerge from the change, the reasons should be communicated in a specially chosen manner. On the macro-level, the semiotics of the foresight information is defined, but the meaning, and the point of view, could be slightly different for every stakeholder. The transfer from the context to the semiotics should involve personalizing rather than simplifying and principle-focusing rather than moralizing.

7.4 Special Questions of Methodology

We pick three main attributes of the methodology to further clarify the foresight process: complexity, uncertainty, and time. Note that these attributes should be discussed in detail when specifying the methods to be applied in the foresight process. The study of complexity determines the manner of information processing. The study of uncertainty determines generality and flexibility. Time determines on which level the method is suitable: qualitative and quantitative methods, for instance. On the meso-level, both qualitative and quantitative methods are used to assess the knowledge. On the micro-level, these two types of knowledge are not integrated, but rather they are split or factored. Integration is possible on the macro-level by qualitative tools.

7.4.1 Complexity

Complexity determines the manner of connection between the levels. We distinguish two kinds of complexity (Alács 2006):

- The top-down approach, which implies that a certain goal could be achieved in a number of similar ways that makes a mathematical optimization problem have several equal solutions.

- The bottom-up approach, which means that the special pattern of interaction among the elements makes the system as a whole work strangely; a new phenomenon emerges that cannot be explained by the characteristics of the separate elements.

In the foresight procedure, we meet both types of complexity. Moving from the meso-level to the micro-level, we meet the top-down approach to complexity. In the first stage of the foresight process, several meso-levels are created. In the second stage, however, we need only, as an input, a comprehensive set of meso-level information. This means that the different meso-levels are compared, extended, and optimized.

Moving from the micro-level to the macro-level, we meet the bottom-up approach to complexity. Choosing the dimensions of key factors means emergence studies and the application of methods that facilitate this emergence.

7.4.2 Uncertainty

Uncertainty represents not only lack of knowledge in the process of foresight but also lack of trust and reliability. A successful creation of the first stage of foresight, however, should involve only the former because, at this level, the different meso-levels are dealt with independently.

On the micro-level, when approaching the consensus about the key future factors, at least a weak connection between the meso-levels is developed. Here uncertainty is utilized to enable flexibility for the development of these connections. On the macro-level, uncertainty arises with the distributed knowledge. The efficiency of the foresight process cannot be directly measured, and failures of the methods or misinterpretations could be revealed only later.

7.4.3 Time

Time can be interpreted in many different ways. The physical approach has been well developed, especially after the elaboration of general relativity. However, interpretations besides the most accepted astronomical interpretation have also emerged in philosophy, economics, or in the social sciences (Sorokin and Merton 1937; Lewis and Weigert 1981).

The common ground, in every interpretation of time, is that the notion of time cannot be defined or understood without understanding the dynamics: reasons and consequences make the states of the system follow a specific order or law. Therefore, after defining the states of the system, usually a measure of this order, time, is defined. In this way, the relationship between time and the evolution of the system is possible, and in our terms, the meso-level is created. Forecasts must first determine the time horizon before further studying the system. For example, the notion of 'short/long term' is sometimes more convenient than the choice of astronomical time. Note that this can be very different according to the field in which it is used: The long term, when considering inflation, is still very much shorter than the short term in demographics.

The foresight activity does not construct time. The objectives of foresight are timeless (not to be confused with eternal), and therefore on each level of the ladder, a certain timelessness is applied to the notions we use.

The meso-level is the level which focuses on the interaction between the elements, and all information can be unified by comprehensive dynamics. On this level, the dynamics is the valid unifying order that defines time, and so time is unique. This must hold at least for each stakeholder but does not necessarily mean that at the basis of the foresight process, a unique time would exist (consider the cases of small, innovative, and dynamic firms compared with government bureaucracy).

The objective of the micro-level is informational enrichment, focusing on the elements. On the micro-level, the informational enrichment of the system overloads the dynamic logical order. Thus, on this level, several projections of the micro-information, as well as several interpretations of time, not only coexist but also interact with each other in a comprehensive manner.

The macro-level concentrates on the phenomenon, the points of view, or an information axis in a descriptive way. Its historical perspective is, however, not to be confused with its relation to time. By revealing the effects and clarifying the phenomenon, we can ensure that this level is targeting the stable, and hence timeless, properties. The goal here is to find the coding in which we can talk about the system in order to understand it. This coding must therefore be stable, and so the macro-level is timeless.

7.5 Efficient Foresight Activity

Through the methodological construction of foresight, we aim to increase the efficiency of the process. Here efficiency means setting up clear conditions and goals for each stage of the process in order to increase and to make the information flow more reliable. The results of the foresight process are not measurable because they are distributed and shared among the stakeholders' attitudes and world views. The methodological construction means that we achieve this through careful selection and in-depth methodological study of the applied tools and methods.

Different foresight processes could be constructed by different methods, but there are some typical methods on each level of the process.

In the first stage, we apply the method for the construction of foresight-specific time dimensions, taking into account the compatibility with the scientific meso-levels. According to our experience, the future cannot be revealed in its whole complexity, but it can only be understood in some special, suitable, well-chosen dimensions. In science, the construction of one specific meso-level is targeted to find special aspects where better understanding is feasible. The meso-level is the most suitable point for scientific method to enter the foresight process. On the meso-level, foresight studies explore the available information and knowledge to provide sufficient information for further decisions. If there is a narrow space for available time dimensions, further knowledge should be created by soft methods. Microsimulations or forecasting with Delphi questionnaires could be applied to explore the future and find the relationship between the stakeholders' decisions and time. These methods should neither replace nor compete with scientific achievements because this would decrease reliability and efficiency. On the other hand, these methods should help to expand scientific achievements, even though they are obviously not supposed to create a comprehensive knowledge base for the wide-ranging needs of foresight.

In the second stage of the foresight process, consensus about the key factors of the individual desirable futures should be achieved. This means:

- Contact with the scientific community to reveal the state-of-the-art meso-levels in science. (Note that the meso-levels of science could be defined as the consensus of the scientific community, but in some cases this is not easy to distinguish from external (non-scientific) interests that are not pure scientific constructions.)
- Contact with the stakeholders of the foresight process to gain access to their point of view and way of understanding to reveal the meso-level of the practice (the individual definition of time). These meso-levels must, however, be in accordance with each other and the scientifically proven meso-levels.
- Expand the revealed meso-levels with additional dimensions to support comprehensiveness.

Without a consensus about the key factors, no main message may emerge. On the micro-level, efficient methods should be developed to help the choice of key factors that represent an emerging meso-level message. It is not necessary that a single meso-level could be determined on the basis of the chosen key factors. A special study of the non-scientific meso-level knowledge is required. Here data mining and expert panels are both applicable to access specific micro-level information in this sense and seem to be suitable choices of methods at this stage.

The foresight process can be considered successful if the stakeholders' decisions to achieve their own desirable future are sufficiently supported. Therefore, the chosen messages at the micro-level should be disseminated to the stakeholders in a way that they can agree and accept changing their attitude and world view if necessary. Because of distributed knowledge, the chosen meso-level cannot be transferred directly. Therefore, macro-level foresight methods should set the factors of importance for all the stakeholder groups for targeting and choose the channels of communication. Typical methods like visioning and networking use a number of such channels to facilitate the personalization of the foresight message.

Finally, let us show how different approaches to foresight appear in the various aspects of its methodological construction in terms of data, time, uncertainty, and complexity.

7.5.1 The Methodological Construction of Foresight Through Social Constructivism

Social constructivism (Prawat and Floden 1994) is the theory behind learning and interpreting the world by individuals exposed to different contextual and cultural environments. With the individual learning process in focus, this theory seems to be a good candidate for the theoretical underpinning of foresight.

Data. Data is what stakeholders or participants reveal from their own, specific world view: expressed as interpretations of the world. The way of expression is usually communication (and not observed behaviour or human experiments), that is, meso-level knowledge with respect to the individual.

Time. From the upcoming dialogue, a consensus is reached, that is, according to some authors, a must in the foresight process (Barnes et al. 1996). Consensus on time in this sense is also considered as a social construct similar to the whole foresight process itself.

Uncertainty. It depends heavily on the tool or method used to facilitate participants' expressions and communication on how conflicts and opposing opinions are handled. Consensus should not aim to determine at the meso-level how stakeholders should think as this would simply be a way of manipulating opinions. As the process advances, some opinions can be suppressed, while others could be overemphasized. In this manner, not only the comprehensive information support but also the direction of the dialogue is exposed to uncertainty. Methods based on social constructivism must deal with these uncertainties on the meso-level.

Complexity. Consensus usually does not mean a common world view among the participants. It rather means a suitable 'greatest common denominator' among individuals' different world views. This kind of consensus is likely to be irrelevant with respect to real future change. Consensus can suppress signals in participants' opinions, or conflicting opinions can be disregarded. On the other hand, without any guided simplification, it is hard to imagine that any social construct as a result of the foresight activity could emerge. If dialogues can be guided in different ways, methods based on social constructivism must deal with this complexity on the macro-level.

The methodological construction of foresight helps to clarify tools rooted in social constructivism in order to develop methods that (1) confine consensus on meso-level communication, usually at the beginning of the activity in order to find the goal of the foresight programme, and (2) encourage emergent thoughts and foster the participants' macro-level learning to develop better understanding of the present and their new behavioural patterns for the future.

7.5.2 Grounded Theory in the Methodological Construction of Foresight

Glaser and Strauss's (1967) original idea of grounded theory could also be a very helpful basis to underpin the methodological construction of foresight between the micro- and the macro-level. The main idea of the theory is that the researcher studying the available data (micro-level) will also address the orientation of the theory (macro-level).

Data. In foresight, grounded theory might be useful to deal with the stakeholders' different interpretations of reality. Communities and differences should be recorded in such a way that the largest variety of possible emerging theories could fit the data.

Time. Grounded theory is timeless: Several categories emerge but do not create unique, comprehensive dynamics within their borders, although the constructed theories can serve as a background for communication and interaction at a later stage. Note that there can be several laws, rules, and dynamics that fit the data but only a few that are also accepted as an explanation. This is the main difference in how grounded theory is applied in social sciences and in foresight: The former focus on what possible theories may emerge and forget about the orientation of the data for the discovery; the latter focuses on the orientation which supplies this macro-level knowledge to the stakeholders and lets them work with possible emergencies in their individual world view.

Uncertainty. 'Theoretical sensitivity' is the researcher's subjective point of view of how to study the data. Induction can be heuristic.

Complexity. Micro-level definition and collection of the data: A 'pure' micro-level does not exist, as all data have been collected on the basis of a certain theory (i.e. theory-ladenness). Also a newer theory could support other aspects of data collection, which means that the available data are likely to be insufficient for new approaches.

Methods developed on the basis of grounded theory are useful in the methodological construction of foresight because it underpins the tools used to address all available information to set up a 'ground' for the discourse. Even when the foresight programme is thematically well-structured, in order to discover the true driving forces of the present, a foresight programme should be able to restructure the current structures of thought. Grounded theory deliberatively chooses possible 'grounds' independently from the meso-level, based on available data (micro-level). The newly constructed direction of thinking causes new types of meso-level behaviour of the participants to emerge, and so there is now a new platform of communication and individual learning.

7.5.3 Evolutionary Theory

Evolutionary theory (László 1996) in foresight is a theory that does not dynamically construct the future but reveals the elements of the future. In this theory, the simplified, but general concepts of mutation, heredity, and selection help to construct meso-elements that can be further developed into a dynamic system.

Data. Evolutionary theory works with micro-level data: entities with full-depth complexity of their properties and usually also the complex properties of the environment.

Time. In the heart of evolutionary theory, we find the questions: What are the possible states of the system that fit a given environment? And which states can be achieved in a gradual, continuous manner? This means that evolutionary theory constructs the meso-level based on the micro-level. Note that, according to different definitions of the entities (micro-level), different meso-levels can be constructed. This does not, however, mean that the theory would consider the macro-level in depth.

Uncertainty. Concerning the purpose of the model, evolutionary theory does not operate on the macro-level. But this purpose can be defined only on the macro-level. What is the phenomenon that we are attempting to model? In foresight studies, evolutionary theory should be applied to help stakeholders to develop their world views and improve their interpretation of micro-level data. Uncertainty arises because the definition of the micro-level may differ among the stakeholders.

Complexity. Concerning the complexity of data, in foresight, evolutionary models should be applied to show which features are decisive. This would help to reduce the complexity of each stakeholder's world view, but may increase the complexity of the whole foresight process, because misunderstanding about the importance of different features would probably increase.

Evolutionary theory can be useful to build the micro-level, but with its general concepts, the theory does not construct the meso-level. Therefore, in the methodological construction of foresight, we would rather consider it as a tool that is less ambitious than, for example, the grounded theory but, on the other hand, more efficient in practice.

7.6 Conclusions

We regard our concept of methodological construction as a first step towards a theoretical underpinning of foresight. In its current stage, it is still rather a philosophical underpinning, as it defines a framework for theories rather than pointing out the desirable theory. The aim of this framework is to define efficiency in the practice of foresight, that is, to ensure clarity of methods: If foresight is focused to support the individual learning process of the participants, they should be able to exploit the foresight programme to the utmost.

The methodological construction is an epistemological approach. The future is complex and needs a holistic understanding of the present. The participants include experts from various fields with limited knowledge. Having individual learning in focus, communication serves only for interactions, and there is no need for consensus to be forged. Other roles also appear clearer in this approach: The role of participation is to support the intrinsic need of the participants to understand the recent change and prepare for future possibilities, and it also ensures that the foresight programme has a more comprehensive basis of sources of information. The role of experts in various fields is to support the communication tools with scientific and more grounded information in order to avoid identifiable contradictions.

Our framework can support several theories through a three-level thinking. The micro-level is a philosophical term that defines the nuclear elements of information. Data are collected on the basis of a theory or a specially defined process that makes information theory-ladenned. The macro-level consists of the ideas, maxims, and axioms of those aspects of the change we would like to study. Complexity and limits of knowledge usually prevent a general theory of everything, but several aspects can be studied in various depths. The logic or understanding of these aspects emerges on the meso-level.

Tools at different levels of the foresight activity may have different theoretical underpinnings. For instance, evolutionary theory provides a substantial foundation for tools with definite uncertainty, moderate complexity, and fully constructed time on the meso-level. On the micro-level, this could be extended by a generalized use of the data based on grounded theory. This is conceptually different from methods designed for the macro-level. On the macro-level, for instance, social constructivism could support principles of shared knowledge for specific methods. The foresight activity itself has diverse goals, but on each level, we define clear objectives for the methods. This can make the foresight activity more reliable, more efficient, and even more successful.

Acknowledgements This work was supported by the Hungarian Science Foundation TÁMOP-4.2.1/B-09/1/KMR-2010-0005 project.

References

Alács, P. (2006). *Information – Model – Society*, Budapest: MTA-BCE (in Hungarian).

Barnes, B., Bloor, D., & Henry, J. (1996). *Scientific knowledge: A sociological analysis*. Chicago: University of Chicago Press.

Dopfer, K., Foster, J., & Potts, J. (2006). Micro-meso-macro. *Journal of Evolutionary Economics, 14*, 263–279.

Glaser, B., & Strauss, A. (1967). *The discovery of grounded theory: Strategies for qualitative research*. London: Weidenfeld and Nicholson.

Horton, A. (1999). A simple guide to successful foresight. *Foresight, 1*(1), 5–9.

László, E. (1996). *Evolution: The general theory* (2nd ed.). New Jersey: Hampton Press.

Lewis, J. D., & Weigert, A. J. (1981). The structures and meanings of social time. *Social Forces, 60*(2), 432–462. Oxford University Press, Oxford.

Prawat, R. S., & Floden, R. E. (1994). Philosophical perspectives on constructivist views of learning. *Educational Psychologist, 29*(1), 37–48.

Slaughter, R. (2003). *Futures beyond dystopia, creating social foresight*. London: Routledge-Flamer.

Sorokin, P. A., & Merton, R. K. (1937). Social time: A methodological and functional analysis. *The American Journal of Sociology, XLII*(5), 615–629. The University of Chicago Press, Chicago.

Wilson, E. B. (1952). *An introduction to scientific research*. New York: McGraw-Hill.

Chapter 8
Going from Narrative to Number: Indicator-Driven Scenario Quantification

Eric Kemp-Benedict

8.1 Introduction

Scenario analysis has more than a half-century of history behind it (Glenn and The Futures Group International 2003), and a wide range of scenario methods and techniques are now available (Bishop et al. 2007). While the term "scenario" refers to a story about the future – that is, a narrative – many scenario exercises include a quantitative analysis. This is particularly true in the environmental realm, and recent important examples include the Special Report on Emissions Scenarios for the Intergovernmental Panel on Climate Change (Nakićenović and Swart 2000), the United Nations Environment Programme's Global Environment Outlook (UNEP 2007), the Millennium Ecosystem Assessment (Carpenter et al. 2005) and the Comprehensive Assessment of Water Management in Agriculture (CA 2007).

Capturing some of the lessons provided by certain of these exercises, a methodology has been developed for combining qualitative and quantitative components of a scenario, the story-and-simulation (SAS) approach (Alcamo 2001, 2008). However, the SAS methodology appears to have been designed with a particular kind of scenario exercise in mind, in which one or more models already exist, and the goal of the exercise is to match the assumptions driving the model with the storyline developed by a team of scenario narrative writers.

While this is the dominant approach at the global level, at smaller scales, another type of scenario-development process predominates, in which a quantitative analysis is to be carried out, but the model does not already exist, or at any rate will not

This chapter is substantially revised version of an earlier paper "From Narrative to Number: A Role for Quantitative Models in Scenario Analysis" (Kemp-Benedict 2004).

E. Kemp-Benedict (✉)
Stockholm Environment Institute,
11 Curtis Avenue, Somerville, MA 02144, USA
e-mail: eric.kemp-benedict@sei-international.org

M. Giaoutzi and B. Sapio (eds.), *Recent Developments in Foresight Methodologies*, Complex Networks and Dynamic Systems 1, DOI 10.1007/978-1-4614-5215-7_8,
© Springer Science+Business Media New York 2013

be identified prior to the scenario narrative development. This chapter presents an approach to quantifying a scenario narrative for the second type of exercise.

The particular focus in this chapter is on scenarios for sustainability assessment[1], and so the system being studied is a socioecological system (Berkes and Folke 1998). Scenario modelling for socioecological systems will always have an element of art to it, so that an automatic procedure is unlikely to ever be achieved. Nevertheless, a process can be put in place that makes the quantification exercise more coherent and manageable. The approach presented in this chapter has been developed over time in the course of their scenario modelling work by staff of the Stockholm Environment Institute and is called "indicator-driven development".

8.2 Fundamental Challenges in Scenario Modelling

Depending on the audience, it is either self-evident that models should be applied to problems of environmental and social sustainability or that the use of models lies somewhere between the questionable and the highly problematic. The author is a modeller and is naturally inclined towards the happy view that models usually do what they are intended to do – to represent a system sufficiently well that they can be used to draw conclusions about the potential future functioning of that system. However, the critiques of models, especially within the environmental domain, are impressive and concerning. Before we present an approach to scenario modelling later in Sect. 8.3, this section offers a defence of the activity.

Socioecological systems, the focus of this chapter, present considerable challenges for modelling. Some authors (Pilkey and Pilkey-Jarvis 2006) have gone so far as to claim that models are nearly always dangerously misleading when used to make policy decisions. Even without considering social dynamics, natural systems are both open and complex (Oreskes 1998, 2003; Beck 2002; Pilkey and Pilkey-Jarvis 2006), while within the subcategory of natural systems, even when the mechanisms are thought to be well understood, a relative scarcity of data compared with the number of free parameters creates fundamental challenges for model calibration (Beven 2001; Beck 2002).

When societies and the actions of prominent individuals are added to the system, the difficulty is even greater. While the issue of complexity in socioecological systems may eventually be brought under some control (Costanza et al. 1993; Kohler 2000; Johnson 2008), especially in very localized studies , where such techniques as "companion modeling" (Bousquet et al., 1999) bring stakeholders directly into

[1] The method described in this chapter shares many features with the "XLRM" method of Lempert et al. (2003) for robust decision making (RDM). The RDM process is distinct from XLRM, which is an approach to building scenario models. While RDM can be run without an attendant narrative, we view the narrative as an essential aspect of scenario development, and see Lempert et al.'s RDM (and related techniques) as very useful ways to use models to give insights to a narrative team.

model building, evaluation, and use. It is so far proving to be elusive for much of policy modelling, and in any case, the question of openness remains. Proceeding to use quantitative models without recognition of the uncertainties has led to what one author called a "credibility crisis" in the use of models for evaluating the future (van der Sluijs 2002).

It is worth noting that there was an earlier debate over the relative merits of qualitative and quantitative approaches to futures in the 1970s and 1980s (Moll 2005). Then, as is also the case today, modelling had been pre-eminent, but its shortcomings were becoming apparent, and practitioners were recommending a shift towards greater participation by non-modellers in scenario exercises. Today's scenario models are more sophisticated than in the 1970s and 1980s, and modellers are not making the same mistakes. Today's critiques of models are likewise more sophisticated: they point to fundamental limitations of models in their application to complex, open systems, in particular socioecological systems.

This chapter starts with the premise that a combination of creative scenario narrative development and quantitative modelling can lead to greater insight than if either method were carried out in isolation, an argument that has been well presented elsewhere (Rotmans 1999; Smil 2000; van Asselt et al. 2001; van der Sluijs 2002).

Where models fall short, conventional scenario analysis is at its most useful. Scenario analysis is called for when there are driving forces that are both highly uncertain and highly important (Schwartz 1996). In such a situation, there is no reason to expect that point forecasts from models will be of any use – the only certainty is that the forecast will be wrong (Smil 2000). Instead, scenario methods (and futures methods in general) direct the analysis towards a multiplicity of futures, any one of which might come to pass. The question for policy then becomes how to develop strategies that are resilient in the face of different possibilities (Schwartz 1996), or how to make a desirable future more likely (Höjer and Mattsson 2000).

This response to the critique of modelling socioecological systems lends urgency to the goal of this chapter, which is to contribute a method for combining qualitative and quantitative analysis. If the combination of a scenario narrative and a quantitative analysis is better than either in isolation, then it is important to develop techniques for carrying out joint qualitative and quantitative exercises effectively.

8.3 An Approach to Scenario Modelling

The response to the critique of scenario models presented in the previous section suggests that there can be a division of labour between the qualitative and quantitative aspects of a scenario exercise. The somewhat provocative claim of this chapter is that the main role for the narrative is to capture complexity and openness, while the main role for quantitative analysis is to capture what might be called "complicatedness", where, following Rotmans (1999), complexity is identified as arising from the interconnections between different parts of a system, such as feedback loops, while complicatedness arises from having a large number of components in a system.

Having made this claim, two caveats should be borne in mind. First, the allocation of complexity to the narrative portion of a scenario exercise is not appropriate in all cases. Some systems, such as the climate, cannot be understood clearly without the aid of computer models that represent the climate as a complex system. Also, complex systems models can provide unique and useful insights into the workings of natural systems. However, in many – perhaps most – instances, the problems that arise from the use of models in policy because of complexity and openness can be ameliorated through the insights of a group of people using foresight techniques. People are capable of imagining diverse possibilities for the future that take into account different layers of social relevance and meaning (Inayatullah 2002; Slaughter 2004). They are furthermore capable of learning through a scenario process in a way that enhances the value of the envisioned futures (Chermack and van der Merwe 2003). Thus, while to some degree quantitative models can capture the uncertainties of complex systems, to the extent that these uncertainties represent highly uncertain and important driving forces, they can be well captured within a scenario narrative exercise. Furthermore, uncertainties due to external influences on an open system can be identified in a thoughtful narrative.

The second caveat is that people are not naturally gifted with the ability to imagine and anticipate departures from normality, and cognitive science has made clear that people can be badly misled by subjective feelings of certainty (Burton 2008). However, many futures techniques, such as those enumerated in *Futures Research Methodology* (Glenn and Gordon 2003) and the *Knowledge Base of Futures Studies* (Slaughter 2005), are designed precisely to assist people in moving beyond their limited view of what is possible and to render their certainty problematic. The approach to scenario quantification proposed in this chapter starts from the notion that the best method available today to carry out policy analysis for socioecological systems is a combination of quantitative modelling and qualitative insights gained from foresight activities.

8.3.1 Narrative First

In the discussion below, the task of building a combined narrative and quantitative scenario is broken into two subtasks: narrative writing and mathematical analysis. Although the same person or group of people may do both subtasks, more often they are carried out by different people with different sets of skills. In this chapter, the two groups will be called the "narrative team" and the "modelling team". In the approach put forward in this chapter, the narrative drives scenario development, while the modelling team follows the narrative team's lead. However, the process is not all one-way: the quantitative analysis also informs the narrative scenario development. Taking the reciprocal influence into account, there are five main roles that quantitative scenario development can play when implemented in response to a narrative:

1. Force a clarification of terms and mechanisms.
2. Expose contradictions in mental models.
3. Provide a feel for the scope of possible outcomes within a narrative framework.
4. Illustrate a particular scenario narrative.
5. Make a study replicable, extensible and transferable.

Most of these roles for quantitative models are straightforward. Regarding the first role, when a narrative is translated into a formal structure, many potentially ambiguous points must be nailed down, and key decisions must be made. This process sharpens the narrative analysis, as the narrative team is forced to address its ambiguous goals and statements. Note that this positive outcome is not reached when the quantitative model drives the analysis, and the narrative follows from it. In this case, the mathematical model has been built by people (the modelling team) who have already encountered ambiguities and resolved them in ways that may, or may not be, acceptable to the people using the quantitative outputs (van der Sluijs 2002). The decisions are not made jointly between the narrative and the modelling teams, so they do not provoke discussion.

The second item – exposing contradictions in mental models – is perhaps less clear, and it highlights a key role that scenarios play, that of fostering cognitive development and learning (Chermack and van der Merwe 2003; Robinson 2003). Constructivist theories of cognition and learning posit that people actively construct mental models through which they filter their experiences. Those mental models are remarkably resilient and are relinquished only when they are shown (repeatedly) to be inconsistent – either internally inconsistent or inconsistent with external reality (Yankelovich 1991; Kempton et al. 1997). Narratives reflect the mental models of their authors, and by translating them into formal terms, contradiction can be exposed, either through the process of developing the formal model or through manipulating the model.

The final three roles for quantitative models are perhaps the main roles that are envisioned when a modelling study is commissioned. The model lets the narrative team "play" with options in a what-if mode in order to gain insight. Once a set of quantitative scenarios is settled upon, model outputs can be used in the narrative text and in graphs to illustrate concretely one way in which the scenario could unfold. Finally, models are usually generalizable beyond the study for which they were designed, and can be modified and extended for further studies.

8.3.2 Indicator-Driven Development

Indicator-driven development is an approach to developing scenario models starting with a scenario narrative. The key idea behind indicator-driven development is that models should be developed in order to provide quantitative indicators that serve to illustrate an existing narrative, while offering the narrative team the "levers" that they need to differentiate one scenario from another. By organizing the modelling effort around the indicators, the focus of the effort remains on the output of most interest to the users of the scenarios. As a side benefit, it provides a convenient metric to track progress on the model, by recording the number of indicators quantified over time.

The entire indicator-driven development process follows these steps:

1. Specify the boundaries of the study.
2. Select indicators and levers.
3. Decide on a model structure.
4. Estimate the time, decide on a schedule and revise the scope if necessary.
5. Iteratively develop, test, document and release the sub-models.
6. Release the final set of quantitative scenarios.

The study boundaries – in time, space and thematic content – should already be specified when the narrative is developed. The narrative will also have some implicit indicators and levers, but further indicators may be proposed by the narrative team when the quantitative work is initiated. The modelling team may propose its own indicators and levers and may decide not to model some of the indicators requested by the narrative team, either because they cannot be modelled, because data are lacking, because of time and budget constraints, or for some other reason. The modelling team should decide the model structure but should also communicate the structure to the narrative team and receive its feedback. Once the structure and set of indicators are decided, the modelling team can estimate the time to complete the modelling activity and, if necessary, decide with the narrative team to revise the scope of the analysis. Finally, the model is ready to be iteratively developed and documented, and then released.

This approach has some significant differences from the story-and-simulation (SAS) methodology proposed by Alcamo (2001, 2008). The SAS methodology asks the narrative team to provide qualitative values for driving variables, which are then quantified by the modelling team. The quantified drivers are then used as inputs to an existing model, the outputs of which are reviewed by the narrative team. In contrast, the indicator-driven approach assumes that no model has yet been identified. The narrative team identifies the indicators and levers – the outputs from and inputs to the model – that it wishes to see. The task of the modelling team is then to fill in the intervening calculations by either finding an appropriate model or, if one does not exist, creating a model.

The basic strategy in indicator-driven development is shown schematically in Fig. 8.1. As shown in the figure, any of the modules in the full model can make use of a common set of driving variables as well as any results from upstream modules (including values calculated in a prior time step). The module is then responsible for providing indicators required by the modelling team as well as inputs to downstream modules.

The schematic in Fig. 8.1 is intentionally non-specific about the modelling approach. The emphasis is on the constraints to the module design: what the module can make use of, and what it must produce. There are further constraints that the modelling team must consider, but these cannot be specified in general terms. Additional constraints fall into at least the following categories:

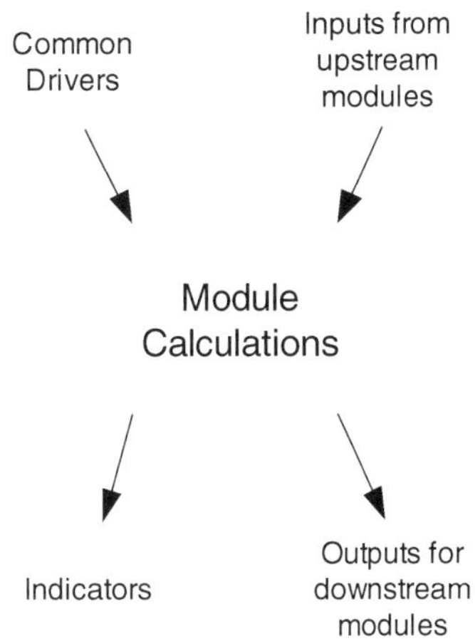

Fig. 8.1 The starting point for module design in indicator-driven development

- Staffing, time and budget
- Spatial and temporal scale
- Level of detail required for outputs
- Standards and state of the art in the field relevant to the study and the model

Further constraints may also apply, such as political or institutional constraints. Within these constraints, the team constructs a model, drawing upon its own knowledge and experience, as well as the numerous models and algorithms available in the various professional literatures and on the Internet.

8.4 Conclusion

There is substantial consensus that quantitative models are, in themselves, insufficient for exploring the future of socioecological systems, because such systems are both complex and open. This chapter started with the premise, as argued in the scenario modelling literature, that qualitative foresight methods – in particular narrative scenario development – can ameliorate the problems of complexity and openness. In the "indicator-driven" approach introduced in this chapter, the scenario narrative is prepared first, and the indicators and "levers" needed by the narrative team are specified. The narratives, indicators and levers are used to constrain, organize and track a model development process.

The indicator-driven approach does not specify any particular modelling technique or structure. However, this chapter argues that for many scenario analyses, models are best used to handle "complicatedness" while leaving complexity to the narrative. This is admittedly a provocative claim and is in any case subject to caveats. However, the modelling effort can be simplified and streamlined by starting with this division of labour between the narrative and quantitative aspects of a scenario analysis and, in combination with an indicator-driven approach, can make scenario quantification more coherent and manageable.

References

Alcamo, J. (2001). *Scenarios as tools for international environmental assessments* (Environmental Issue Report No. 24). Copenhagen: European Environment Agency.

Alcamo, J. (2008). The SAS approach: Combining qualitative and quantitative knowledge in environmental scenarios. In J. Alcamo (Ed.), *Environmental futures*. Amsterdam: Elsevier.

Bousquet, F., Barreteau, O., Le Page, C., Mullon, C., Weber, J., 1999. An environmental modelling approach. The use of multi-agents simulations. In: Blasco, F., Weill, A. (Eds.), Advances in Environmental and Ecological Modelling. Elsevier, Paris, pp. 113–122.

Beck, M. B. (Ed.). (2002). *Environmental foresight and models: A manifesto*. Amsterdam: Elsevier.

Berkes, F., & Folke, C. (1998). Linking social and ecological systems for resilience and sustainability. In F. Berkes & C. Folke (Eds.), *Linking social and ecological systems: Management practices and social mechanisms for building resilience*. Cambridge: Cambridge University Press.

Beven, K. (2001). How far can we go in distributed hydrological modelling? *Hydrology and Earth System Science, 5*(1), 1–12.

Bishop, P., Hines, A., & Collins, T. (2007). The current state of scenario development: An overview of techniques. *Foresight, 9*(1), 5–25.

Burton, R. A. (2008). *On being certain: Believing you are right when you're not*. New York: St. Martin's Press.

Carpenter, S. R., Pingali, P. L., Bennett, E. M., & Zurek, M. B. (Eds.). (2005). *Ecosystems and human well-being scenarios: Findings of the scenarios working group of the millennium ecosystem assessment*. Washington: Island Press.

Chermack, T. J., & van der Merwe, L. (2003). The role of constructivist learning in scenario planning. *Futures, 35*, 445–460.

Comprehensive Assessment of Water Management in Agriculture (CA). (2007). *Water for food, water for life: Comprehensive assessment of water management in agriculture*. London: Earthscan.

Costanza, R., Wainger, L., Folke, C., & Mäler, K.-G. (1993). Modelling complex ecological economic systems: Toward an evolutionary, dynamic understanding of people and nature. *BioScience, 43*(8), 545–555.

Glenn, J. C., & Gordon, T. J. (Eds.). (2003). *Futures research methodology*, V.2.0 (CD-ROM). Washington, DC: American Council for the UNU.

Glenn, J. C., & The Futures Group International (2003). Scenarios. In J. C. Glenn, & T. J. Gordon (Eds.), *Futures research methodology*, V. 2.0 (CD-ROM). Washington, DC: American Council for the UNU.

Höjer, M., & Mattsson, L. G. (2000). Determinism and backcasting in future studies. *Futures, 32*, 613–634.

Inayatullah, S. (2002). Layered methodology: Meanings, epistemes, and the politics of knowledge. *Futures, 34*, 479–491.

Johnson, J. (2008). Science and policy in designing complex futures. *Futures, 40*(6), 515–596.

Kemp-Benedict, E. (2004). *From narrative to number: A role for quantitative models in scenario analysis, iEMSs 2004*. Osnabrück,: International Environmental Modelling and Software Society.

Kempton, W., Boster, J. S., & Hartley, J. A. (1997). *Environmental values in American culture*. Cambridge, MA: MIT Press.

Kohler, T. A. (2000). Putting social sciences back together again: An introduction to the volume. In T. A. Kohler & G. J. Gumerman (Eds.), *Dynamics in human and primate societies: Agent-based modeling of social and spatial processes*. New York: Oxford University Press.

Lempert, R.J., Popper, S.W., Bankes, S.C., 2003. Shaping the Next One Hundred Years: New Methods for Quantitative, Long-term Policy Analysis. Rand Pardee Center, Santa Monica, CA.

Moll, P. (2005). The thirst for certainty: Futures studies in Europe and the United States. In R. Slaughter (Ed.), *Knowledge base of futures studies* (CD-ROM) (Vol.1). Queensland: Foresight International.

Nakićenović, N., & Swart, R. (Eds.). (2000). *Emissions scenarios: Special report of the intergovernmental panel on climate change*. Cambridge: Cambridge University Press.

Oreskes, N. (1998). Evaluation (not validation) of quantitative models. *Environmental Health Perspectives, 106*(Suppl. 6), 1453–1460.

Oreskes, N. (2003). The role of quantitative models in science. In C. D. Canham, J. J. Cole, & W. K. Lauenroth (Eds.), *Models in ecosystem science*. Princeton: Princeton University Press.

Pilkey, O. H., & Pilkey-Jarvis, L. (2006). *Useless arithmetic: Why environmental scientists can't predict the future*. New York: Columbia University Press.

Robinson, J. (2003). Future subjunctive: Backcasting as social learning. *Future, 35*, 839–856.

Rotmans, J. (1999). *Integrated assessment models: Uncertainty, quality and use* (Working Paper No.199-E005). Maastricht: ICIS.

Schwartz, P. (1996). *The art of the long view: Planning for the future in an uncertain world*. New York: Currency Doubleday.

Slaughter, R. (2004). *Futures beyond Dystopia: Creating social foresight*. London: Routledge Falmer.

Slaughter, R. (Ed.). (2005). *Knowledge base of futures studies (CD-ROM)*. Queensland: Foresight International.

Smil, V. (2000). Perils of long-range energy forecasting: Reflections on looking far ahead. *Technological Forecasting and Social Change, 65*, 251–264.

United Nations Environment Programme (UNEP). (2007). *Global environment outlook GEO4: Environment for development*. Valletta: Progress Press.

van Asselt, M., Mellors, J., Rijkens-Klomp, N., Greeuw, S., Molendijk, K., Beers, P., & van Notten, P. (2001). *Building blocks for participation in integrated assessment: A review of participatory method*. Maastricht: International Centre for Integrative Studies.

van der Sluijs, J. P. (2002). A way out of the credibility crisis of models used in integrated environmental assessment. *Futures, 34*, 133–146.

Yankelovich, D. (1991). *Coming to public judgement: Making democracy work in a complex world*. Syracuse: Syracuse University Press.

Chapter 9
On Foresight Design and Management: A Classification Framework for Foresight Exercises

Totti Könnölä, Toni Ahlqvist, Annele Eerola, Sirkku Kivisaari, and Raija Koivisto

9.1 Introduction

The gradual paradigm shift in innovation research and policy from linear to systemic innovation models has challenged also the conventional technocratic technology-driven forecasting practices and called for new participatory and systemic foresight approaches (Smits and Kuhlmann 2004). In the 1980s, publicly funded foresight activities were commonly seen as an instrument for assisting in the development of priorities for research and development (R&D) resource allocation (Irvine and Martin 1984). Later on, stakeholder participation and networking have been regarded as increasingly important elements of foresight activities for 'wiring up' the multilayered innovation systems both in the public (Martin and Johnston 1999) and private sectors (e.g. Salmenkaita and Salo 2004). Reports from recent foresight projects have, in turn, emphasized the importance of common vision building as a

T. Könnölä (✉)
Impetu Solutions, Palacio de Miraflores, Carrera de San Jerónimo, 15-2, 28014 Madrid, Spain
e-mail: totti.konnola@impetusolutions.com

T. Ahlqvist
VTT Technical Research Centre of Finland, P.O. Box 106, 20521 Turku, Finland
e-mail: Toni.Ahlqvist@vtt.fi

A. Eerola • S. Kivisaari
VTT Technical Research Centre of Finland, P.O. Box 1000, FI-02044 VTT, Espoo Finland
e-mail: annele.eerola@vtt.fi; sirkku.kivisaari@vtt.fi

R. Koivisto
VTT Technical Research Centre of Finland, Tekniikankatu 1, P.O. Box 1300, FI-33101 Tampere, Finland
e-mail: koivisto@vtt.fi

M. Giaoutzi and B. Sapio (eds.), *Recent Developments in Foresight Methodologies*, Complex Networks and Dynamic Systems 1, DOI 10.1007/978-1-4614-5215-7_9,
© Springer Science+Business Media New York 2013

step towards the synchronization of the innovation system (Cuhls 2003). In these developments, the locus of foresight activities has tended to shift from positivist and rationalist technology-focused approaches to the recognition of broader concerns that encompass the entire innovation system, including its environmental, social and economic perspectives. The High Level Expert Group appointed by the European Commission crystallized these trends by defining foresight as follows (European Commission 2002): 'A systematic, participatory, future intelligence gathering and medium-to-long-term vision-building process aimed at present-day decisions and mobilizing joint action'.

While the expansion of foresight scope has provided significant opportunities for learning and synchronized action between different business units and/or policy fields, it may also have caused digression and ambiguity in the practice and theory of the management of foresight processes (Könnölä 2006). Managing a larger set of foresight activities and designing and managing individual, even strikingly different foresight projects, requires profound understanding of the varying nature of the foresight tasks and expectations. For this purpose, this chapter first develops a conceptual framework for the classification of foresight projects. The framework is then applied to examine a portfolio of foresight projects, in which a horizontal foresight expert team at the VTT Technical Research Centre of Finland (which is responsible of coordinating foresights projects involving experts from different VTT units and external stakeholder organizations) has recently been either a coordinator or a participant. The classification framework is expected to support the coherent and effective management of individual projects as well as the strategic management of the portfolio of foresight projects. The findings of the analysis also point out to the need for flexible modular design in the management of foresight projects.

The remainder of this chapter is organized as follows. Section 9.2 defines and discusses the key design dimensions to be considered in the planning and management of foresight projects. Section 9.3 elaborates a coherent classification framework for foresight projects, making use of the four key dimensions identified. The classification framework is then applied to the analysis of a number of foresight processes in which VTT has recently been engaged. Section 9.4 reflects the findings from the empirical analysis and, in particular, examines the needs and reasons for the modular design of foresight projects. Section 9.5 concludes this chapter.

9.2 Key Dimensions of Foresight Project Design and Management

The systemic understanding of innovation processes has challenged conventional technology-driven forecasting practices and called for new participatory foresight approaches that also address the consideration of diverse perspectives, the formation of shared knowledge and the examination of alternative futures. Foresight activities are also increasingly seen as crucial functions to prepare for the future: not only to identify promising technological pathways but also to engage relevant stakeholders and translate common visions into action (Eerola and Jørgensen 2002; Ahlqvist 2005; Dannemand Andersen et al. 2007; Könnölä et al. 2007a). Furthermore, foresight processes can often be seen as a pertinent design phase for the creation of new value networks that are based on novel combinations of technologies, organizational partnerships and institutional arrangements. The design and management of such processes are likely to face major challenges in responding to the diverse expectations of the client(s) and other stakeholders and in supporting efficient R&D resource allocation, enhanced networking and operational visions for joint action.

Thus, the design of foresight activities can benefit from structured approaches, which help to identify the expectations and challenges concerning the management of the process and the final outcomes. Towards this end, we will discuss and define the key design dimensions in the management of foresight processes. The key design dimensions, including *outcomes, chosen future perspectives, management and stakeholder engagement*, are first defined and elaborated and then applied to the analysis of selected foresight processes in which the VVT foresight experts have recently been engaged.

9.2.1 Informative Versus Instrumental and Informative Outcomes

Outcomes consist of the outputs, results and impacts of the foresight exercise. Outputs refer to the products and services, both tangible and intangible. The results, in turn, refer to the advantage (or disadvantage) that the beneficiaries obtain soon after the end of their participation in the foresight project, and the impacts refer to consequences affecting beneficiaries during and after the project.

For the purpose of defining the key design dimensions, the nature of the outcomes of foresight activities can be divided into instrumental and informative outcomes. *Informative outcomes* refer to the use of foresight to improve the understanding of present and future challenges of the innovation system and its parts. Thus, the informative outcomes do not refer to the expectations that foresight activity would necessarily lead directly to specific actions, although informative outcomes may increase the preparedness to act in some unspecified future situations (e.g. making it easier to recognize emerging risks and discover new windows of opportunities). *Instrumental and informative outcomes* refer not only to informative outcomes but also to the use of foresight to support the specific foreseen decision-making situation: for example, related to resource allocation or to the formation of strategic partnerships/joint actions.

Foresight processes not only elaborate but also produce shared new knowledge for the stakeholders. New knowledge in the foresight project is essentially produced through collaborative work, which benefits the strategic thinking and work of each participant. A useful framework for analysing such processes is provided by the SECI model of the dynamics of shared knowledge creation (Nonaka 1994; Nonaka and Takeuchi 1995; Eerola and Jørgensen 2002, 2008; Eerola and Väyrynen 2002).

In the SECI model, shared knowledge creation is envisaged as a spiral process in which tacit and explicit knowledge, as well as the different modes of knowledge conversion – that is, socialization, externalization, combination and internalization – plays a central role. Here, 'shared knowledge creation' refers to *generation of new knowledge* that the stakeholders can share, without necessarily agreeing about the exact meaning when applying it to specific problems and goals collectively or individually.

9.2.2 Consensual Versus Diverse Future Perspectives

The nature of future perspectives needs to be addressed in order to define the ways and methods by which the project develops an understanding of the future. Foresight activities often focus on the production of *consensual future perspectives* that refer to the creation of a common understanding on priorities, relevant collaborative networks and future actions. For example, the taxonomy of Barré (2002) reflects this trend by distinguishing between objectives for (1) setting scientific and technological priorities, (2) developing the connectivity and efficiency of the innovation system and (3) creating a shared awareness of future technologies. Alternatively, foresight activities can be seen as useful tools to identify diverse future perspectives with a view to understanding diverse ideas, opinions and perspectives in priority setting, identifying and fostering alternative and competing coalitions and value networks, as well as exploring alternative futures and generating rivalling visions (Könnölä et al. 2007a). The value of the communication of diverse perspectives and their inclusion in the decision making has been commonly recognized, for instance, in the field of risk analysis (e.g. Koivisto et al. 1997, 2002) and in the societal embedding of innovations (Kivisaari et al. 2004).

In more specific terms, diversity can be linked to widely discussed foresight objectives (i.e. priority setting, networking and common vision building) as follows (Könnölä et al. 2007a):

- Priority setting supports the identification of common future actions and the efficient allocation of resources (Irvine and Martin 2004), but may decrease the diversity of options that could challenge conventional approaches and dominant designs (Arthur 1989) and escape from techno-institutional lock-ins (David 1985; Arthur 1994; Jacobsson and Johnson 2000; Unruh 2000). Here, foresight can generate ideas on alternatives and recognize the diverse perspectives in priority setting (Salo et al. 2003; Keenan 2003).
- Networking enhances the connectivity of the innovation system and can improve its performance (Lundvall 1992; Martin and Johnston 1999) but may lead to the excessive strengthening of *existing* networks (e.g. Grabher and Stark 1997), creating path dependencies and locking out alternative technological options (Unruh 2000). Thus, foresight should also contribute to the creative restructuring and even the destruction of lock-in conditions by engaging different stakeholders in

the proactive generation of rivalling visions for competing coalitions based on different value networks with different architectures, configurations, features and standards (Tushman and O'Reilly 1997; Könnölä et al. 2011).
- Building a consensual vision of the future reduces uncertainties and helps synchronize the strategies and joint actions of different stakeholders (e.g. Cuhls 2003), but may lead to conservative and abstract results (Keenan 2003), to the effect that existing path dependencies are further strengthened. Nor are general abstractions readily actionable, especially if responsibilities are not clearly identified (Salmenkaita and Salo 2004). The search for a consensual vision of the future should therefore be complemented with – or even replaced by – the exploration of alternative futures and respective techno-institutional arrangements (Könnölä et al. 2007b). It might then also be easier to recognize the relevant key actors and their responsibilities.

The attempt to accommodate diverse perspectives on the future is also central to the methods for the scanning of weak signals defined as 'imprecise early indications about impending significant events'. Later on, this definition has been expanded to accommodate additional characteristics, such as *new, surprising, uncertain, irrational, not credible, difficult to track down, related to a substantial time-lag before maturing and becoming mainstream* (Ansoff 1975; Coffman 1997; Harris and Zeisler 2002; Mendoça et al. 2004).

9.2.3 From Fixed to Autonomous Management

The foresight process can be taken up with different kinds of management approaches, which is often driven by the diverse expectations laid on the project. Those in charge of the foresight process are likely to benefit from the sharp definition of their role and approach to the management of the foresight process. This makes it easier to design the process in a coherent way and to communicate the responsibilities of different stakeholders so that there are also good conditions for carrying out a high-quality foresight process. Here, two extreme approaches can be identified with respect to the classification purposes. *Fixed management* can be characterized as a centralized approach, in which coordinators fix the scope and methods of the exercise at the outset and control the process, which is often the case, for example, in Delphi exercises (Helmer 1983). *Autonomous management*, in turn, refers to the process intermediated by the coordinators, who facilitate an autonomous and evolving participant-led continuum of meetings and other activities, which may be the case, for example, in expert panel work (Salo et al. 2004).

The creation of new, especially shared, knowledge is challenging, in particular when the people participating in the foresight process have heterogeneous backgrounds, which occurs when various interest groups (industry, academia, government, NGOs, etc.) and different geographical areas (countries, regions, etc.) are engaging in the foresight process. Therefore, special attention must be paid to the organization of the process and to the appropriate use of formal tools and procedures. In these circumstances, the SECI model helps us to understand the dynamics of the knowledge conversion, also providing a tool for designing well-functioning autonomous foresight exercises that emphasize the learning process in the foresight exercise (Eerola et al. 2004; Eerola and Jørgensen 2008).

Foresight processes are increasingly designed as 'open source' and 'wisdom of crowds' types of open creative ideation processes, rather than centralized and consensual processes driven by one key actor. These processes may produce an increased number of novel ideas that can be utilized in the search for means to improve strategic intelligence and competitiveness. So, the complexity of the actual process management is likely to increase hand in hand with the amount, and potentially also the quality, of information. However, in such more 'people-intensive' processes, the process management and knowledge 'logistics' may become difficult if state-of-the-art practices and techniques, such as Internet-based decision support systems and computer-supported workshop tools, are not adequately applied.

9.2.4 From Exclusive to Extensive Stakeholder Engagement

Developments in the networking in foresight activities can be characterized in terms of the extensiveness of the stakeholder engagement, referring to the set of qualitative and quantitative factors including the number of stakeholders involved, openness to participate (who are invited/allowed to participate) and diversity of stakeholders involved (diverse disciplines, policy and industrial sectors, NGOs, etc.). Thus, *extensive stakeholder engagement* refers to the approach in which the actual number of participants is high. The stakeholder participation is encouraged and open for all the interested stakeholders, and many kinds of stakeholders are invited to participate in the process. *Exclusive stakeholder engagement*, in turn, means that stakeholder participation is limited and is, thus, not open for all the stakeholders interested. The exclusive stakeholder engagement can provide opportunities for intensive stakeholder communication and for dealing with sensitive issues related to intellectual property rights and competitiveness, for example.

Building on Barré (2002) and Meulen et al. (2003), Salo et al. (2004) have distinguished three levels of stakeholder intensity of engagement with respective objectives:

- *Low engagement:* Stakeholders exchange ideas on, and perceptions of, future challenges and comment on foresight deliverables, thus contributing inputs to the exercise. This does not, however, necessarily lead to notable changes in their perceptions.
- *Medium engagement:* Stakeholders engage in collaborative learning processes and proactive development of innovative options, to the effect that the perceptions of individual stakeholders are shaped by these processes.

- *High engagement:* Stakeholders are intensively involved in the collaborative management of the foresight exercise and also assume responsibilities in contributing to the development and realization of jointly approved action plans.

In the implementation of stakeholder engagement, it can be difficult to pursue both extensive and high engagement, because high engagement requires continuous and transparent processes of learning (Cruickshank and Susskind 1987). Such processes are likely to become time-consuming and expensive when the number of foresight panel participants increases (Hjelt et al. 2001; Meulen et al. 2003). However, without a sufficient number and diversity of participants, the activity may fail to produce innovative results or to reach sufficient media attention, or difficulties may be encountered during the response to the results.

Voss and Kemp (2006) discuss a parallel problem in reflexive governance. The problem consists of the contradicting requirements of opening up (towards the consideration of an extensive set of stakeholder perspectives) and closing down (towards focused analyses and intensive engagement): 'opening up is necessary to adequately grasp the factual embedding of decision making and problem solving in systemic contexts which comprise complex dynamics, heterogeneous values and distributed power. Closing down is necessary to reduce complexity in order to avoid anomy and keep up the ability to act'. One important aspect of opening up refers to the number and heterogeneity of actors who are involved in problem analysis, goal formulation or strategy development. Eventually, opening up needs to be linked in one way or the other with extended participation, since the knowledge about different problem aspects and values as well as resources is distributed among different actors. Ultimately, it is the diversity of perceptions held by different actors which is the key trigger for opening up governance processes. At the same time, however, it is also the key trigger for controversy and possible misunderstanding, which makes governance often difficult and seemingly ineffective.

Within these premises, in the foresight processes, recurring workshops can be functional in building networks, trust and shared understanding through exploring experiments and/or sequential opening up and closing down of stakeholder engagement. Transparency of the process and the maintenance of a proper balance between the qualitative and quantitative elements facilitate trust building. When there is sufficient time between the events for stakeholder interactions, for instance, between workshops, there is also an opportunity to genuinely refine the collective knowledge base with some further back-office work by a core group intensively engaged in the process. On the other hand, too long intervals between the events for stakeholder interaction may affect the intensity of the process and engagement as a whole (Eerola et al. 2005; Kivisaari et al. 2004).

In the subsequent sections, the design dimensions of foresight management discussed above are used to characterize a number of foresight activities in which the VTT horizontal foresight expert team has coordinated or participated. The positioning of individual projects in the framework clarifies the methodological decisions and the rationales of stakeholder engagement. Once the projects are positioned in the framework, they provide an overview of the whole portfolio of foresight projects that supports building the holistic view of the activities.

Table 9.1 The design dimensions of foresight management

Instrumental versus informative outcomes
Informative outcomes refer to the use of foresight to improve the understanding of the present and future challenges of the innovation system and its parts. Informative outcomes are not expected to lead to specific actions, although they can improve the preparedness to act in some unspecified forthcoming situations
Instrumental and instrumental *outcomes* refer not only to informative outcomes but also to the use of foresight to support the specific foreseen decision-making situation, for example, in relation to resource allocation or the formation of strategic partnerships/joint actions
Consensual versus diverse future perspectives
Consensual future perspectives refer to the creation of a common understanding on priorities, relevant collaborative networks and future actions
Diverse future perspectives refer to understanding diverse ideas, opinions and perspectives in priority setting and identifying and fostering alternative and competing coalitions and value networks, as well as exploring alternative futures and generating rivalling visions
Fixed versus autonomous management
Fixed management can be characterized as a centralized approach in which coordinators fix the scope and methods of the exercise at the outset and control the process, which is often the case, for example, in Delphi exercises
Autonomous management, in turn, refers to the process intermediated by the coordinators, who facilitate an autonomous and evolving participant-led continuum of meetings and other activities, which may be the case, for example, in expert panel work
Extensive versus exclusive stakeholder engagement
Extensive stakeholder engagement refers to the approach in which the actual number of participants is high, the stakeholder participation is encouraged and open for all the interested stakeholders, and many kinds of stakeholders are invited to participate in the process
Exclusive stakeholder engagement means that stakeholder participation is limited, and thus not open for all the stakeholders interested

9.3 Empirical Analysis of Foresight Projects in a Contract Research Organization

This section builds on the key dimensions of foresight project design discussed in Sect. 9.2 (see also Table 9.1). In particular, the classification framework is developed and applied to study the recent foresight activities in which the VTT foresight experts have recently been engaged, representing different regional, organizational and sectoral contexts. Thereby, the section intends to create an improved understanding of the methodological choices made during these projects and clarify how different foresight dimensions are linked to one another and some methodological choices.

The defined conceptual dichotomies of the foresight dimensions provide a useful structure for the analysis, assuming that foresight activities consist of identifiable elements for the purposes of classification. In practice, foresight activities often consist of some elements of both sides of these dichotomies, and altogether they are integrated in a case-specific process design. This is manifested in the summary of selected VTT foresight activities in Annex 1.

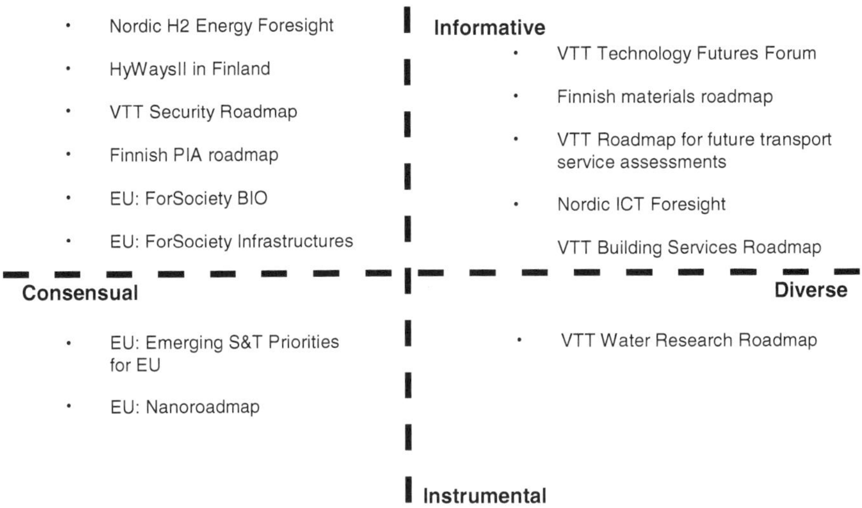

Fig. 9.1 Foresight projects positioned with respect to the dimensions of outcomes (informative vs. instrumental and informative) and future perspectives (consensual vs. diverse)

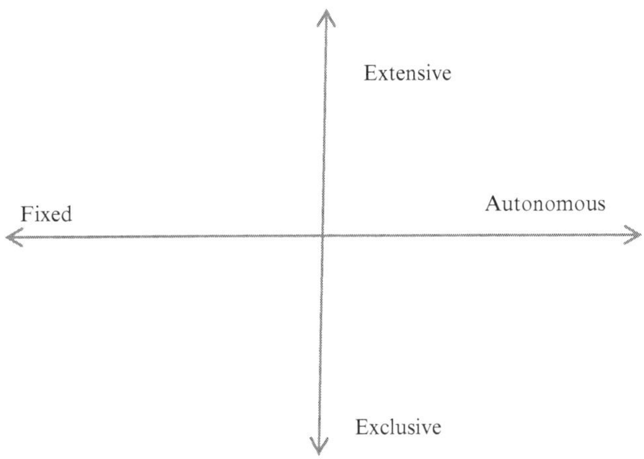

Fig. 9.2 Process management (from fixed to autonomous) and stakeholder engagement (from exclusive to extensive) dimensions in a coordinate system

Foresight projects can be further classified according to their design dimensions. When the dimensions of outcomes (informative vs. instrumental) and future perspectives (consensual vs. diverse) correspond, respectively, to the horizontal and vertical lines, the projects (described in Annex 1) can be positioned in four different quadrants (consensual and informative, consensual and instrumental, diverse and informative, and diverse and instrumental) (Fig. 9.1).

In parallel, the projects can also be positioned in relation to process-oriented dimensions. When the process management (autonomous vs. fixed) and stakeholder

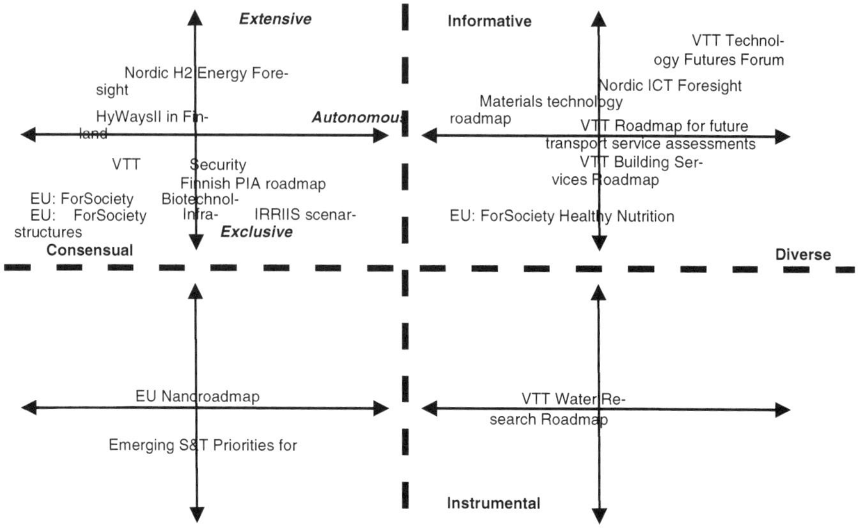

Fig. 9.3 Foresight projects positioned with respect to the dimensions of outcomes (informative vs. instrumental) and future perspectives (consensual vs. diverse) and in the coordinate system of stakeholder engagement (extensive vs. exclusive) and management (autonomous vs. fixed)

engagement (extensive vs. exclusive) dimensions are considered to correspond, respectively, to the horizontal and vertical axes, together they produce a coordinate system (see Fig. 9.2). Here, the horizontal axis represents the qualitative continuum from fixed to autonomous management, and the vertical axis the continuum from extensive to exclusive stakeholder engagement.

Further on, if the coordinate system (Fig. 9.2) is positioned in relation to each quadrant of Fig. 9.1, the foresight projects can be positioned in relation to the coordinates to provide detailed information on the nature of the outcomes and process of each project (Fig. 9.3).

Hence, once the project is in one of the four quadrants (according to consensual vs. diverse and informative vs. instrumental), the exact position of the project can be defined in the coordinates (from fixed to autonomous and from exclusive to extensive).

Positioning the projects in the quadrants, and in the coordinate systems within the quadrants, provides important insight for further analysis of their characteristics and the methodological choices of the projects. Subsequently, we discuss the projects in each quadrant and discuss their positioning in the coordinate systems.

9.3.1 Consensual and Informative Processes

Consensual, informative processes create a common understanding on priorities, relevant collaborative networks and/or future actions. They are expected to improve the understanding of present and future challenges of the innovation system and its parts. However, specific short-term actions are not necessarily expected after the projects. This setting relieves the participants partly from claiming value and partly from the pressures of policymaking and lobbying and, hence, may also enable otherwise adversarial parties to learn together and search for common ground for long-term agendas.

Two hydrogen-related foresight projects included visioning and modelling of future energy systems. These projects were characterized by fixed or partly fixed management, consensual perspectives and rather extensive stakeholder engagement. In the *Nordic H2 Energy Foresight*, the major challenge was to create shared understandings between different stakeholder groups representing five different countries. This required getting the right balance between quantitative and qualitative approaches. Much of the efforts were directed towards the creation of a common language and understanding between technical experts and modelling people and foresight experts and the various groups of stakeholders (industry, research institutes, public organizations and associations) in the five Nordic countries. The project applied the combination of interactive workshops, modelling, analytic back-office work and a small-scale Delphi-type enquiry. A quantitative model of the Nordic hydrogen energy system was built during the project. The stakeholders who participated in the foresight process also provided the inputs for the model. *HyWays II in Finland*, in turn, was conducted as part of the European integrated project that looked at the alternative scenarios of hydrogen demand up till 2050 in selected member states. MARKAL modelling with the focus on the demand side of the development of energy systems and KCAM stakeholder analysis methodology enabled the production of relevant inputs for European analysis, but meant a strict structure and focus for backcasting and forecasting workshops. This created few opportunities for general discussion on the role of hydrogen solutions in Finland, which might have been even more productive for national hydrogen developments.

The consensual and rather fixed approach was also applied in the less extensive stakeholder process for the *VTT Security roadmap*. This process management was challenged to address, on the one hand, the expectations on linking different units and operational project plans and, on the other hand, the strategic needs to construct roadmaps and identify VTT priorities. Towards this end, a questionnaire was circulated on future technologies, challenges, opportunities and VTT expertise. Furthermore, workshops were organized for brainstorming, roadmapping and priority setting, aiming at consensual statements on the future and corresponding R&D needs.

The IRRIIS scenario work was a more exclusive consensual and informative process on future developments in energy and communication technology and their interactions. It was expected that the project results would describe the future scenarios in detail. This was considered very challenging by the participants who examined diverse uncertainties in such scenarios. This challenge was dealt with in the brainstorming workshops, intensive e-mail communication, commenting and co-writing.

Two foresight exercises within the framework of the EU *ForSociety project – Future Dialogues on Biotechnology and Transnational Infrastructures* – represent a joint European expert work with rather exclusive participation. The major challenge was how to form a small expert group representative enough to cover diverse stakeholder perspectives for the pre-planned panel work in the workshops; these results were further complemented with experts' commentaries. In the Finnish *PIA roadmap*, the implementation challenges consisted especially of a tight schedule and the management of a large amount of background material. In addition to the core group meetings, stakeholders were engaged through workshops, interviews and an e-mail questionnaire.

From the methodological and management viewpoint, *HyWays II in Finland* was carried out as a part of the European energy system modelling project, which required a rigid structure for workshops and the consensual fixed process design. The two *ForSociety* dialogues, in turn, were guided by a general-level fixed process, but creative autonomous stakeholder panel work still offered considerable freedom for the invited participants to express their opinions to form consensual statements on the future. While informative processes supported visionary work which produced consensual understanding on possible future directions, the participation of decision makers was scant (an aspect revealed by examining the participation dimension). In the *Nordic H2 Energy Foresight and HyWays II in Finland*, specific efforts were made to engage policymakers but with limited immediate success. This may be partly due to the initial positioning of the projects as informative rather than instrumental and thus not considered as policymaking processes (Könnölä et al. 2007b). At best, indirect and diffuse policy links during and after such projects may, however, be influential in the long run.

9.3.2 Consensual and Instrumental and Informative Processes

Consensual and instrumental and informative processes create a common understanding on priorities, networks and/or future actions, as well as support the specific foreseen decision-making situation. Among decision makers, this is likely to lead to interest in the results. However, policy interests may also enter in the foresight process and create rigidities and difficulties to provide new and fresh perspectives for change. The potential problem may be mitigated by ensuring extensive stakeholder participation through the diversity and high number of participants.

The EU foresight project on *Emerging S&T Priorities in public research policies* engaged a large number of S&T stakeholders from industry, research institutes and funding organizations. The project applied both fixed and autonomous methods to try and balance the transparency of the process, tight deadlines and budget, and the genuine involvement of the client (DG Research). Towards this end, the methods used were enquiries by email, PESTE-type document analyses, structured stakeholder interviews and interactive workshops. In the *EU Nanoroadmap*, the major challenges included how to get government officers, business leaders and academics engaged in the roadmap process and not only wait for the final results. This was pursued via a questionnaire and workshops on roadmapping and priority setting.

9.3.3 Informative Processes with Diverse Future Perspectives

Informative processes with diverse future perspectives take into account diverse ideas, opinions and perspectives in priority setting and the identification and fostering of alternative and competing coalitions and value networks, as well as the exploration of alternative futures and the generation rival visions. This relieves participants of the intensive search for consensus and direct support for decision making, which provides opportunities for creative thinking and the inclusion of diverse and alternative viewpoints that can challenge incumbent and path-dependent approaches which hinder – especially radical – changes in the innovation system (Könnölä 2006).

Technology Futures Forum was planned as a free and open forum engaging VTT stakeholders in future-oriented thinking and dialogue. This enables the application of innovative methods to enhance creative thinking and the formulation of possible, even radically different, futures. Extensive stakeholder engagement of foresight practitioners, government officers, business leaders and academics creates the challenge of how to take into account the different backgrounds in the seminars and roadmapping workshops.

The Finnish material technologies roadmap supported the work of the Federation of Finnish Technology Industries and of the Finnish Funding Agency for Technology and Innovation (Tekes) and fostered the networking of stakeholders from many different organizations with different backgrounds. In spite of the variety of views, the project managed to construct roadmaps with a rather fixed process structure and rigid management, including a questionnaire on future technologies, challenges, opportunities and Finnish competitive advantages, and workshops for brainstorming, roadmapping and priority setting.

Dominating views, as well as institutional and political reasons, may sometimes downplay the consideration of diverse perspectives. This is particularly true in national- and organizational-level foresight settings, whereas the presentation of diverse perspectives may be easier in cross-border foresight exercises where the participants are not aware of – and need not obey – organizational- and national-level 'musts' and 'taboos' (Eerola et al. 2005). Good example of such a process is the European *ForSociety Future Dialogue on Healthy Nutrition*, in which invited stakeholders from different member states produced alternative future scenarios to support policymaking on the national and the European level. Another example is the *Nordic ICT Foresight*, which was designed using a similar modular process structure to that used in the *Nordic H2 Energy Foresight*. However, the project laid particular emphasis on incorporating diverse perspectives in the vision, scenario working and roadmapping. The major challenge in the process was to facilitate communication among stakeholders from various knowledge fields. ICT (information and communication technologies) is a broad theme, which made it difficult to form common understanding about the visions and highlighted ICT applications, especially when there were some personal changes in the stakeholder engagement along the way. Many subsequent phases – including desktop analyses; SWOT workshops and questionnaires; scenario and vision work, facilitated by brainstorming, clustering and assessments; and construction of structured socio-technical roadmaps – were needed to generate the resulting action proposals and policy recommendations.

The *VTT Roadmap for future transport service assessments* and the *VTT Building Services Roadmap* followed a similar process design – with an emphasis on multi-faceted communication among stakeholders, forming a common understanding about the objectives of the project, the creation of roadmaps in a robust yet simple enough fashion and intensive core group engagement. Various participatory foresight methods were applied in the roadmapping workshops towards this end.

9.3.4 Instrumental Processes with Diverse Future Perspectives

Instrumental processes with diverse future perspectives generate diverse ideas, opinions and perspectives, which support the specific foreseen decision-making situation or the formation of strategic partnerships/joint actions. The VTT *Water Research Roadmap* was initially considered rather similar to the VTT *Security roadmap* designed to develop consensual and informative roadmaps. However, the key foci of the VTT *Water Research Roadmap* were the creative combination of wide-ranging water-related issues at the VTT and the generation of new R&D initiatives. Questionnaires and mind-mapping and brainstorming workshops were used to engage VTT experts in roadmapping and the formulation of innovation ideas. During the project, it appeared that there were diverse opinions on VTT technology expertise and future market opportunities in water research. This called for mapping these differences and identifying diverse project plans instead of producing consensus statements on common future priorities. The roadmaps were finally constructed through co-writing with different intensities of participant engagement.

9.3.5 Conclusions on the Framework Analysis

The framework was regarded as one way to position even strikingly different foresight projects in a common framework in order to provide a holistic understanding of the foresight activities. The constructed portfolio of the VTT horizontal foresight expert team illustrates the clear emphasis on informing rather than providing instrumental support for decision making. This provides relevant input for the strategic discussion on the role of foresight activities, especially the positioning of foresight with respect to decision making. Furthermore, the quadrant of the instrumental outcomes with diverse perspectives (see Fig. 9.3) includes only one roadmap project. Is this a natural consequence of the inherent difficulties in communicating diverse perspectives to decision makers? And should this be a crucial development area for the VTT foresight team? With regard to the consensual and diverse perspectives and the process management, it appears that consensus-oriented projects put the emphasis on fixed processes and the diverse perspectives oriented projects, leading, in turn, towards autonomous processes.

9.4 Discussion

9.4.1 Towards Modular Foresight Design

Section 9.3 introduced a general framework for the classification of foresight projects. It illustrated, with examples from VTT foresight activities, how different design dimensions, which focus on outcomes, future perspectives, management and stakeholder involvement, interact with each other. Conceptual work, theory-aided process design and systematic practices are, however, needed for the successful integration of various qualitative and quantitative methods. In addition, it is important to consider case-specific conditions and the flexible application of different methods, for instance, building on interactive workshop methods, online generation, elaboration and assessment of stakeholder ideas and opinions, and analytic back-office methods.

At the contract research organization, foresight projects are carried out for a number of different clients and in different roles: as initiator, coordinator, cooperation partner and/or invited participant of a foresight process. This also means specific requirements with regard to the contract research organization's preparedness to meet their clients' and other stakeholders' varying expectations. Furthermore, the experience of the VTT foresight processes summarized in Table 9.1 indicates that the foresight management in a contract research organization is prone to experience tensions produced by diverse policy and business-driven expectations concerning the foresight process and the results. For example, the need to formulate relevant instrumental outcomes for decision makers that lead to direct actions may create barriers to exploring alternative futures, especially those with radical changes. On the other hand, it might be difficult for the processes that are strictly informative to engage policymakers and other decision makers, who could often be valuable contributors in the foresight project. In many cases, the foresight projects seem to be best designed when they balance the different dimensions of foresight management.

In view of these empirical observations, we agree with Salo et al. (2004) that *responsiveness* – which refers to purposely instituted managerial mechanisms for making warranted midcourse adaptations to foresight objectives and implementation plans – should be regarded relevant in the management of foresight activities. In effect, responsiveness requires *receptivity*, vis-à-vis the interests and expectations of participating stakeholders, and *flexibility* in planning and implementation, achieved through the ability to envision and execute even radical changes in the foresight process (Salo et al. 2004). The need to adapt the foresight process to the requirements of the policy context and decision-making phase has also been discussed by other authors (Eriksson and Weber 2006; Da Costa et al. 2008).

The responsiveness in foresight management can be introduced through *modular foresight design*, inherent in the categorized foresight VTT projects and processes described in Sect. 9.3. Modularity refers to process design where analogous subprocesses – or modules – can be enacted relatively independently from the other subprocesses (Könnölä et al. 2011). Thus, the modular foresight design refers to the planning and execution of foresight processes as modules that can be linked together to form tailored processes for different needs. Modules themselves consist of certain standard elements (e.g. method), but they can also be somewhat tailored (e.g. target, scope, variety). As such, modular foresight design enhances the contract research organization's capabilities for expected quick response to unique client needs. In addition, the modular foresight design can also be useful to address the foresight design dimensions:

- *Outcomes.* Accumulation of foresight knowledge for informative and instrumental outcomes can be enhanced via modular design that supports the commensurable and comparable process flows and results. This provides a strong basis to create linkages between different issues and actors, as well as detailed elaboration of outcomes with respect to stakeholder needs. Moreover, the modular foresight design can make the assessment and comparison of the foresight processes and outcomes easier, more systematic and more transparent. For example, the SECI model referred to in Sect. 9.2 of this chapter allows the comparison of the foresight processes by setting them in a common conceptual framework. This setting makes the processes commensurable and provides scope for critical reflection and further development.
- *Chosen future perspectives.* The modular design supports the application of different methods to address diverse perspectives on the future and to develop solid action plans. Hence, modularity helps to achieve a balance between the expectations on outcomes with regard to the emphasis on consensus or diverse perspectives.
- *Management.* The modular design allows the utilization of the synergies between different foresight processes, and it makes it easier to link the results of subprocesses. Thus, it supports the attainment of scalability by enabling organizing parallel processes. Within the modules, even very rigid structures can be applied to ensure that the objectives are met, while between the modules the coordinators can reflect and decide about even radical changes for forthcoming modules. On the other hand, the modularity can support the attainment of objectives through a rather rigid process structure, keeping in mind that the results of one module should contribute to another. However, inside the modules, there can be considerable flexibility to respond to stakeholder expectations and foster creativity during the process.
- *Stakeholder engagement.* Modular design makes it easier for the different stakeholders to enter and contribute to the knowledge accumulation in the different project phases. Modular design thus supports a reasonable division of work between foresight experts of different organizations and between experts of different technical fields and backgrounds. Furthermore, the wide consultative processes and stakeholder engagement can be included without losing the control and strategic focus of the project.

At VTT, there seems to be an increasing demand for roadmapping the future with varying foci and objectives. Roadmapping is considered both as a line of strategic thought and as a process methodology. Firstly, roadmapping as a line of strategic thought aims to combine different modes of knowledge with specific activity layers (Kostoff and Schaller 2001; Phaal et al. 2004). In other words, roadmaps are strategic tools for the crystallization and combination of organizational knowledge that might seem 'unlinkable' with other strategic methods. Secondly, as a process methodology, roadmapping consists of several modules that can be flexibly linked together. Modularization allows researchers to form a tailored 'response chain' to answer different kinds of research and development problems in different technology areas (e.g. Lee and Park 2005). Modularization also makes space for the combination of different research methods (Ahlqvist et al. 2007a, b). Furthermore, modularization enables the tailoring of the roadmapping process to suit the needs of the different actors and the different tasks in the innovation network. On many occasions, VTT roadmaps can be considered as informative processes with diverse perspectives. However, roadmapping studies at VTT are usually connected to some instrumental procedure that embeds the outcomes of the roadmaps in the strategic management and priority setting in the client organization. Generally, the roadmapping process in VTT is formed from different modules that combine the external future potentials with the internal strategic activities.

Modular design has also been applied in cross-border foresight processes, such as the *Nordic ICT Foresight* project (Ahlqvist et al. 2007a, b) and the *Nordic H2 Energy Foresight* (Dannemand Andersen et al. 2005). In these contexts, the modular design offered possibilities to apply different methods, to engage different stakeholders in different phases and to develop country-specific approaches to better respond to different national conditions. The successful implementation of such a modular process called for careful consideration of dependability between the phases. Responsive management throughout the process supported adequate changes in stakeholder participation and communication when completing the complex scenario-based process.

Table 9.2 illustrates the meaning of the foresight management dimensions (expected outputs, future perspectives, methods and stakeholder participation) in the context of a VTT roadmapping process. In particular, the generic key modules of a VTT roadmapping process are related to the choices within the various dimensions. Module 1 '*Drivers and bottlenecks*' refers to societal development trajectories. In the roadmapping process, the outcomes of the first module are at the same time specified for a certain theme, that is, instrumental, and yet simultaneously they form a kind of general view of the changes in the societal drivers and bottlenecks. The future perspectives of the module are based on a mixture of consensus on the most important trajectories and sensitivity to the potential disruptive phenomena that, if realized, could transform the picture drastically. The module can be managed either as a fixed process by means of, for example, questionnaires or interviews or as an autonomous – and hence a more closed – workshop process. Depending on the management preference, the stakeholder engagement can be more open or closed. Module 2 '*Markets*' is basically quite similar to the first module. However, the topic of analysis focuses on the market dynamics and business environment of the topic area. This module can also be completed in a more open or closed fashion, depending on the preferences and the topic area.

Table 9.2 Examples of generic roadmap modules and their foresight management dimensions

Roadmapping modules	Outcomes	Future perspectives	Management	Stakeholder engagement
Module 1: Drivers and bottlenecks	*Instrumental*	*Consensual*	*Fixed*	*Extensive*
	Analysis of the drivers and bottlenecks in a specific topic area	Analysis usually based on the formation of consensus	Questionnaires, thematic interviews, Delphi	Evaluation with large stakeholder participation
	Informative	*Diverse*	*Autonomous*	*Exclusive*
	General understanding of the emerging trajectories	Understanding the potential disruptive phenomena	Workshop dialogue; matrices and tables	Evaluation in a closed workshop set up for a specific purpose
Module 2: Markets	*Instrumental*	*Consensual*	*Fixed*	*Extensive*
	Market trajectories in a specific topic area	Market trends, for example, in potential consumer preferences	Questionnaires, thematic interviews, Delphi	Evaluation with large stakeholder participation
	Informative	*Diverse*	*Autonomous*	*Exclusive*
	General understanding of the market dynamics	Understanding market dynamics in specific fields, counter trends, disruptive phenomena	Workshop dialogue, matrices and tables	Evaluation in a closed workshop set up for a specific purpose
Module 3: Business concepts	*Instrumental*	*Consensual*	*Autonomous*	*Exclusive*
	Visionary business concepts in a specific topic area	Requires a consensus of the directions of the business concepts	Workshop dialogue, matrices and tables; visioning business concepts requires expert knowledge and multiple iterations	Visionary business concepts are mostly created for a specific purpose and in a specific topic area; requires mostly exclusive engagement of the key experts in the topic area
	Informative	*Diverse*		
	Generic ideas for the construction of business concepts	Requires diversifying ideas, understanding the markets and business environment		

Module 4: Applications/ services	*Instrumental* Visionary applications or services in a specific topic area	*Consensual* Requires a consensus of the directions of the applications and/or services *Diverse* Requires contextual understanding and diversifying ideas	*Autonomous* Workshop dialogue, matrices and tables; visioning business concepts requires expert knowledge and multiple iterations	*Exclusive* Visionary applications are mostly created for a specific purpose and in a specific topic area; requires mostly exclusive engagement of the key experts in the topic area
Module 5: Enabling technologies	*Instrumental* Technologies that enable applications, services and business concepts in a specific topic area	*Consensual* Consensus on the most important enabling technologies	*Fixed* Questionnaires, thematic interviews, Delphi	*Extensive* Utilization of a large pool of technology experts and citizens
	Informative General enabling technologies that could be applicable in multiple applications	*Diverse* Alternative technological trends, technological disruptions	*Autonomous* Workshop dialogue, matrices and tables	*Exclusive* Closed evaluation with technology experts in some specific topic area, R&D and noncommercial-ized technologies

Modules 3 '*Business concepts*' and 4 '*Applications/services*' of Table 9.2 aim to characterize visionary business concepts and applications that could proactively respond to, or even change the whole landscape of, the emerging challenges identified in Modules 1 and 2. Modules 3 and 4 also differ from Modules 1 and 2 in the sense that they are more exclusive and autonomous, expert-driven modules. The aims of these modules are topically more contextualized, that is, dependent on the institutional and organizational setting of the roadmapping process. Therefore, constructing visionary business concepts or ideas for applications and services requires mainly on organizational consensus on the general directions of the development, that is, where to focus and where not to focus. Beyond that, the ideation and construction of business concepts, services and applications is an expert-driven and organizationally contextualized process which requires space for open dialogue and more systematic iterations.

Module 5 '*Enabling technologies*' is again a module that builds on both consensus and diversity. Enabling technologies develop both along linear development paths and through radical shifts that can be identified quite feasibly by combining fixed and autonomous management of the process. In this module, it is therefore useful to combine extensive and exclusive participation in order to obtain the most feasible results. However, the stakeholder engagement might be dependent on the roadmap topic: if the roadmap is dealing with technologies that are critical for the client and are still in the R&D phase, then extensive participation might not come into question.

To sum up, Table 9.2 illustrates how the design and management dimensions differ in different modules of the VTT roadmapping process. Such a modular approach helps to address case-specific conditions including the diverse expectations that exist among stakeholders.

9.5 Conclusions

In this chapter, the conceptual framework was developed and introduced for the classification of foresight projects. The application of the framework to examples from the VTT foresight portfolio illustrated how the different dimensions of foresight management influence the methodological choices and stakeholder participation.

Further to the management of foresight activities on the portfolio level, the application of a framework seemed to provide relevant inputs to the strategic discussion on the role of foresight activities in the decision-making processes and on the management of foresight projects from methodological and process facilitation perspectives. However, the role of foresight activities is strongly related to the advances in the whole innovation system. This raises further questions on the understanding of the context in which the foresight activities are implemented. Because the framework does not explicitly consist of such perspectives, it is not sufficient for the purposes of the evaluation of foresight activities. Rather, it is appropriate as a tool for providing complementary support for the design and management of foresight projects and the portfolio of projects.

The framework presented in this chapter has value for both the design and management of both individual foresight projects and a portfolio of foresight activities. For example, building on the application of the framework in the analysis of VTT foresight activities, we identified a coherent and modular application of foresight methods and the responsive engagement of stakeholders, as relevant approaches when balancing the expectations and objectives related to different dimensions of foresight management. In support of building a broader view of the management of foresight activities, the framework was regarded as one way to position even strikingly different foresight projects in the common framework of analysis. The framework was tested as a useful tool in the context of a contract research organization. The test makes us confident to recommend its further elaboration. The elaboration should include both its further theoretical refinement, taking into account other strands of foresight and futures research, and its further application in contract research organizations, as well as in other policy and organizational contexts.

9.6 Annex 1: VTT Foresight Projects, According to the Dimensions of Foresight Management

Project	Outcomes	Future perspectives	Management	Stakeholder engagement
VTT Technology Futures Forum[a]	*Informative* Inform stakeholders on VTT foresight activities. To foster and provide forum for future-oriented thinking and dialogue among VTT stakeholders	*Diverse* Present foresight results at VTT and among its stakeholders, offering visions on alternative futures	*Autonomous* Participant empowerment. Future-oriented work-shops on selected themes facilitated by VTT experts	*Exclusive* Paid memberships for VTT clients interested in foresight partnership (planned option for deeper involvement)
				Extensive Seminar and workshop participation open and free for all stakeholders
VTT Roadmap for future transport service assessments[b]	*Instrumental* Support the development of R&D and assessment activities in VTT and in Finnish transport planning	*Consensual* Visionary creation of roadmaps and view on new potential technology services	*Fixed* Structured visionary mind mapping; thematic and combinatory matrices; roadmap templates	*Exclusive* Stakeholder engagement in core group and workshops
	Informative Visionary roadmaps of the potential transport system technology services and assessment knowledge needed in their development	*Diverse* Identification of potential new technology services and new assessment factors for the future transport system	*Autonomous* Flexible utilization of open discussion and brainstorming	*Extensive* Cooperative idea and concept creation among VTT's experts and stakeholders from other key transport organizations; supported networking

VTT Building Services Roadmap[c]	*Instrumental* Endorse the R&D activities in VTT and, in more general, support the development of Finnish construction and building services	*Consensual* Identification of technological and business development trajectories in building services	*Fixed* Structured driver evaluation; thematic and combinatory matrices; roadmap templates	*Exclusive* Cooperative idea and concept creation among VTT's experts; supported networking
	Informative Visionary roadmaps of the key development trajectories in building services	*Diverse* Identification of potential new technology services and new assessment factors for future systems	*Autonomous* Flexible utilization of open discussion and brainstorming	*Extensive* Stakeholder engagement in core group and workshops to enable intensive communication
VTT Water Research Roadmap 2006[d]	*Informative* Create a common understanding on future challenges and VTT expertise	*Consensual* The identification of key action areas for VTT water research and their priority setting	*Fixed* Structured questionnaire; defined agenda for workshops and structured priority setting	*Exclusive* VTT expert engagement in steering group and workshops to enable intensive communication
	Instrumental Support the formation of common R&D actions within VTT	*Diverse* Inclusion of alternative views on priority setting	*Autonomous* Flexible use of methods in working groups	*Extensive* Networking among VTT experts on water-related R&D through questionnaires, co-writing

(continued)

(continued)

Project	Outcomes	Future perspectives	Management	Stakeholder engagement
VTT Security roadmap[e]	*Instrumental* Present ideas for new VTT R&D Programmes on security technologies	*Consensual* Different themes presented with the same framework Shared understandings were sought for in order to find the key developments	*Fixed* Structured questionnaire; defined agenda for workshops and structured priority setting	*Exclusive* Participation by invitation only
	Informative To foster and provide forum for future-oriented thinking and dialogue among VTT researchers	*Diverse* A variety of views and opinions were considered and debated during the process	*Autonomous* There was still a significant degree of freedom to adapt to the perceived needs during the process	*Extensive* The participants were from different parts of VTT
Finnish material technologies roadmap[f]	*Instrumental* Present ideas for new R&D Programmes on materials technologies	*Consensual* Different themes presented with the same framework Shared understandings were sought in order to find key developments	*Fixed* Structured questionnaire; defined agenda for workshops and structured priority setting	*Exclusive* Participation by invitation only
	Informative To foster and provide forum for materials technology development and dialogue among Finnish stakeholders	*Diverse* A variety of views and opinions were considered and debated during the process	*Autonomous* There was still a significant degree of freedom to adapt to the perceived needs during the process	*Extensive* The participants were from research institutes, industry, associations and public organizations throughout Finland

Finnish PIA roadmap	*Informative* To produce a vision and roadmap for an industry branch	*Consensual* Consensus among a certain group of stakeholders	*Fixed* A procedure was planned in the beginning, including back-office work, different types of workshops, interviews and a questionnaire	*Exclusive* For invited participants only
	To inform national decision makers		*Autonomous* Along during the course of the work the procedure had to be changed to be efficient	
Nordic ICT Foresight[g]	*Instrumental* Action proposals and policy recommendations	*Consensual* Identification of ICT applications with development potential in Nordic region; future-oriented elaboration of factors affecting the Nordic business and development environment in ICT	*Fixed* Structured discussion and the generation of new ideas in the workshops	*Exclusive* Intensive stakeholder engagement in core group
	Informative Evaluations of key ICT applications, Nordic scenario set in context of ICT development, scenario-based visionary roadmaps Building views of the Nordic potential in ICT development among key actors	*Diverse* Linking the perspectives of alternative scenarios to the research process	*Autonomous* Creative brainstorming and ideation in the different workshops	*Extensive* Cooperative idea and concept creation among stakeholders from different Nordic organizations and firms; networking

(continued)

(continued)

Project	Outcomes	Future perspectives	Management	Stakeholder engagement
Nordic H2 Energy Foresight[h]	*Instrumental* A clear action plan for the Nordic key actors – without a direct link to any decision process	*Consensual* Shared understandings were sought in order to be able to give action recommendations for the Nordic key actors	*Fixed* The overall design of the process was already determined when planning the project. The model and modelling techniques in use guided the data gathering of the system analysis part	*Extensive* The participants were from research institutes, industry, associations and public organizations of the five Nordic countries
	Informative Awareness raising and deepening the overall understanding of the entire value chain (hydrogen production, storage, distribution, stationary hydrogen uses and hydrogen uses in transport)	*Diverse* A variety of views and opinions were considered and debated during the process	*Autonomous* There was still a significant degree of freedom to adapt to the perceived needs during the process	
EU: Nanoroadmap[i]	*Instrumental* Clear guidance and recommendations for the new Framework Programmes were expected by the EC/DG Research	*Consensual* Shared understandings were sought in order to find the key developments	*Fixed* Structured questionnaire; defined agenda for workshops and structured priority setting	*Exclusive* Participation by invitation only
	Informative Present the key development paths and visions of nanotechnologies with respect to energy applications	*Diverse* Different themes presented with the same framework A variety of views and opinions were considered and debated during the process	*Autonomous* There was still a significant degree of freedom to adapt to the perceived needs during the process	*Extensive* The participants were from research institutes, industry, associations and public organizations; both EU and non-EU views were collected

EU: Emerging S&T Priorities in public research policies[j]	*Instrumental* Clear guidance and recommendations for the new Framework Programmes were expected by the EC/DG Research	*Consensual* Shared understandings were sought between the project group and the client. For that purpose, it was also important to get a balanced European picture of the various topics	*Fixed* The process consisted of extensive excel-sheet inquiries, systematic PESTE-type SEEP analyses of socio-economic S&T environments, structured stakeholder interviews and interactive workshops, together with the client's representatives	*Exclusive* Direct participation limited to European S&T experts, some key persons from DG Research + the foresight experts of the project group
	Informative In addition to compilation of action recommendations, the process covered identification of new emerging technologies, their prioritization, country-specific and regional analyses of socio-economic S&T environments and construction of alternative action scenarios	*Diverse* A variety of views and opinions were considered during the process. Three European scenarios were produced on the basis of different sets of assumptions	*Autonomous* But there was still a significant degree of freedom in the analytic back-office work and in the compilation of action scenarios and recommendations	*Extensive* In total, a considerable number of scientific and technical stakeholders from various European countries + the USA and Japan were consulted. Representatives of DG Research participated in the intermediate workshops and commented on the intermediate reports
EU: HyWays II in Finland[k]	*Informative* As part of HyWays II integrated project, to produce policy-relevant information for the Commission and member states	*Consensual* National-level common vision on hydrogen demand up to 2050 *Diverse* Formulation of alternative hydrogen-based energy systems	*Fixed* Energy system modelling Stakeholder analysis *Autonomous* National workshops for the identification of national target levels	*Exclusive* Selected energy experts invited to the workshops. Intensive back-office work *Extensive* Energy stakeholders from different sectors and organizations

(continued)

(continued)

Project	Outcomes	Future perspectives	Management	Stakeholder engagement
EU: IRRIIS scenario work[l]	*Informative* Identification of emerging safety and security issues in an EU project to ensure the safety of critical infrastructures	*Consensual* A project level consensus on the future developments *Diverse* Different scenarios for the future development	*Fixed* A fixed procedure and selected methods *Autonomous* The content of the work resulted from autonomous work among the stakeholders. The experience of stakeholders 'overrode' the methodological rigidity in some points	*Exclusive* The work was carried out among the project partners *Extensive* The results were tested against available expertise outside the project consortium
EU: ForSociety Future Dialogue on Biotechnology[m]	*Informative* To produce information for EU decision making	*Consensual* Consensus among the invited expert groups	*Fixed* Pre-planned process *Autonomous* Some room for creative panel work	*Exclusive* Invited participants only
EU: ForSociety Dialogue on Transnational Infrastructures[n]	*Informative* To produce information for EU decision making and transnational projects at the EU level	*Consensual* Consensus among the invited expert group, also included sharing of different views on the topics	*Fixed* Pre-planned process *Autonomous* Some room for creative panel work	*Exclusive* Invited participants only

EU: ForSociety Future Dialogue on Healthy Nutrition[o]	*Informative* To build a roadmap for research policy institutions To contribute to national, EU and ETP discussions on future research and innovation	*Diverse* Creation of alternative scenarios for future development	*Fixed* Pre-planned process *Autonomous* Some room for creative panel work	*Exclusive* Invited participants only

[a]http://www.vtt.fi/proj/tff/?lang=en
[b]http://www.vtt.fi/inf/pdf/tiedotteet/2007/T2390.pdf
[c]http://www.vtt.fi/inf/pdf/tiedotteet/2007/T2379.pdf
[d]Unpublished
[e]http://www.vtt.fi/inf/pdf/tiedotteet/2007/T2368.pdf
[f]http://www.teknologiateollisuus.fi
[g]http://www.vtt.fi/inf/pdf/publications/2007/P653.pdf
[h]http://www.h2foresight.info
[i]http://www.nanoroadmap.it/
[j]http://ec.europa.eu/research/foresight/pdf/21960.pdf; http://www.efmn.info/kb/efmn-brief42.pdf
[k]http://www.hyways.de/
[l]http://www.irriis.org/?lang=en&nav=218; http://www.irriis.org/File.aspx?lang=2&oiid=8661&pid=572
[m]http://www.eranet-forsociety.net//resources/future_dialogues/
[n]http://www.eranet-forsociety.net/ForSociety/resources/future_dialogues/
[o]http://www.eranet-forsociety.net/ForSociety/resources/future_dialogues/nutrition.html?dbX_sid=f82615f5016

References

Ahlqvist, T. (2005). From information society to biosociety? On societal waves, developing key technologies, and new professions. *Technological Forecasting and Social Change, 72*, 501–519.

Ahlqvist, T., Carlsen, H., Iversen, J., & Kristiansen, E. (2007a). *Nordic ICT foresight: Futures of the ICT environment and applications on the Nordic level*. Helsinki: VTT Publications 653, Edita Prima Oy.

Ahlqvist, T., Carlsen, H., Iversen, J. S., & Kristiansen, E. (2007b). *Nordic ICT foresight. Futures of the ICT environments and applications on the Nordic level* (Summary Report). Publication of the Nordic Innovation Centre Series.

Ansoff, I. (1975). Managing strategic surprise by response to weak signals. *California Management Review, 17*(2), 21–33.

Arthur, W. B. (1989). Competing technologies, increasing returns and lock-in by historical events. *Economic Journal, 99*(394), 116–131.

Arthur, W. B. (1994). *Increasing returns and path dependence in the economy*. Ann Arbor: University of Michigan Press.

Barré, R. (2002). Synthesis of technology foresight. In A. Tübke, K. Ducatel, J. Gavigan, & P. Moncada (Eds.), *Strategic policy intelligence: Current trends, the state of play and perspectives* (Technical Report EUR-20137-EN). Seville: Institute for Prospective Technological Studies (IPTS).

Coffman, B. S. (1997). *Weak signal research, Part III: Sampling, uncertainty and phase shifts in weak signal evolution*. MG Taylor Corporation, http://www.mgtaylor.com/mgtaylor/jotm/winter97/wsrsampl.htm.

Cruickshank, J., & Susskind, L. (1987). *Breaking the impasse: Consensual approaches to resolving public disputes*. U.S.: Basic Books.

Cuhls, K. (2003). From forecasting to foresight processes – New participative foresight activities in Germany. *Journal of Forecasting, 22*(2–3), 93–111.

Da Costa, O., Warnke, P., Cagnin, C., & Scapolo, F. (2008). The impact of foresight on policy-making: Insights from the FORLEARN mutual learning process. *Technology Analysis and Strategic Management, 20*(3), 369–387.

Dannemand Andersen, P., Jørgensen, B. H., Eerola, A., Koljonen, T., Loikkanen, T., & Eriksson, E. A. (2005). *Building the Nordic research and innovation area in hydrogen* (Summary Report). Nordic Innovation Centre and Nordic Energy Research, http://www.h2foresight.info/

Dannemand Andersen, P., Borup, M., Borch, K., Kaivo-oja, J., Eerola, A., Finnbjörnsson, T., Øverland, E., Eriksson, A., Malmér, T., & Mölleryd, B. (2007). *Foresight in Nordic innovation systems* (Summary Report of the Nordic Foresight Forum funded by the Nordic Innovation Centre in 2005–2007), http://www.nordicinnovation.net/_img/nordic_foresight_forum_final_report.pdf.

David, P. A. (1985). Clio and the economics of QWERTY. *The American Economic Review, 75*(2), 332–337.

Eerola, A., & Jørgensen, B. H. (2002). *Technology foresight in the Nordic countries*. A Report to the Nordic Industrial Fund, Center for Innovation and Commercial Development, Risoe-R-1362(EN).

Eerola, A., & Väyrynen, E. (2002). Developing TF and TA practices on the basis of European experience, VTT Tiedotteita – Research Notes 2174 (in Finnish, abstract in English), http://www.inf.vtt.fi/pdf/tiedotteet/2002/T2174.pdf.

Eerola, A., & Jørgensen, B. H. (2008). Foresight in the Nordic countries. In L. Georghiou, J. Cassingena-Harper, M. Keenan, I. Miles, & R. Popper (Eds.), *The handbook of technology foresight – Concepts and practice*. Cheltenham and Northampton: Edward Elgar.

Eerola, A., Loikkanen, T., Jørgensen, B. H., Dannemand Andersen, P., & Eriksson, E. A. (2004, May 13–14). *Nordic H2 energy foresight – Complementary contribution of expert views and formal analyses*. A paper presented in the EU-US Scientific Seminar "New Technology

Foresight, Forecasting and Assessment Methods", Seville, http://www.h2foresight.info/Publications/Publications_Conference.htm.
Eerola, A., Loikkanen, T., Koljonen, T., Joergensen, B. H., Dannemand Andersen, P., & Eriksson, E. A. (2005, June 9–10). Nordic H2 energy foresight – Managerial challengers of the interactive process. In *Proceedings of the 17th annual international conference "Foresight management in corporations and public organizations – New visions for sustainability"*, Helsinki, http://www.tukkk.fi/tutu/conference2005/default.htm.
Eriksson E. A., & Weber M. (2006, September 28–29). Adaptive foresight: Navigating the complex landscape of policy strategies. In *Second international Seville seminar on future-oriented technology analysis*, Available at http://forera.jrc.es/documents/papers/FTA_EAE_MW%20 Final%20060807.pdf.
European Commission (2002, April 24). *Thinking, debating and shaping the future: Foresight for Europe* (Final Report of the High Level Expert Group for the European Commission). Brussels: European Commission.
Grabher, G., & Stark, D. (1997). Organizing diversity: Evolutionary theory, network analysis and postsocialism. *Regional Studies, 31*(4), 411–423.
Harris, D., & Zeisler, S. (2002). Weak signals: Detecting the next big thing. *The Futurist, 36*, 21–28.
Helmer, O. (1983). *Looking forward: A guide to futures research*. Beverly Hills: Sage.
Hjelt, M., Luoma, P., van de Linde, E., Ligvoet, A., Vader, J., & Kahan, J. (2001). Experiences with National technology, *Foresight Studies* (Sitra Report Series 4/2001). Helsinki.
Irvine, J., & Martin, B. R. (1984). *Foresight in science, picking the winners*. 166. Pinter Pub Ltd. London: Dover.
Jacobsson, S., & Johnson, A. (2000). The diffusion of renewable energy technology: An analytical framework and key issues for research. *Energy Policy, 28*(9), 625–640.
Keenan, T. (2003). Identifying emerging generic technologies at the national level: The UK experience. *Journal of Forecasting, 22*(2–3), 129–160.
Kivisaari, S., Lovio, R., & Väyrynen, E. (2004). Managing experiments for transition: Examples of societal embedding in energy and health care sectors. In B. Elzen, F. W. Geels, & K. Green (Eds.), *System innovation and the transition to sustainability: Theory, evidence and policy*. Cheltenham: Edward Elgar.
Koivisto, R. A., Reunanen, M. O. E., & Nissilä, M. H. (1997). Safety reports in Finnish process industries: Experiences based on some case studies. *Journal of loss prevention in the process industries, 10*(2), 77–81.
Koivisto, R. A., Törmäkangas, K. M., & Kauppinen, V. S. (2002). Hazard identification and risk assessment procedure for genetically modified plants in the field – GMHAZID, ESPR. *Environmental Science and Pollution Research, 9*(2), 110–116.
Könnölä, T. (2006). *Escaping path dependence – Essays on foresight and environmental management*. Dissertation for the degree of Doctor of Science in Technology, Helsinki University of Technology.
Könnölä, T., Brummer, V., & Salo, A. (2007a). Diversity in foresight – Insights from the fostering of innovation ideas. *Technological Forecasting and Social Change, 74*, 608–626.
Könnölä, T., Unruh, G., & Carrillo-Hermosilla, J. (2007b). Toward prospective voluntary agreements: Reflections from a hydrogen foresight project. *Journal of Cleaner Production, 15*, 259–265.
Könnölä, T., Salo, A., & Brummer, V. (2011). Foresight for European coordination: Developing national priorities for the forest based sector technology platform. *International Journal of Technology Management, 54*(4), 438–459. Special Issue on Technology Management.
Kostoff, R. N., & Schaller, R. R. (2001). Science and technology roadmaps. *IEEE Transactions on Engineering Management, 48*(2), 132–143.
Lee, S., & Park, Y. (2005). Customization of technology roadmaps according to roadmapping purposes: Overall process and detailed modules. *Technological Forecasting and Social Change, 72*, 567–583.
Lundvall, B.-Å. (Ed.). (1992). *National systems of innovation: Towards a theory of innovation and interactive learning*. London: Pinter.

Martin, B. R., & Johnston, R. (1999). Technology foresight for wiring up the national innovation system. Experiences in Britain, Austria, and New Zealand. *Technological Forecasting and Social Change, 60*, 37–54.

Mendoça, S., Pina e Cunha, M., Kaivo-oja, J., & Ruff, F. (2004). Wild cards, weak signals and organisational improvisation. *Futures, 36*(2), 201–218.

Meulen, B., de Wilt, J., & Rutten, H. (2003). Developing futures for agriculture in the Netherlands: A systematic exploration of the strategic value of foresight. *Journal of Forecasting, 22*(2–3), 219–233.

Nonaka, I. (1994). A dynamic theory of organisational knowledge creation. *Organisation Science, 5*(1), 14–37.

Nonaka, I., & Takeuchi, H. (1995). *The knowledge-creating company*. New York: Oxford University Press.

Phaal, R., Farrukh, C. J. P., & Probert, D. R. (2004). Technology roadmapping – A planning framework for revolution and revolution. *Technological Forecasting and Social Change, 71*, 5–26.

Salmenkaita, J.-P., & Salo, A. (2004). Emergent foresight processes: Industrial activities in wireless communications. *Technological Forecasting and Social Change, 71*(9), 897–912.

Salo, A., Gustafsson, T., & Ramanathan, R. (2003). Multicriteria methods for technology foresight. *Journal of Forecasting, 22*(2–3), 235–255.

Salo, A., Könnölä, T., & Hjelt, M. (2004). Responsiveness in foresight management: Reflections from the Finnish food and drink industry. *International Journal of Foresight and Innovation Policy, 1*(1–2), 70–88.

Smits, R., & Kuhlmann, S. (2004). The rise of systemic instruments in innovation policy. *International Journal of Foresight and Innovation Policy, 1*(1), 4–32.

Tushman, M. L., & O'Reilly, C. A. (1997). *Winning through innovation*. Boston: Harvard Business School Press.

Unruh, G. C. (2000). Understanding carbon lock-in. *Energy Policy, 28*(12), 817–830.

Voss, J.-P., & Kemp, R. (2006). Sustainability and reflexive governance: Introduction. In J.-P. Voss, D. Bauknecht, & R. Kemp (Eds.), *Reflexive governance for sustainable development* (pp. 3–27). Cheltenham: Edward Elgar.

Chapter 10
Will Entrepreneurship, Knowledge Management and Foresight Emerge in a System?

Arturs Puga

10.1 Introduction

This chapter discusses a theme related to methodological settings, learning and knowledge production in the realm of futures studies and foresight. The author focuses on *the synergy that comes from combining an entrepreneurial mindset and transdisciplinary research with organizational and personal knowledge management activities in the context of foresight initiatives and projects*. Evolving ideas and concepts to develop research, which deals with the integration of entrepreneurship, foresight and knowledge management, have been put forward in several national-level initiatives, projects and higher education modules in Latvia (2003–2009).

Can we perceive already executed or new foresight projects as drivers for entrepreneurial and innovative thinking that enhance the capabilities of research organizations? The answer might be 'yes' in one case and 'no' in the others. And what would we say about foresight and knowledge management (KM) interrelations? Perhaps not very much for many reasons. In our world, KM has many faces, interpretations and applications, and this field of enquiry deserves a particular place in the foresight area.

This chapter describes a number of interrelated elements and processes and tracks how they develop synergies for foresight. The study is structured in the following sections.

Section 10.2 informs on the knowledge base of this chapter. The first steps are concerned with explaining the synergy, providing the definitions applied and explaining their value for our research and projects. Section 10.3 looks at the work of foresight organizations (teams) and presents the author's reflections on their achievements and failures, considering the synergy issues of Latvian foresight activities, as well as experience obtained in European projects. Section 10.4 refers

A. Puga (✉)
Forward Studies Unit, 13-4, Virsu Street, Riga, Latvia
e-mail: arturs_puga@yahoo.com

M. Giaoutzi and B. Sapio (eds.), *Recent Developments in Foresight Methodologies*, Complex Networks and Dynamic Systems 1, DOI 10.1007/978-1-4614-5215-7_10,
© Springer Science+Business Media New York 2013

to entrepreneurial KM activities integrated into foresight research and presents a KM case that we can design for the task concerned. The outcomes of these endeavours are identified at the national level. The conclusions to this chapter, in Sect. 10.5, contain thoughts based on the identified challenges concerning possible ways to work more effectively and efficiently in the proposed direction.

10.2 Definitions to Enlighten What Is in the Task Concerned

10.2.1 The Knowledge Base for This Chapter

The exploratory studies for this chapter, its theoretical and problem-solving issues, are mostly related to the project 'Latvia towards Knowledge Societies of Europe: new options for entrepreneurship and employment achieving the goals of the Lisbon strategy' (the LNELS project) (Puga 2007). This initiative, intended for the period 2003–2010, was launched by the Forward Studies Unit, an independent research body located in Riga. The aim and main task for this project is to explore theoretical issues and offer conceptual and practical ways of how to promote entrepreneurship in R&D activities by finding new options for knowledge workers to gain employment related to forward-looking intelligence, including foresight on technology and social issues.

The core problem of the LNELS initiative was how to promote the understanding/incremental recognition of the futures and foresight area as a particular field of inquiry in Latvia and some other countries. Therefore, we need to approach both individual mindsets and organizational structures in political, academic, business and other areas.

In 2004–2009, some basic ideas and frameworks of the LNELS project were transferred to foresight activities and projects conducted by the Latvian Technological Centre, the Latvian Academy of Sciences, the Latvian Academy of Agricultural and Forestry Sciences, the Riga Technical University and the University of Latvia and to students then at the School of Business Administration Turiba in Riga (Puga 2007; Puga 2008). The LNELS researchers have grounded their research on methodological studies, projects and exercises performed in the EU and other countries in recent decades, and they themselves have contributed inputs to more than ten European and regional foresights in the last 5 years (e.g. Knowledge Society Foresight, the European Foresight Monitoring Network, FISTERA, ForeIntegra-RI) (Puga 2007, 2008).

10.2.2 How to Initiate Synergy

The LNELS approach proposes that in launching a foresight project/exercise, many introductory and explanatory activities should be set up in order to improve the

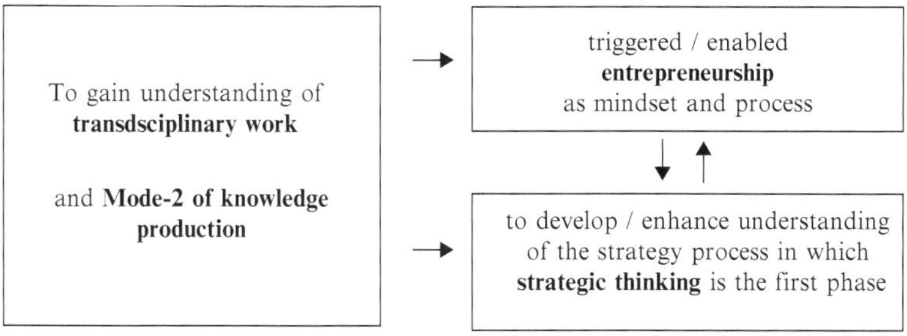

Fig. 10.1 Set in motion the foresight organization: Enabling entrepreneurship by and for transdisciplinary work and strategic thinking

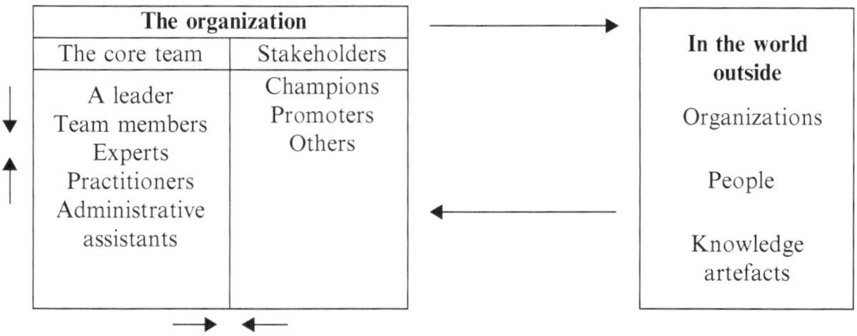

Fig. 10.2 The foresight project organization and its interactions to obtain knowledge (the *arrows* indicate the directions of interaction and knowledge flow)

understanding of the project team and stakeholders about the particularities of the foresight study concerned.

The perception of the intangibles and research processes in which the project team and stakeholders would/should be engaged, of activities of different kinds and their outputs and deliverables and of the synergetic potential of the people involved can all be important in shaping the foresight team's philosophy, research culture and particular ways of knowledge production. European foresight initiatives and exercises and the European foresight area itself are 'young children' unknown or obscure for many people in S&T policymaking, academia and other groups and communities in European countries.

Knowledge on transdisciplinarity and mode 2 of knowledge production (see definitions in Sect. 10.2.3) can help the foresight organization to adapt more easily to the multidimensional scope of knowledge work (see Figs. 10.1 and 10.2 in this section, and Fig. 10.3 in Sect. 10.3.1). Such work requires willingness and entrepreneurial efforts leading to evolving strategic thinking for new products and applications and also to the self-assurance of those people who might try to interpret their own individual foresight in the project team activities.

To begin with, it is also significant to consider selected examples for the development of a better understanding of different parts of the strategy process: strategic thinking and foresight as its basic elements, strategic development or decision making, strategic planning and ultimately strategic implementation (Voros 2003).

In the case of success, a variety of ideas, concepts and discussed examples of transdisciplinarity and mode 2 of knowledge production interacting with individuals' tacit and explicit knowledge evolve for the tasks concerned, and this can also trigger and increase the entrepreneurial abilities of the team members and their strategic thinking, while setting up step by step a future-friendly mental and physical environment for futures research and knowledge management.

In such a way, synergy can be created for futures research processes. This can mitigate the inevitable cognitive and organizational gaps between what we know and what we should learn and perform in dealing with futures.

Opposite to this approach might be the attitude and activities of some individuals, developments in the foresight organization characterized by terms like 'unreadiness', 'unsociability', 'apathy' and 'conflict'.

People's mindsets, endeavours and capabilities motivated towards outputs and deliverables, involved in individual and organizational knowledge production using technological options and intertwined with individual and organizational values *are often challenged and specifically 'addressed' by the foresight* to which they contribute.

10.2.3 Definitions for the Foresight Organization's Work

The author argues that a good understanding of the definitions (selected by the team or individual) that are the main building blocks for knowledge work and provide a platform for the task concerned can be one of the key issues of success when dealing with futures and KM activities. The aim is to enter in depth into the complexity of definitions and coherently integrate evolving outputs: that is, to enhance and orientate our mental work in an entrepreneurial direction.

In adopting this approach to futures studies, the LNELS project team selected the following definitions and explanations which seemed to be most appropriate for reasons of both cognitive and practical application:

Foresight – thinking, debating and shaping the future – can increase strategic intelligence capabilities for policies, industry and innovation, research, education, etc. (Busquin 2002). 'Foresight is a systematic, participatory process, collecting future intelligence and building medium- to long-term visions, aimed at influencing present-day decisions and mobilizing joint actions. It helps in making choices in complex situations by discussing alternative options, bringing together different communities with their complementary knowledge and experience' (UNIDO 2007).

Entrepreneurship – 'the mindset and process to create and develop economic activity by blending risk-taking, creativity and/or innovation with sound management, within a new or an existing organisation' (European Commission 2003).

Mode 2 of knowledge production (a set of cognitive, research and social practices – ideas, methods, values, norms, etc.) is characterized by:

- Knowledge produced in the context of application
- Transdisciplinarity
- Heterogeneity and organizational diversity
- Enhanced social accountability
- More broadly based system of quality control (Gibbons 1997)

Transdisciplinarity – Enables the development of a distinct but evolving framework to guide problem-solving efforts. This is generated and sustained in the context of application and not developed first and then applied to that context later by a different group of practitioners.

Transdisciplinary knowledge develops its own distinct theoretical structures, research methods and modes of practice, though they may not be located on the prevailing disciplinary map.

The diffusion of the results is initially accomplished in the process of their production. Subsequent diffusion occurs primarily when the original practitioners move to new problem contexts, rather than through reporting results in professional journals or at conferences. Communication links are maintained partly through formal and partly through informal channels.

Transdisciplinarity is dynamic. It is problem-solving capability on the move. A particular solution can become the cognitive site from which further advances can be made, but where this knowledge will be used next and how it will develop are as difficult to predict as are the possible applications that might arise from discipline-based research (Gibbons 1997). 'Rather, it is in the context of application that new lines of intellectual endeavour emerge and develop, so that one set of conversations and instrumentation in the context of application leads to another, and another, again and again' (Nowotny 2003).

Knowledge worker – 'An individual whose primary contribution is through the knowledge that they possess or process. This contrasts with workers whose work is predominantly manual or following highly specified procedures with little scope for individual thought' (David Skyrme Associates 2008).

Organizational knowledge management (OKM) – Knowledge management is the name given to the set of systematic and disciplined actions that an organization can take to obtain the greatest value from the knowledge available to it. 'Knowledge' in this context includes both the experience and understanding of the people in the organization and the information artefacts, such as documents and reports, available within the organization and in the world outside. Effective knowledge management typically requires an appropriate combination of organizational, social and managerial initiatives along with, in many cases, deployment of appropriate technology (Marwick 2001).

Personal knowledge management (PKM) – 'A set of concepts, disciplines and tools for organizing often previously unstructured knowledge, to help individuals take responsibility for what they know and who they know' (European Committee for Standardization 2004b). 'KM only makes sense if knowledge is important for the job at hand and when the individual possesses and/or needs knowledge to reach his or her objectives' (European Committee for Standardization 2004a).

The LNELS approach aims at horizontal application and adaptation of the designed and tested frameworks and examples of its teams' activities that have roots in processes defined above. These cases are intended for a broad spectrum of applicants in different areas of knowledge production about futures.

10.3 The Foresight Organization and Its Interactions to Stimulate Mindsets and Obtain Knowledge

10.3.1 The Project Team in the World of Knowledge

A general model of the foresight project organization (FO) can be perceived to be adaptable to a wide variety of applications to design, plan and manage new and ongoing interactions. In this case, *organization is an evolving set of the processes involved in handling futures-related knowledge and managing knowledge work by a group of people* (e.g. *the LNELS team) who have a shared purpose and interests.* The FO and its members interact with organizations and individuals in the world outside and apply a combination of organizational, social and managerial actions to reach research targets using appropriate technologies. They should interact with the project stakeholders throughout the entire project time and, what is important, also after the job is completed.

We can consider FO as a mechanism which ensures the relationship between the project researchers and their environment, people and organizations and bodies of knowledge. In this context, one should think about KM issues.

In global knowledge exchange and trade, the FO actions for the task concerned can be twofold and interrelated. They should support:

- A set of organizational, social, managerial and technological processes which ensure innovative ways and appropriate methods for knowledge production and execution of the project as a whole
- The foresight process itself – to generate ideas, use methods and work with alternative futures, and manage the foresight knowledge production which results in the project deliverables

Following transdisciplinarity, LNELS applies the ideas, conceptual findings and results of each of its phases to new tasks, and KM activities contribute to the incremental progress towards the goals. The combination of integrative, exploratory, interactive, participative, experimental and skills-building processes has been developed by applying and rethinking knowledge and experience gained from European and other international projects and workshops and knowledge artefacts (Puga 2007).

<table>
<tr><td>having commitment</td><td>supporting</td><td>learning</td><td>exploring</td></tr>
<tr><td>networking</td><td colspan="2" rowspan="2">performing knowledge, building capabilities, producing deliverables</td><td>conversing, having dialogue</td></tr>
<tr><td>collaborating</td><td>understanding</td></tr>
<tr><td>achieving objectives and purpose</td><td>innovating</td><td>changing</td><td>searching</td></tr>
</table>

Fig. 10.3 The foresight project organization (developed processes): entrepreneurship and KM intertwined with personal values for the task concerned

The richness of the LNELS expanding knowledge repository, the application of the explicit knowledge for increasing understanding about the methodology and practice of foresight programmes and exercises by the LNELS team and the use of possibilities to obtain experience and new impressions by working together with the leading foresight specialists of EU and other countries in both real and virtual environment – all these factors, coupled with knowledge-intensive work – were essential to ensure a sustained LNELS process. Entrepreneurship was the key to exploit the new opportunities the researchers began to develop for themselves and others.

In this case, values play an important role – to understand and promote foresight to be a knowledge domain like history, literature and management, and many intangible and other values have an impact on it.

10.3.2 Different Approaches to Foresight: Will Mode 1 or Mode 2 Be the Leader?

This section refers to five national-level projects and exercises related to foresight and supported by the Latvian Council of Science, the Latvian Academy of Sciences and the EU Structural Funds in 2003–2008. The first (the LNELS foresight organization) explored and translated into research activities a great deal of knowledge on KM topics. The LNELS outputs of 2004–2007 include a methodological framing for the practical application, enhancement and further development of KM approaches to foresight. The results indicate that understanding of KM terminology and processes enables the active participation of researchers in foresight projects at the national and the European level (Puga 2007, 2008).

We look at four projects that had a different approach than LNELS. The next FO comprised about 15 scientists and researchers, almost all of whom were from academia. They implemented two national-level projects dealing with agricultural and rural foresight.

Another research team worked on the elaboration of a conceptual module of the KM system for corporate foresight in SMEs. This organization included about ten researchers and young scientists mainly from a technical university.

Evidence for this was also collected from the teaching and learning processes of a Master's studies module at the University of Latvia in 2008. In this course, futures methodology and synergy between foresight and knowledge management activities had been discussed for application in the natural sciences and in building capabilities to promote processes of sustainable development (Puga 2008).

The above-mentioned four projects were conducted (over the course of 1 or 2 years) by academic organizations. Those core teams consisted of top-level scientists and professors, and they mostly emphasized a disciplinary way of thinking and practices (what is termed as mode 1 of knowledge production) in both organizational and research activities. The teams tended to maintain well-known and routine ways of management while running foresight. The research methods and results were accepted by clients from government bodies. However, the projects have not attracted much attention from the scientific or business communities: for instance, to discuss the foresight methodology and consider the foresight role for R&D policy or societal issues. It is now difficult to find information on the application of the foresight projects. In Latvia, as in some other countries, S&T policy and research funding continue to operate predominantly within disciplinary constraints.

At the same time, research practices of foresight projects are associated with mode 2. Considering another domain of thought that challenges and often embarrasses a part of the academic world – KM – we also find it associated with mode 2 of knowledge production.

In the view of the author, possible ways to generate the synergy of entrepreneurial thinking, KM and foresight activities can be learned and discussed by the team members and, as far as possible, by individual stakeholders. That requires commitment from the project champions, team and other participants – to think out of the box, to be open to other mindsets and innovative ideas and to try to understand, analyse and synthesize those ideas – producing individually and collectively new knowledge for the following phases of the project and eventually for applications of the results that the public can freely evaluate and develop.

10.4 Entrepreneurship and KM Support to Develop Foresight at Different Levels

10.4.1 PKM for and in the FO

We need to address a great deal of intangibles and material elements, tools and assets to integrate entrepreneurship, knowledge management and foresight in order to develop a system of knowledge production in the FO. A case of personal knowledge management to promote individual foresight (IF) capabilities can be considered as a useful framework while performing the task concerned. How could such a case be mentally constructed and really shaped to become adapted to changing realities? Let us think about the following parts:

- To understand the objectives and the goals of the foresight project/exercise, careful and often time-consuming knowledge work should be done. In their personal and organizational activities, individuals should be motivated and interested in enhancing their ownership of knowledge in order to reach new objectives by orienting their work in an entrepreneurial direction.
- Knowledge means both assets and processes that benefit from and give benefit to IF. New assets are produced and accumulated during transdisciplinary work within the FO. Processes of socialization, externalization, combination and internalization of knowledge are in progress. Sets of ideas, concepts, approaches, methodologies and frameworks are evolving for the task concerned and far beyond it.
- Working with the FO and people in the outside world, the knowledge worker can identify needs and critical skills deficiencies. This process leads him/her to a better understanding and acquisition of new/improved skills in some disciplines and research areas, which are to be integrated in producing knowledge and applied in transdisciplinary work.
- A part of the framework of PKM for IF relates to tools for foresight research and knowledge management activities. We need both instruments and skills to search for and capture information, to create, organize and share knowledge. This leads to an evolving virtual workplace available around the clock from 'anywhere in the world'. To ensure access to information artefacts and knowledge owners, a set of emails, software programs, online dictionaries, social media, communications and other tools are applied for the individual and FO work. Databases and knowledge repositories support researchers in all phases of the foresight process (Puga 2007).

10.4.2 The Improved Dialogue Results in the Dissemination and Recognition of Foresight at the National Level

In 2006–2008, some methodological settings, methods and techniques through dialogues and the dissemination of international experience and LNELS activities were transferred to, and later implemented in, several projects supported by the Latvian Council of Sciences, the Latvian Academy of Sciences and ministries and universities in Riga. Top-level policymakers from the Ministry of Education and Science and the Ministry of Agriculture, as well as two academies, signed an agreement on the cooperation within the network of foresight specialists and practitioners for Latvian agricultural research (Zinātnes Vēstnesis 2007). In 2007, foresight was recognized as a new research area by the Latvian Academy of Sciences. In that year, the Terminology Commission of the Latvian Academy of Sciences also made a decision to introduce new terms that apply to this discipline (Latvijas Zinātnu akadēmijas gadagrāmata 2008).

10.4.3 To Integrate Entrepreneurial Activities, Strategic Thinking and KM with Learning New Transdisciplinarities (Options of Horizontal Application)

As illustrated in Figs. 10.1 and 10.2, frameworks and processes that stem from and develop further synergy of entrepreneurship, knowledge management and futures research have been applied during the teaching and learning of the course on lobbying (interest representation) techniques of the School of Business Administration Turiba in Riga, Latvia. This course is a module of a programme to obtain a professional qualification and the Master's degree in public relations. Interest representation, similarly to KM and foresight, is a relatively new area for both education and research in some European universities.

All of the following support the entrepreneurial endeavour and students' ways of obtaining knowledge on and practical skills associated with lobbying topics: the introduction to transdisciplinary work and shift into fields of knowledge that are important for understanding the complexity of the lobbying area; the description of this area as very necessary for a society in which everyone individually and/or through the organization/s contacts (business, R&D, education, non-profit/NGO, local and other communities) with parliamentary or government institutions might be involved; and insights into lobbying cases with a focus on strategy development and its aspect – strategic thinking that plays a significant role for lobbying.

The author's experience as a lobbying teacher indicates that cases of PKM are useful and applicable for lecturers and may be also suggested to students. During the lobbying course, a virtual learning and knowledge-sharing space (website of LTehno group on SCRIBD) was developed by more than 30 Master's students organized in a knowledge community (LTehno group 2009).

In a future-friendly environment, the majority of students will have shown commitment that leads to higher levels of knowledge and skills when studying the course subjects. Members of the group have collaborated, networked, searched, explored, supported each other and step by step contributed with inputs to develop a critical mass for this new area of knowledge in Latvia.

How to deal with futures and innovate in order to change citizens' attitude to be at ease with lobbying rather than to be hostile to it – these are important issues that students have tried to evaluate and discuss. They produced individual and team outputs resulting in deliverables, for instance, presentations placed in the team's virtual space; shared knowledge with people outside the student group at work, their family; and designed lobbying cases for organizations and others.

The LNELS project and its research on the elaboration of specific frameworks and approaches, their application and lessons about relationships based on knowledge, trust and commitment point to the usefulness of theoretical work for developing the synergy of entrepreneurship, KM and foresight. The author of this chapter proposes to continue research and extend its potential for adaptation (for different foresight tasks, institutions, areas).

10.5 About Consequences

Discussing, researching and evaluating ideas/good practice/obstacles for synergies of entrepreneurship, foresight and knowledge management can be a constructive and perhaps particularly fruitful way to attract the attention of people to each of these domains and to their natural unity.

The futures activities and methodological framing described support the opinion that entrepreneurship can be promoted by creating, converting and sharing knowledge in a favourable mental and physical environment and by approaching in skilled way knowledge sources in and outside the organization.

People live in communities – self-organizing and adapting to a previously established environment and to economic, political and ideological meta-frameworks. National and supranational powers and interconnected and overlapping communities have their own 'systems of values'. The author believes that 'knowledge' and 'intelligence' are not prioritized enough in existing and self-developing 'systems of values' across the world. Self-organizing change in companies, research organizations and communities, supported by a deeper approach to knowledge for foresight culture, remains a challenging theme for European research.

References

Busquin, P. (2002). The foresight dimension of the European research area. In The role of foresight in the selection of research policy priorities. *Conference Proceedings*, EC, JRC-IPTS, pp. 19–23.

David Skyrme Associates (2008). http://www.skyrme.com/resource/glossary.htm accessed on 30 sep 2012

European Commission (2003) *Green Paper: Entrepreneurship in Europe*, Brussels, Document based on COM 27 final.

European Committee for Standardization. (2004a). *European guide to good practice in knowledge management – Part 1: Knowledge management framework*. Brussels: CEN.

European Committee for Standardization. (2004b). *European guide to good practice in knowledge management – Part 5: KM terminology*. Brussels: CEN.

Gibbons, M. (1997). What kind of University? *Research and Teaching in the 21st Century*, 1997 Beanland Lecture, Melbourne: Victoria University of Technology.

Latvijas Zinātņu akadēmijas gadagrāmata (2008). Riga, 199.lpp.

LTehno group (2009). *Lobbying technologies*. Course for the Masters Program in PR, BAT. http://www.scribd.com/group/70099-ltehno Accessed on 18 April 2012.
Marwick, A. (2001). Knowledge management technology. *IBM Systems Journal, 40*(4), 814–830.
Nowotny, H. (2003). The potential of transdisciplinarity. Rethinking interdisciplinarity. http://w7.ens-lyon.fr/amrieu/IMG/pdf/Nowotny_Interdisciplinarity_5-08.pdf Accessed on 30 sep 2012.
Puga, A. (2007). A Latvian experience addressing issues of the foresight innovation. *International Journal of Foresight and Innovation Policy, 3*(4), 369–387. accessed on 30 sep 2012.
Puga, A. (2008, April 23–25). Pursuing innovation: The foresight issues in Latvian universities. In *TII 2008 annual conference*, Valencia. http://www.tii.org Accessed on 30 sep 2012.
UNIDO (2007, September 27–29). Technology Foresight Summit 2007, Documentation, Budapest, p. viii.
Voros, J. (2003). A generic foresight process framework. *Foresight, 5*(3), 10–21.
Zinātnes Vēstnesis (2007). *Scientific Bulletin*: 3, 5 February, Association of Latvian Scientists, Riga, Latvia. http://www.lza.lv/ZV/zv070300.htm Accessed on 30 sep 2012.

Chapter 11
Scenario Transfer Methodology and Technology

Bartolomeo Sapio and Enrico Nicolò

11.1 Introduction

Many research activities have been carried out over the years in the area of *scenario methodologies* (see, e.g. Godet 1987; Georgantzas and Acar 1995), but no great efforts have been devoted to facilitating the fruition, by end-users, of the results obtained through the application of these methodologies. That is, enormous attention has been paid over time to the theoretical aspects of formal scenario methods, but the gap between the analytical details of the ensuing findings in various application fields and the necessity of easy-to-learn knowledge by decision makers and strategic planners has not been adequately bridged. In other words, yet again, the availability of only complex mathematical outputs has often discouraged top managers from adopting suggestions derived from the utilization of the relevant methods and has frustrated the precious potentialities of their conceptual frameworks and computerized tools.

Therefore, methodological and technical efforts have to be made in order to both make scenario methodologies more effective and help key decision takers with their work. It is possible to make these attempts by exploiting the tools provided today by technology, in particular by the 'networked hypermultimedia' epoch which we are entering, characterized by the evolving global digital superhighway infrastructure and dominated by the huge planetary diffusion of the networked hypermedia, the World Wide Web. Scenario modellers should integrate the capabilities offered by technology within the process of scenario creation itself, from conception to delivery, in order to develop their methodologies and, specifically, to *transfer* them

B. Sapio (✉) • E. Nicolò
Fondazione Ugo Bordoni,
Viale del Policlinico 147, 00161 Rome, Italy
e-mail: bsapio@fub.it; nic@fub.it

M. Giaoutzi and B. Sapio (eds.), *Recent Developments in Foresight Methodologies*, Complex Networks and Dynamic Systems 1, DOI 10.1007/978-1-4614-5215-7_11,

© Springer Science+Business Media New York 2013

effectively and efficiently to strategic end-users. In this way, the comprehension, interpretation, acceptability and usability of scenario methods and their results can be facilitated and increased.

In sum, both a theoretical framework and a practical context are necessary to provide methods with a conceptual and instrumental 'interface' capable of adequately *transferring* 'scenario knowledge' to strategic managers who need easy- and ready-to-use highly aggregated and concise elements. To this purpose, we introduce here some fundamentals of *scenario transfer methodology and technology* (Nicolò and Sapio 1999), which are both developed within the logical framework of *scenario engineering*.

11.2 Scenario Analysis: History and Definitions

The term 'scenario' was officially introduced in futurology by Herman Kahn (1967), but the word has generally retained primarily literary connotations, concerning either ideal or apocalyptic prophecies about the future. This was the theme of the work of many authors, such as Verne, Huxley and Orwell. A scenario, in these circumstances, could be defined as an all-inclusive planetary vision, sketching the panorama that mankind might face in the foreseeable future. However, modern scenario analysis requires the consideration of more 'concrete' current phenomena in the real world in order to predict future outcomes.

Before proceeding further, it is necessary, in the context of this work, to define the term 'scenario' and to comment on traditional methods of scenario analysis now being used.

Schnaars (1987) distinguishes three classes of methods for scenario analysis. Firstly, we have empirical studies, which focus on particular issues that aim to emphasize the validity of the method. Secondly, there are applications within specific firm realities, with interesting strategic implications. Finally, we find the stream of futurology, along which many methods for scenario construction flourish.

Many interpretations of the method of creating scenarios have been advanced, but the meaning of the term can vary greatly with different authors. Kahn (1967) defines a scenario as a hypothetical sequence of events with the aim of focusing attention on causal processes and decisional moments. However, some authors use the term 'scenario' to denote any series of multiple forecasts (Carlson and Umble 1980), whilst, according to others, a scenario is nothing more than a set of visionary forecasting and the work of imagination (Chambers et al. 1971).

A series of definitions that have appeared in the literature are purely generic and variations on a theme. Scenarios are general descriptions of future conditions and events (Gershuny 1976); a group of possible but not absolutely certain future conditions (Becker 1983); a set of plausible futures, each one possible but not indubitable (Schnaars and Berenson 1986); an internally coherent vision of what the future might be (Porter 1985); hypothetical futures expressed by means of a sequence of temporal images (Georgoff and Murdick 1986); and a technique to analyse alternative futures and develop firm strategies (Mobasheri et al. 1989).

Other definitions refer to the methodologies employed to build scenarios. Julien et al. (1975) define 'scenario' as a methodology that simulates, step by step and in a plausible and coherent way, a series of circumstances leading a system to a future situation and shows an overall picture of this situation. According to the working group that formulated ASP (Alternate Scenario Planning), a scenario is a series of events, trends and developments that are listed in a logical and sufficiently chronological sequence (Vanston et al. 1977). SRI International defines scenarios as a means of organizing one's perceptions about alternative environments where decisions are to be taken (Huss and Honton 1987). According to Huss (1988), a scenario is a narrative description of a consistent set of factors, defining, in a probabilistic way, alternative sets of future economic conditions.

There are some absolutely specific definitions. Nair and Sarin (1979) define 'scenario' as a description of a certain aspect of the state of the world: namely, a combination of occurrence or non-occurrence of events. According to the American Medical Association, scenario analysis supplies a method to deal with elusive factors without imposing the need for unrealistic data, yet based upon a predetermined methodology that can maintain a notable rigour (Balfe 1985). Battelle Columbus Division considers scenarios as descriptions of interrelated conditions based on cause-effect relationships and logically consistent groups of likely future conditions (Millett and Randles 1986).

However, beyond definitions, there are some fundamental characteristics that make scenarios an essentially different tool from traditional forecasting techniques, which are founded on the hypothesis that the future world will be broadly similar to the present one, so they cannot anticipate primary changes in the environment (Wack 1985a, b). Cause-effect relationships might permit rebuilding the past and extrapolating the future from the present, in a way we can call 'deterministic', provided that we have adequate information about the current state, highly sophisticated models and unlimited power of calculation.

In these times, however, the environment is increasingly turbulent under the effect of powerful technological thrusts. It has often been stressed that the modern economic world is moving ever faster towards chaos, getting further and further from a predictable order (Gleick 1987; Waterman 1987). According to Godet (1983), the year 1973 represents the turning point, after which the future has ceased to resemble the past. High-tech industrial sectors, such as telecommunications or biotechnologies, are inherently uncertain and difficult to predict.

11.3 Scenario Engineering in Brief

Scenario engineering (Nicolò and Sapio 1996) includes a general corpus of scenario methods and techniques for strategic planning and pre-planning. These methodologies may be both interpretative and project-oriented. They are quantitative and/or qualitative, and they may be both analytical and simulative. Scenario engineering can be regarded as a pragmatic approach which has arisen in the area of scenario methodologies and which has been conceived and developed for the construction of scenarios to be produced by exploiting multisectoral technology resources. Broadly speaking, scenario engineering could actually also be considered as a subdiscipline of engineering, related to management science, operations research, mathematics, information technology, telecommunications, strategic planning, forecasting and project management.

11.4 On the Component Phases of Scenario Transfer

Five phases for scenario engineering can be distinguished: *scenario generation*, *scenario analysis*, *scenario presentation*, *scenario communication* and *scenario documentation* (Nicolò and Sapio1998).

Scenario presentation, scenario communication and many elements of scenario documentation constitute a subset of scenario engineering, which can be referred to as *scenario transfer*. Their brief descriptions follow below.

Scenario presentation is concerned with the practical arrangement (for presentation purposes) of the results of the entire scenario modelling process. The description of the scenarios can be made by exploiting traditional tools (e.g. magnetic mass memories, transparencies, slides and paper) and using textual and graphical facilities. Audio, video and photographic facilities can be used to add value to presentation outputs, thus producing scientific documentation supported by, for example, video cassettes and multimedia CD-ROMs including sound, full-motion pictures and digital still images.

Scenario communication mainly deals with the possibility of sending documentation on scenarios (e.g. results, findings and final outputs) to the scientific community over the Internet. From a logical point of view, the activities regarding the communication with experts via the World Wide Web to gather opinions and filled-in questionnaires can be referred to this phase of scenario engineering, even though they are carried out in previous stages. Interactive multimedia capabilities can be used for scenario communication on the Internet.

Scenario documentation includes all the activities needed to keep a record of the entire process of scenario modelling, in order to add both the possibility of accessing at any time the data and hypotheses that led to the generation of specific outputs and the possibility for any reader and modeller to reconstruct the entire creative path. Documentation can use traditional media (e.g. text) and also exploit multimedia capabilities to build hypermedia links.

Scenario transfer is intended to serve as a multifaceted methodology that we name *scenario transfer methodology*.

It is a technological, user-oriented approach to scenario diffusion for strategic management purposes. In this context, effective, user-friendly scientific divulgation can be mainly performed by adequate use of multimedia devices. The scenario presentation phase is here heavily involved, together with scenario documentation and scenario communication.

It is also a technological, expert-oriented approach for scenario documentation interchange. In this case, efficient communication of information can mainly be reached through the utilization of local and wide area network devices. Scenario documentation and communication are mainly involved here.

Scenario transfer is actually a technological modeller-oriented approach for documentation purposes. It assures the possibility of re-performing the whole scenario modelling process, from conception to delivery. This task is accomplished by the scenario documentation phase.

11.5 More Considerations About Scenario Transfer

Both the usability and the practical success of scenario methodologies depend on the effective transfer of information about application hypotheses and methodological outlines, as well as about results and findings to strategic end-users. Clear (intrinsically), comprehensible (for non-specialists), concise and correct (as to formal aspects) information can assure immediate and time-saving catching of the key aspects of scenario inputs and outputs by managers and planners.

We can call this important requirement the 'four "C"s rule' (based on the initial letters of the words: clear, comprehensible, concise and correct), and we consider it as a basic principle for efficient and effective scenario transfer.

In this context, multimedia technology can provide powerful capabilities that should be suitably employed in order to increase the acceptability and usefulness of scenario applications. The exploitation of sound and images (still images and full-motion pictures) can significantly contribute to delivering easily understood, synoptical studies and therefore to facilitating their utilization. Furthermore, scenario documentation, presentation and communication are more effective if they are research-engineers-driven (i.e. modellers-, analysts-, simulationists-, methodologists-driven) through the use of multimedia facilities. As a matter of fact, such capabilities provide modellers with flexible technological tools which are suitable to calibrate and modulate the complexity level of the *transfer contents* according to the circumstances and the category of end-users. In the meantime, the key role played by modellers in the scenario transfer phases assures the accurate maintenance and diffusion of scenarios. In other words, their active presence in all the phases of scenario engineering ensures that the activities concerned with scenario spreading are in compliance with the scientific rigour required. Apart from research engineers, also technicians belonging to the fields of informatics and audio-video production and post-production can obviously contribute to carrying out the technical activities concerned with scenario transfer. Thus, a team of experts could act as a scenario-engineering task force which operates a scenario science lab applied to strategic studies on technology evolution.

It is worth noting that scenario transfer can be oriented to education objectives in the field of scenario methodologies. To this purpose, the instructional pattern for learners designed by research engineers may include advanced navigation hypermedia tools and allow distance learning over the global network. Instructional projects concerned with scenario transfer methodology could be addressed to scenario modellers.

11.6 Some Principles of Scenario Transfer Methodology

Some basic principles and guidelines of scenario transfer methodology are integration, digitalization, personalization, simplification, transduction, complementation, aggregation, linkage, openness, guidance and control.

Integration is obtained through the combined utilization of several types of apparatus which differ in the technology of their components. Computers, audio-video equipment and photographic devices are used co-operatively.

Digitalization is the fundamental means to put technical integration into practice. It allows one to actually handle diverse media in a simultaneous and very efficient way.

Personalization is employed to dimension and tune scenario information according to the type of end-users. It is a key aspect of successful scenario transfer.

Simplification must be used to deliver scenario information to strategic decision makers who are not interested in complex and formal details.

Transduction is a very basic principle of effective scenario transfer methodology and technology. It lies in using multimedia, mainly images and sound, as a vehicle for conveying – if it is feasible – information which may have different and unfamiliar formats, for example, mathematical ones. As an example, the expressive power of video images may be used to describe variables involved in the scenarios under consideration. Transduction is important with particular reference to the above-mentioned simplification principle.

Complementation concerns the opportunity to add to scenario presentations any other multimedia element which could further help end-users, thanks to the exploitation of familiar formats (e.g. photographic images). Such elements, not directly implied by the scenario under consideration, can be more easily introduced by scenario modellers if they are also experts in the specific application field which is the current object of the scenario method considered. Complementation activities may be regarded as an auxiliary or complementary scenario creation task which can add value to scenario transfer, with particular reference to the comprehensibility of scenario information. From an application-related point of view, complementation can be regarded as an extension of the transduction driving principle.

Aggregation is important for simplification, too, and exploits transduction. It lies in presenting scenario information in a synthetic and clustered way, while discarding inessential aspects.

Linkage mainly refers to the opportunity of using hypermedia technology to facilitate the users' task of identifying connections between scenario aspects.

Openness concerns the possibility for users to obtain scenario information from remote sites with respect to the opportunity of performing *networked scenario transfer* over the Internet. Openness also refers to the possibility for scenario modellers to gather data from remote experts through the Net.

Guidance is concerned with the guided learning paths introduced by scenario research engineers within the specific contents of the types of technical support chosen for transfer (e.g. multimedia CDs, video cassettes and files transmitted over the Internet).

Control by scenario research engineers deals with scientific supervising and guarantees conformity with formal aspects.

11.7 On Format Transduction

It is worth underlining that transduction (as well as complementation) is not an automated process. It deals with the further utilization of the creative potentialities of scenario modellers, which can be orientated towards the improvement and transformation of traditional scenario presentation and delivery. Thus, transduction mainly concerns the substitution of conventional, formal output formats (e.g. analytical ones) – and the related media (e.g. paper) – with easy-to-understand, figurative and audio formats (and media). Such a process must be capable of preserving the precise meaning of formal aspects, not introducing spurious or deviant elements and not conditioning users.

Transduction exploits the huge capabilities of still-image and full-motion picture formats, which can quickly convey enormous quantities of both explicit and symbolic simultaneous information. Such information can be also conveyed by audio formats. In particular, vocal comments can efficiently guide users towards the scenario findings. Transduction may also represent a revolutionary approach to scenario documentation, with particular reference to the possibility of storing multimedia information about scenario inputs.

Transduction and complementation constitute both driving principles and practical processes addressed to the effective scientific diffusion of scenarios. A key point is that the people responsible for scenario transfer are the same scientists and researchers who generate and analyse the scenarios under consideration. This careful supervision by the experts themselves assures rigour and continuity for the entire scenario creation process.

As mentioned above, also technicians who are experts in computing, video production and digital photography can efficiently contribute to transduction and complementation activities under the supervision of scenario research engineers.

In sum, the format transduction process can be considered as a media-type *virtual transducer* or a conceptual transformation block (i.e. media transposition box), which allows scenario makers to operate creative multimedia scenario transfer functions in compliance with current constraints fixed by scenario scientists according to the circumstances.

However, some general *transduction criteria* to be met can be identified. They are listed briefly below. It is worth considering that they are often interrelated.

Speech mediation is concerned with the opportunity of providing scenario delivery support with user-oriented oral explanations of contents by scenario scientists.

Graphics prioritization refers to the preferability of simple, graphic, synoptical schemes with respect to any other type of formal representation.

Image primacy deals with the pre-eminence of still and full-motion pictures as vehicles to convey huge quantities of explicit, as well as symbolic, messages which interpret formal and/or abstract contents.

Metaphoric transposition specifically concerns the figurative translation of elements of scenario contents into images, which may also exploit graphical representations and audio support. It is the core of format transduction and, to a certain extent, even of transfer methodology. Metaphoric transposition can make use of the association of ideas which may be induced by scenario researchers between a conceptual and/or formal element (i.e. represented object) and the relevant figurative sublimation (i.e. representative subject). Of course, simple, clear, formal elements may be as such directly transferred.

Transduction deviance minimization concerns the great attention which must be paid by modellers in carrying out metaphoric transposition activities in order to limit as much as possible distortion phenomena (with respect to scientific evidence from scenario models) and to avoid ensuing conditioning effects which could damage end-users.

Semeiotic universalization is concerned with the use of visual images and, in general, multimedia descriptors, which are fundamentally common-sense-driven and common-experience-based in order to assure adequate comprehensibility for as many users as possible.

11.8 Conclusions

In sum, scenario transfer methodology and technology can be considered as a conceptual and practical interface between methods and users, which is oriented to translate complex data into comprehensible information through familiar languages. This approach can be considered as a new research topic in scenario methodologies.

It offers two main advantages. From the point of view of methodologists, it can limit the risk of complex methods proving to be useless. From the point of view of decision makers, it helps them identify problems and possible solutions in their own terms rather than in unfamiliar professional languages.

It is worth considering that the goal of scenario methodologies is not only to provide outputs but also to give opportunities for structured thinking and intelligent information exchange about a fixed problem. The fundamental aspect of a perspective study is not the report which derives from it but what takes place in the minds of people involved in the thinking process. Scenario transfer methodology is intended to serve as a contribution to orientate scenario research towards end-users and ultimate strategic planning objectives, taking into account that the 'process of planning is more important than the plan itself' (Godet 1987).

Therefore, scenario transfer methodology and technology allow scenario research engineers not only to give traditional presentations more easily but also to document, present and communicate scenarios using format transduction, as well as to carry out pilot experiments while setting up a methodological rationale for scenario science laboratories which could be involved in scenario transfer.

References

Balfe, B. (1985). *Environmental analysis at the AMA.* Paper presented at the Midwest Planning Association, Chicago.

Becker, H. S. (1983). Scenarios: A tool of growing importance to policy analysts in government and industry. *Technological Forecasting and Social Change, 23*, 95–120.

Carlson, R. L., & Umble, M. M. (1980). Statistical demand functions for automobiles and their use for forecasting in an energy crisis. *Journal of Business, 53*, 193–204.

Chambers, J. C., Satindev, K. M., & Donald, D. S. (1971). How to choose the right forecasting technique. *Harvard Business Review, 49*(4), 45–74.

Georgantzas, N. C., & Acar, W. (1995). *Scenario-driven planning: Learning to manage strategic uncertainty*. Westport, CT: Quorum Books.

Georgoff, D. M., & Murdick, R. G. (1986). Manager's guide to forecasting. *Harvard Business Review, 64*, 110–120.

Gershuny, J. (1976). The choice of scenarios. *Futures, 8*, 496–508.

Gleick, J. (1987). *Chaos: Making a new science.* New York: Viking.

Godet, M. (1983). Reducing the blunders in forecasting. *Futures, 15*, 181–192.

Godet, M. (1987). *Scenarios and strategic management.* London: Butterworths.

Huss, W. R. (1988). A move toward scenario analysis. *International Journal of Forecasting, 4*, 377–388.

Huss, W. R., & Honton, E. J. (1987). Scenario planning – What style should you use? *Long Range Planning, 20*, 21–29.

Julien, P. A., Lamonde, P., & Latouche, D. (1975). La Méthode des Scénarios. *L'actualité économique, 51*(2), 253–281.

Kahn, H. (1967). *The year 2000.* New York: Macmillan.

Millett, S. M., & Randles, F. (1986). Scenarios for strategic business planning: A case history for aerospace and defence companies. *Interfaces, 16*, 64–72.

Mobasheri, F., Orren, L. H., & Sioshansi, F. P. (1989). Scenario planning at Southern California Edison. *Interfaces, 19*, 31–44.

Nair, K., & Sarin, R. K. (1979). Generating future scenarios – Their use in strategic planning. *Long Range Planning, 12*, 57–61.

Nicolò, E., & Sapio, B. (1996, June). SIMULAB (Scenario-engineering Interactive MUltimedia LABoratory): An integrated environment to forecast and plan for possible futures, ISF '96. In *The Sixteenth Annual International Symposium on Forecasting "Financial markets and forecasting"*. Istanbul.

Nicolò, E., & Sapio, B. (1998, June 10–13). NET-SIMULAB: A scenario-engineering laboratory in the networked hypermultimedia era, ISF '98. In *18th International Symposium on Forecasting "Forecasting social, technological & economic change"*. Edinburgh.

Nicolò, E., & Sapio, B. (1999, March 15–17). Fundamentals of scenario transfer methodology and technology. *Proceedings of IAMOT '99, Eighth International Conference on Management of Technology* (Vol. 2, pp. 755–763). Cairo.

Porter, M. J. (1985). *Competitive advantage.* New York: Macmillan.

Schnaars, S. P. (1987). How to develop and use scenarios. *Long Range Planning, 20*, 105–114.

Schnaars, S. P., & Berenson, C. (1986). Growth market forecasting revisited: A look back at a look forward. *California Management Review, 28*, 71–88.

Vanston, J. H., Jr., Frisbie, W. P., Lopreato, S. C., & Poston, D. L., Jr. (1977). Alternate scenario planning. *Technological Forecasting and Social Change, 10*, 159–180.
Wack, P. (1985a). Scenarios: Uncharted waters ahead. *Harvard Business Review, 62*, 73–89.
Wack, P. (1985b). Scenarios: Shooting the rapids. *Harvard Business Review, 63*, 139–150.
Waterman, R. H. (1987). Strategy in a more volatile world. *Fortune, 116*, 181–182.

Part IV
Foresight Tools and Approaches

Chapter 12
Willingness of Stakeholders to Use Models for Climate Policy: The Delft Process

Serge Stalpers and Carolien Kroeze

12.1 Introduction

Participatory integrated assessments (PIAs) can be defined as 'an IA approach in which social stakeholders… contribute their knowledge and policy preferences to the assessment of complex policy problems' (Schlumpf et al. 1999: p. 2). PIAs often involve dialogues between scientists, decision-makers and other stakeholders. Participatory research is increasingly used in integrated assessments (IAs) of climate change (Dahinden et al. 2000; Kloprogge and van der Sluijs 2006). PIAs differ with respect to their degree of involvement of stakeholders (Van de Kerkhof 2004). Here, we focus on PIAs with co-productive participation, where the IA is carried out in co-production between stakeholders and scientists (Van de Kerkhof 2004). In co-productive PIAs, participants decide what information to use and therefore also decide what models they are willing to use for producing the integrated insights in the PIA.

Climate change, being characterized by large uncertainties and high decision stakes, is an example of a post-normal problem on which multiple legitimate perspectives on the problem exist (Funtowicz and Ravetz 1993; Van de Kerkhof and Leroy 2000). For such problems, Funtowicz and Ravetz (1993) propose a post-normal science approach, which, as opposed to the routine puzzle-solving approach of 'normal' science, explicitly manages uncertainties and spells out values. In a

S. Stalpers
Environmental Systems Analysis Group, Wageningen UR, Wageningen, P.O. Box 47 6700 AA, The Netherlands

C. Kroeze (✉)
School of Science, Open University of the Netherlands, Heerlen, The Netherlands
e-mail: Carolien.Kroeze@wur.nl

M. Giaoutzi and B. Sapio (eds.), *Recent Developments in Foresight Methodologies*, Complex Networks and Dynamic Systems 1, DOI 10.1007/978-1-4614-5215-7_12,
© Springer Science+Business Media New York 2013

post-normal approach, an 'extended peer community' of stakeholders evaluates the quality of scientific inputs to decision-making, so that the multiple legitimate perspectives on the problem are taken into account. Therefore, PIAs can help find solutions for climate policy that are scientifically credible, as well as socially legitimate.

In addition, stakeholder participation is often used in IAs because it can foster learning – both by including stakeholder knowledge in the IA and by generating new knowledge through exchange of ideas between participants – and because it can improve implementation by involving actors who can implement policy options (Fiorino 1990; Stalpers et al. 2009).

Integrated assessment models are promising tools for use in PIAs. Models have been applied extensively in IAs of climate change and are a means of using scientific knowledge as a basis for policymaking (Dowlatabadi 1995; Parker et al. 2002; Schröter et al. 2005b). Using models in PIAs is a promising means of making scientific knowledge accessible to participants, thereby enhancing the use of scientific knowledge in PIAs (Toth and Hizsnyik 1998; Dahinden et al. 2000; Yearley et al. 2001; Siebenhüner and Barth 2005).

However, it can be difficult to use models in a way that meets participants' needs and requirements. This requires effective scientist-stakeholder communication about models. A good understanding by scientists of the willingness of participants to use models is an important precondition for the effective use of a model in a PIA (henceforth, 'participants' refers to non-scientist participants, i.e. all participating stakeholders except those participating solely in their capacity as scientists). We define *willingness to use* (WTU) as the extent to which participants are willing to use models as a source of information for an IA. The explicit or tacit assessment of participants of their WTU models is then called the *WTU assessment*. We distinguish between WTU assessment, which is the participants' assessment of their WTU models, and the integrated assessment (IA) within which the WTU assessment takes place. Understanding the demand and supply of information in WTU assessments can help to find ways to improve the WTU assessment process and improve the supply of scientific information about models in PIAs.

This chapter aims to investigate how participants assessed their WTU models in the Delft Process, a PIA-supporting climate policy appraisal in which the IMAGE model (Integrated Model to Assess the Global Environment) was used with international climate negotiators. This chapter contains two sections. The first section gives an analysis of the Delft Process. The second section presents an outline for a conceptual model of WTU assessments, based on the findings from the Delft Process. The analysis of the Delft Process focuses on the process by which information pertaining to the WTU is requested and supplied and on the criteria that stakeholders apply to assess WTU and is based on workshop reports (Berk et al. 1995; Van Daalen and Grünfeld 1995, 1996) and a publication (Van Daalen et al. 1998). In addition, interviews with the process facilitator, a modeller from the IMAGE team, and three participants were used to confirm the main conclusions of the reports and publication. The study of Stalpers et al. (2009) complements this chapter and investigates the reconciliation of model results and user needs in the Delft Process.

12.2 The WTU Assessment in the Delft Process

There are several examples of PIAs dealing with climate change (Van Daalen et al. 1998; Holman and Loveland 2001; Van Ierland et al. 2001; Gupta et al. 2004; Schröter et al. 2005a). In these PIAs participants produced a part of the IA results in collaboration with scientists and therefore had to consider whether to use information from models. Therefore, we expect that WTU assessments occurred in these PIAs. In this section, we examine how the WTU assessment process worked in practice in the case of the Delft Process.

The Delft Process (Van Daalen et al. 1998) aimed to provide a platform for policy-science dialogue on issues on the international climate agenda and to make the IMAGE 2.1 model more policy-relevant. The Delft Process consisted of five two-day workshops, each attended by about 15 policy makers and senior policy advisors of the Ad Hoc Group of the Berlin Mandate (AGBM)[1] involved in the negotiations leading up to the third Conference of the Parties (COP 3) of the United Nations Framework Convention on Climate Change (UNFCCC), at which the Kyoto Protocol was adopted. The participants were primarily from Europe and developing countries, and most sympathized with an 'environmentally oriented point of view' (Van Daalen et al. 1998: p. 278). The IMAGE model 'aims to contribute to scientific understanding and support decision making by quantifying the relative importance of major processes and interactions in the society-biosphere-climate system' (IMAGE-team 2001). IMAGE is a geographically explicit policy simulation and evaluation model, which integrates a general equilibrium economic model, a population model and models of the energy-industry system (to quantify industrial emissions), a terrestrial environment system (to quantify land-use-change-related emissions and CO_2 exchange with the biosphere) and the atmosphere-ocean system (to quantify atmosphere-ocean interactions). Model outputs include CO_2 concentration, changes in temperature and precipitation, impacts on agriculture and sea level rise.

12.2.1 The WTU Assessment Process

The five workshops consisted of one and a half days of presentation and discussion of model results, followed by a half-day session in which participants prioritized their requested analyses by allocating points with the aid of the Group Decision

[1] The AGBM is a UNFCCC body which was established at the first Conference of the Parties in 1995, aimed to assist the Kyoto negotiations by assessing greenhouse gas emission reduction options, targets and time frames 'in the light of the best available scientific information' (UNFCCC 1995, Art.2,3).

Room (GDR), a computerized meeting facility. Each workshop followed the same general process, so that each workshop was an iteration of the following steps:

- The IMAGE team gave presentations about the model (first workshop) or changes to the model (subsequent workshops) and about the model results.
- Participants discussed the results and how they could use the IMAGE model.
- Participants prioritized requests for new analyses and suggestions for model improvement.
- Each workshop ended with an evaluation questionnaire and discussion.

Facilitators took note of issues that came up during the discussion of the model results. These were then grouped and presented at the beginning of the prioritization session, and participants could add additional requests. These requests were then clustered and scored by giving each participant a number of points (e.g. 20) they could allocate to the requests they felt were most important. The IMAGE team had the time and resources to improve the model and produce new results between workshops so that the participants knew their comments would be used. The prioritized requests included suggestions to change the model to make it more relevant; requests for additional information on, for example, the validity of the model; and requests for running the model using different emission scenarios.

The prioritization of requested analyses is a form of WTU assessment, as participants reflect on how the IMAGE model can be improved to make it more useful to them. Participants of the Delft Process were not asked whether or not they wanted to use the results of the IMAGE model, and as such there was no explicit WTU assessment in the Delft Process. Rather, participants were asked *what* analyses would be useful and how the model could be improved, so that the role of the model was adjusted to what participants were willing to use the model for. This occurred in an iterative process: as the workshops proceeded and participants got better acquainted with IMAGE's possibilities, they could identify for what purposes they could use the model.

12.2.2 Information Demand: WTU Criteria

The analyses requested by the participants reveal the WTU criteria on the basis of which they assessed their WTU. IMAGE is designed to quantify climate impacts, most of which become manifest in the long term, that is, towards the year 2100. In the first workshop, participants requested analyses focusing on shorter-term emission profiles rather than the long-term emission profiles presented by the IMAGE team, reflecting the AGBM's focus on short-term policy action (Van Daalen et al. 1998). Apparently, the participants felt that the IMAGE results would be more relevant if put in a more short-term perspective. The focus on the short term remained throughout the Delft Process workshops. The participants particularly wanted to know what the stabilization levels of greenhouse gasses are and what the associated impacts are for different emission profiles, such as various graduation mechanisms, in which developing countries start reducing emissions at different points in time.

In the second workshop, the focus on short-term emission profiles led to the conceptualization of the 'safe landing analysis' (SLA), an analogy for landing an aircraft safely: reducing emissions too quickly costs too much, and reducing emissions too slowly means risking unacceptable climate impacts. From the third workshop onwards, the participants' attention shifted to the SLA rather than the IMAGE model itself, because the SLA became more relevant as the need arose to link long-term impacts to short-term policy targets. Although the relevance of the IMAGE model became less, it was still needed to lend credibility to the SLA by quantifying the impacts of the long-term climate targets from which the SLA deduces short-term emission profiles.

In the second workshop, the participants requested more information on the impacts associated with the climate change indicators of IMAGE, such as the rate of temperature change per decade, and more information on regional impacts associated with the safe landing goals. Apparently, the model needed to be more comprehensive and the results presented at a greater spatial resolution to better suit the participants' needs.

In the third workshop, the participants discussed the assumptions behind setting a critical value for the indicators of the SLA, from which the future emission concentration targets are derived: 'Selecting critical values of the indicators is a political choice and relates to the amount of risk which is considered acceptable' (Van Daalen et al. 1998, p. 270). From this discussion, it follows that the legitimacy of the process by which critical values were selected is a WTU criterion.

In the first workshop, the participants requested information on model validation and on the uncertainties of the model, such as uncertainty ranges for model output, reflecting a need for information on the credibility of IMAGE. In the second workshop, the IMAGE team was able to present only some of the requested uncertainty information. Even though the demand for uncertainty information was not met, there were only a few more requests for uncertainty-related information in subsequent workshops, and these did not rank high on the priorities list. In fact, comparisons of the IMAGE model and the results from other climate modelling groups presented in the fourth workshop, which gave some insights into the range of uncertainties related to choices of model structure, were not well received by participants. Participants did not appreciate the detailed discussions on model technicalities between the modelling groups. By the fourth workshop, the participants apparently no longer demanded information on model uncertainties. Possible explanations for this change in demand are the following: that the IMAGE model was perceived by participants as being the 'best available knowledge' after peer-reviewed scientific publications on the model were published between the first and second workshop and that the participants had no interest in technical model details (Van Daalen et al. 1998).

The WTU criteria were not explicitly elicited, but became apparent indirectly through the prioritization of requested analyses. Therefore, only the WTU criteria that were not met, or for which participants did not have enough information to assess, were articulated.

12.2.3 Information Supply

The requested analyses prioritized at the end of each workshop were used to decide on the programme for the next workshop. The IMAGE team selected what model analyses to perform between workshops and what information to present based on the requested analyses with the highest priority. The presentations at the first workshop were selected on the basis of the project team's expectations of what participants would like to know. An exception is the presentation of comparisons between the IMAGE model and other climate models in the fourth workshop held at the request of other model teams.

The presentations mainly consisted of descriptions of the model results obtained by running different scenarios, as requested by the participants. The presentations about the IMAGE model could not cover all model components in detail. Therefore, presentations of model structure and model assumptions were based on expert judgement of the IMAGE team of what are the most important model components. Information on uncertainty was gathered using a limited uncertainty analysis based on expert judgement of the most uncertain model parameters and also on investigation of the sensitivity of model results to various parameters, such as climate sensitivity and assumptions about technological change.

12.3 Towards a Conceptual Framework of Willingness-to-Use Assessment

Participants of the Delft Process decided on their WTU models on the basis of at least the information available and on explicit or implicit WTU criteria. The information flows involved in the participants' decision to use the IMAGE model results in the Delft Process can be represented by information demand and supply flows, as shown in Fig. 12.1.

The left-hand side of Fig. 12.1 represents information demand, and the right-hand side represents information supply. The demand for information follows from the criteria by which participants decide whether to use models. The supply side is the information supplied by scientists that is relevant for the WTU assessment by participants. In PIA practice, participants will use various sources of information including news media, information from the participants' network and previous experience with the model or the issue concerned, but we limit the conceptual framework to information supplied by scientists because this is the only information supply which the PIA research team can improve.

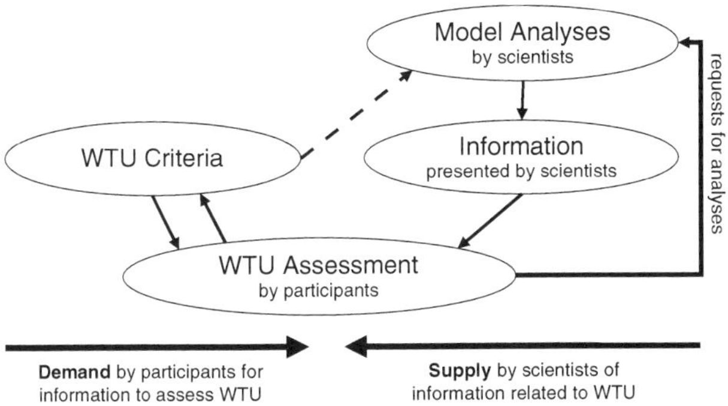

Fig. 12.1 *Conceptual model of WTU assessment.* Information on demand and supply for assessing the WTU IA models. Information on demand and supply lie along the axis of the nature of the information, moving from the stakeholders' WTU criteria (demand) to science (supply). The *solid arrows* in the diagram represent flows of information identified in the Delft Process. The *dashed arrow* suggests an improvement of information supply by taking WTU criteria into account when deciding what model analyses to carry out

Table 12.1 Willingness-to-use concepts

Willingness to use (WTU)
The extent to which participants are willing to use a model as a source of information for their input to the integrated assessment
WTU assessment
The tacit or explicit assessment by participants of their WTU models
WTU criteria
The criteria by which participants, implicitly or explicitly, assess their WTU models
Information for WTU
The information supplied by scientists to participants which is relevant for the WTU assessment
Model analyses
Analyses by scientists of the model and of the problem concerned, in order to generate information for the WTU assessment

12.3.1 The Demand Side of WTU Assessments

There can be various criteria which participants consider relevant for assessing WTU, and we refer to these as *WTU criteria* (see also Table 12.1). The WTU criteria can remain implicit or they can be explicitly articulated by the participants. The WTU assessment can result in a clarification of the WTU criteria, represented in Fig. 12.1 by the arrow going from WTU assessment to WTU criteria.

The WTU criteria by which participants of the Delft Process assessed their willingness to use IMAGE can be classified according to a taxonomy of information attributes borrowed from research on science and technology studies: relevance, credibility and legitimacy (Cash et al. 2003). Although alternative taxonomies exist, this taxonomy is relatively comprehensive and applicable to the information 'value demand' of the decision-makers (McNie 2007) and other stakeholders.

Relevance (termed 'salience' in Cash et al. (2003)) refers to the extent to which information is relevant, given the needs of stakeholders. The relevance of models considers such aspects as the spatial and temporal extent and resolution (Dumont et al. 2008), whether essential components of the studied system are modelled and whether the model includes variables or 'leverage points' (Meadows 1999) of the system that can be managed. In the Delft Process, participants requested that IMAGE results focus more on short-term emission profiles, rather than long-term policy action, which led to the development of the safe landing analysis (SLA). Also, participants requested (and received) more comprehensive impact indicators including rate of temperature change per decade and more detailed regional information on impacts.

Credibility refers to the 'scientific adequacy of technical evidence and arguments' (Cash et al. 2003: p. 8086). Credibility considers aspects such as the scientific and technical believability of models (Tuinstra et al. 2006) and the internal consistency of the model and scenarios used (Van Aardenne 2002). In the Delft Process, participants requested information on the uncertainty of IMAGE in the first workshop. In subsequent workshops, participants had less interest in uncertainty information and may have assessed the credibility of IMAGE indirectly by the reputation of the modelling institute and the standing of the model and derived publications in the scientific community.

Legitimacy refers to the extent to which the process of information production is unbiased and takes into account divergent stakeholder values, beliefs, views and interest (Cash et al. 2003). Legitimacy also includes whether value-laden assumptions (Van der Sluijs et al. 2005) in the model match with users' perspectives or at the least are made explicit. In the Delft Process, an example of the legitimacy criterion is the selecting of critical values for the safe landing analysis.

The accuracy and transparency of a model can pertain to both relevance and credibility, and transparency may be a precondition for assessing the legitimacy of a model. The *accuracy* of a model can determine, on the one hand, if the model is relevant, that is, if the model can answer relevant queries, given its uncertainty. On the other hand, the accuracy of a model can also determine if a model is credible, that is, if conclusions drawn from the modelling exercise are valid, given the uncertainties. Many typologies of uncertainties exist which may serve for subclassification of WTU criteria (e.g. Van Asselt and Rotmans 1996; Van der Sluijs 1997; Gabbert and Kroeze 2003; Sarewitz 2004; IPCC 2005; Peterson 2006). The IPCC guidance notes on addressing uncertainties for the Fourth Assessment Report (IPCC 2005) work well for classifying WTU criteria because it can easily be related to different aspects of using a model: the model inputs, the model itself and the real future. It is therefore expected to be transparent to non-scientific stakeholders. The typology is (1) *value uncertainty* – uncertainty from, for example, missing, inaccurate or non-representative data, inappropriate spatial or temporal resolution, and poorly known or changing model parameters; (2) *structural uncertainty* – uncertainty from, for example, inadequate model form, processes not considered or wrongly specified and ambiguous system boundaries; and (3) *unpredictability* – uncertainty from unknowable future developments, for example, human behaviour and chaotic behaviour of complex systems. The corresponding WTU criteria are (1) accuracy of model inputs and parameters, (2) accuracy of model structure and (3) accuracy in the face of unpredictability.

Transparency is the degree to which a model can be understood by other experts or non-experts. Transparency is a precondition for assessing other WTU criteria. For example, assessing relevance requires understanding of the model structure; assessing credibility requires understanding of the internal consistency of the model; and legitimacy requires accessibility of value-laden assumptions in the model.

The participants' WTU criteria are not static during a PIA. WTU criteria will likely be determined by the participants' perspectives on the role of the model in the PIA, and more generally on participants' perspectives on the role of science in the PIA, which may change as the participants gather more information about the model and the environmental problem. For example, if participants perceive the role of the model in the PIA to be modest because other sources of information are more relevant, they are likely to be less critical than if the model forms the main argument of the PIA. The participants' WTU criteria also depend on their perspective on what constitutes sufficient quality of science for policy. For example, the participants' demand for accuracy depends on whether participants are more risk-seeking, risk-accepting or risk-averse (Van Asselt and Rotmans 1996). It is likely that the act of explicitly discussing user requests and WTU criteria can result in a group learning process, which leads to a different prioritization of WTU criteria than if the WTU assessment remained implicit.[2]

12.3.2 The Supply Side of WTU Assessments

The information that scientists can supply about a model is limited by the analytical methods available to generate information relevant to the WTU assessment. We refer to such analyses as *model analyses*. Model analyses can be broad in scope, including not only analyses of the problem concerned by using the model but also analyses of the model behaviour and uncertainties.

Scientists generally choose what model analyses to perform and what information to present, depending on their perception of the PIA's aim and their assumptions about participant needs. The model's characteristics also constrain what analyses are possible. In the Delft Process, for example, a full global uncertainty analysis was not possible given the large number of parameters in IMAGE.

Model analyses can include, for instance, non-participatory methods such as uncertainty analysis and related methods, scenario analysis and expert judgement,

[2] We are indebted to Dale Rothman, for pointing this out in his review.

and participatory methods such as workshop discussions and interviews on the relevance, credibility and legitimacy of a model, but also more elaborate frameworks combining these, such as the Numeral Unit Spread Assessment Pedigree (NUSAP) system (Van der Sluijs et al. 2005) and the Pluralistic fRamework for Integrated uncertainty Management and risk Analysis (PRIMA) (Van Asselt 2000). In the Delft Process, model analyses included expert judgement of what are important model components, assumptions and uncertain model parameters, limited uncertainty analysis and sensitivity analyses.

Improving a model's transparency is usually dependent on the modellers' skill of judging what model components and assumptions are important for participants. An alternative method to increase the transparency of the model structure, and thereby help users to make judgements on the relevance and credibility of the model, is given by Cuppen et al. (2007). This method uses the Toulmin model of argumentation (Toulmin et al. 1979; Toulmin 2003) to reveal the grounds on which the model is based by identifying what variables are taken into account and which assumptions are (explicitly or implicitly) made, and why.

12.3.3 Matching Supply and Demand

An appropriate choice for model analyses by scientists would be one in which the information supplied matches the demand for information by participants, so that participants have sufficient information to assess, for themselves, what their WTU model is. When scientists select what information about a model they will present to participants, they can base this choice on their assumptions about participants' WTU criteria, but without eliciting WTU criteria directly from the participants, there is no guarantee that the information presented matches participants' needs. An appropriate choice of model analyses can be made by eliciting WTU criteria from participants and basing the choice of model analyses on participants' WTU criteria, as represented in Fig. 12.1 above by the dashed arrow from WTU criteria to model analyses.

In the Delft Process, the supply of information matched demand reasonably well, and the participants evaluated the information positively. Key model analyses which match demand include the comprehensive set of baseline and stabilization scenarios and the development of the SLA (including the regionalized version). A notable mismatch is the model comparison which participants evaluated negatively because they were not interested in being present at discussions between modellers, although they did express interest in the outcomes of the comparison. The iterative process in which the prioritized requests of each workshop guided what information was presented at the next workshop improved the match between supply and demand. However, this process required an iteration of five 2-day workshops. It may be more efficient to base the choice for model analyses on a direct elicitation of WTU criteria. The elicitation of WTU criteria may be done after presenting and discussing the initial model results so that the participants can better understand the model's capabilities.

12.4 Concluding Remarks

The Delft Process illustrates how the demand and supply of information in WTU assessments occur in practice. The WTU assessment in the Delft Process was only partly explicit because participants were not asked whether they wanted to actually use the model, but only for what purpose. In the Delft Process, information for the WTU assessment was supplied by the IMAGE modelling team in the form of model presentations, and the demand for information was structured using a computerized meeting facility for prioritization of participants' requested analyses to make the IMAGE results more useful. The participants in the Delft Process implicitly referred to WTU criteria in their requests for analyses. These WTU criteria, which characterize the information needs of participants, related to relevance, credibility and legitimacy. The Delft Process illustrates how participants' WTU model may change during the assessment, where the IMAGE model was relevant in the first workshop for quantifying long-term climate impacts for setting long-term targets, and in later workshops, the SLA became more relevant as attention shifted to short-term greenhouse gas emission targets. Overall, the supply of information seemed to match the demand for information in the Delft Process because the IMAGE team supplied information on the basis of participants' requested analyses. One exception is the model comparison presented in the fourth workshop which was not well received by participants. Model comparisons are common practice for modellers to investigate model uncertainties and thereby (in part) establish model credibility, but in this case, the participants had no demand for more information on model uncertainty.

We presented a conceptual model of the demand and supply of information in WTU assessments. The conceptual model assumes that participants assess their WTU models on the basis of the WTU criteria of relevance, credibility and legitimacy. On the basis of the conceptual model, we propose that supply of information can match the participants' demand if the WTU criteria of participants are taken into account when scientists select what model analyses to apply to provide information for participants. Stalpers et al. (2007) further elaborates on this conceptual model to facilitate reconciliation of model results and user needs.

We argue that an explicit WTU assessment, as suggested by our conceptual model, can improve scientist-stakeholder communication in a PIA by ensuring that the information provided by scientists better matches the information needs of participating stakeholders for assessing their WTU models. In such an explicit WTU assessment, scientists become more aware of participants' information demands by explicitly eliciting WTU criteria from them and then using these WTU criteria to select what model analyses should be carried out, and what information should be supplied to match participants' information demands.

We conclude that, in the Delft Process, the WTU assessment was only partly explicit. It should be realized that the Delft Process is likely to be the PIA in which most attention was paid to WTU. Our analysis may therefore be an indication that in many PIAs, WTU assessments are to a large extent implicit or even non-existent. It would be interesting to analyse this in more detail in studies of other PIAs.

Acknowledgements The authors would like to acknowledge Dale Rothman, Eefje Cuppen, Matthijs Hisschemöller, Rik Leemans and Ekko van Ierland for comments; members of COST Action 22 for lively and useful discussions; and participants of the 2007 Amsterdam Conference on the Human Dimensions of Global Environmental Change 'Earth System Governance: Theories and Strategies for Sustainability', 24–26 May 2007, Vrije Universiteit, Amsterdam, and the COST A22 Conference 'From Oracles to Dialogue', 9–11 July 2007, National Technical University, Athens, at which a previous version of this chapter was presented.

References

Berk, M., van Daalen, E., & Grünfeld, H. (1995). *First International Workshop on Using the IMAGE 2 Model to Support Climate Negotiations*, Summary Report, Delft, Faculty of Systems Engineering, Policy Analysis and Management, and National Institute of Public Health and Environment.

Cash, D. W., Clark, W. C., Alcock, F., Dickson, N. M., Eckley, N., Guston, D. H., Jäger, J., & Mitchell, R. B. (2003). Knowledge systems for sustainable development. *Proceedings of the National Academy of Sciences of the United States of America, 100*, 8086–8091.

Cuppen, E., Leemans, R., Hisschemöller, M., & Stalpers, S. I. P. (2007, May 24–26). Enhancing the use of scientific models in participatory integrated assessment: Applying a method to articulate argumentations to the IMAGE 2.2 model. In *Amsterdam Conference on the Human Dimensions of Global Environmental Change*, Amsterdam.

Dahinden, U., Querol, C., Jäger, J., & Nilsson, M. (2000). Exploring the use of computer models in participatory integrated assessment – Experiences and recommendations for further steps. *Integrated Assessment, 1*(4), 253–266.

Dowlatabadi, H. (1995). Integrated assessment models of climate change: An incomplete overview. *Energy Policy, 23*(4–5), 289–296.

Dumont, E., Bakker, E. J., Bouwman, A. F., Kroeze, C., Leemans, R., & Stein, A. (2008). A framework to identify appropriate spatial and temporal scales for modelling N flows from watersheds. *Ecological Modelling, 212*(3–4), 256–272.

Fiorino, D. J. (1990). Citizen participation and environmental risk: A survey of institutional mechanisms. *Science, Technology, & Human Values, 15*(2), 226–243.

Funtowicz, S. O., & Ravetz, J. R. (1993). Science for the post-normal age. *Futures, 25*(7), 739–755.

Gabbert, S., & Kroeze, C. (2003). Uncertainty analysis in integrated assessment models for acidification. In J. Wesseler, H.-P. Weikard, & R. D. Weaver (Eds.), *Risk and uncertainty in environmental and natural resource economics* (pp. 189–214). Cheltenham, UK: Edward Elgar.

Gupta, J., van Asselt, H., Amelung, B., Berk, M., Buiteveld, H., Dalenoord, E., Hein, L., Huynen, M., Kuik, O., Leemans, R., Martens, P., Mulder, J., Oost, A., Schaeffer, M., Verbeek, K., & van der Wegen, M. (2004). *Re-evaluation of the Netherlands' long-term climate targets, summary report.* Institute for Environmental Studies, Free University of Amsterdam, Amsterdam.

Holman, I. P., & Loveland, P. J. E. (2001). *Regional climate change impact and response studies in East Anglia and North West England (RegIS).* London, UK: Department for Environment, Food & Rural Affairs.

IMAGE-team (2001). *The IMAGE 2.2 Implementation of the SRES Scenarios: A comprehensive analysis of emissions, climate change and impacts in the 21st century.* RIVM CD-ROM publication 481508018.

IPCC (2005). *Guidance notes for lead authors of the IPCC fourth assessment report on addressing uncertainties.* IPCC, Geneva.

Kloprogge, P., & van der Sluijs, J. P. (2006). The inclusion of stakeholder knowledge and perspectives in integrated assessment of climate change. *Climatic Change, 75*, 359–389.

McNie, E. C. (2007). Reconciling the supply of scientific information with user demands: An analysis of the problem and review of the literature. *Environmental Science & Policy, 10*(1), 17–38.

Meadows, D. (1999). *Leverage points: Places to intervene in a system*. Hartland: The Sustainability Institute.

Parker, P., Letcher, R., Jakeman, A., Beck, M. B., Harris, G., Argent, R. M., Hare, M., Pahl-Wostl, C., Voinov, A., Janssen, M., Sullivan, P., Scoccimarro, M., Friend, A., Sonnenshein, M., Barker, D., Matejicek, L., Odulaja, D., Deadman, P., Lim, K., Larocque, G., Tarikhi, P., Fletcher, C., Put, A., Maxwell, T., Charles, A., Breeze, H., Nakatani, N., Mudgal, S., Naito, W., Osidele, O., Eriksson, I., Kautsky, U., Kautsky, E., Naeslund, B., Kumblad, L., Park, R., Maltagliati, S., Girardin, P., Rizzoli, A., Mauriello, D., Hoch, R., Pelletier, D., Reilly, J., Olafsdottir, R., & Bin, S. (2002). Progress in integrated assessment and modelling. *Environmental Modelling & Software, 17*(3), 209–217.

Peterson, S. (2006). Uncertainty and economic analysis of climate change: A survey of approaches and findings. *Environmental Modelling and Assessment, 11*(1), 1–17.

Sarewitz, D. (2004). How science makes environmental controversies worse. *Environmental Science & Policy, 7*(5), 385.

Schlumpf, C., Behringer, J., Dürrenberger, G., & Pahl-Wostl, C. (1999). The personal CO_2-calculator: A modelling tool for participatory integrated assessment methods. *Environmental Modeling and Assessment, 4*(1), 1–12.

Schröter, D., Cramer, W., Leemans, R., Prentice, I. C., Araujo, M. B., Arnell, N. W., Bondeau, A., Bugmann, H., Carter, T. R., Gracia, C. A., de la Vega-Leinert, A. C., Erhard, M., Ewert, F., Glendining, M., House, J. I., Kankaanpaa, S., Klein, R. J. T., Lavorel, S., Lindner, M., Metzger, M. J., Meyer, J., Mitchell, T. D., Reginster, I., Rounsevell, M., Sabate, S., Sitch, S., Smith, B., Smith, J., Smith, P., Sykes, M. T., Thonicke, K., Thuiller, W., Tuck, G., Zaehle, S., & Zierl, B. (2005). Ecosystem service supply and vulnerability to global change in Europe. *Science, 310*(5752), 1333–1337.

Schröter, D., Polsky, C., & Patt, A. (2005). Assessing vulnerabilities to the effects of global change: An eight step approach. *Mitigation and Adaptation Strategies for Global Change, 10*(4), 573–595.

Siebenhüner, B., & Barth, V. (2005). The role of computer modelling in participatory integrated assessments. *Environmental Impact Assessment Review, 25*(4), 367–389.

Stalpers, S. I. P., van Amstel, A. R., Dellink, R. B., Mulder, I., Werners, S. E., & Kroeze, C. (2007). Lessons learnt from a participatory integrated assessment of greenhouse gas emission reduction options in firms. *Mitigation and Adaptation Strategies for Global Change, 13*(4), 359–378.

Stalpers, S. I. P., van Ierland, E. C., & Kroeze, C. (2009). Reconciling model results with user needs to improve climate policy. *Environmental Science & Policy, 12*(7), 959–969.

Toth, F. L., & Hizsnyik, E. (1998). Integrated environmental assessment methods: Evolution and applications. *Environmental Modeling and Assessment, 3*, 193–207.

Toulmin, S. (2003). *The uses of argument*. Cambridge: Cambridge University Press.

Toulmin, S., Rieke, R., & Janik, A. (1979). *An introduction to reasoning*. New York: Macmillan.

Tuinstra, W., Hordijk, L., & Kroeze, C. (2006). Moving boundaries in transboundary air pollution: Co-production of science and policy under the convention on long range transboundary air pollution. *Global Environmental Change, 16*(4), 349–363.

UNFCCC (1995). *Report of the Conference of the Parties on its First Session*. Berlin, 27 March–7 April 1995, FCCC/CP/1995/7.

Van Aardenne, J. A. (2002). *Uncertainties in emission inventories* (p. 143). Wageningen: Wageningen University.

Van Asselt, M. B. A. (2000). *Perspectives on uncertainty and risk: The PRIMA approach to decision support*. Dordrecht: Kluwer Academic.

Van Asselt, M. B. A., & Rotmans, J. (1996). Uncertainty in perspective. *Global Environmental Change, 6*(2), 121–157.

Van Daalen, E., & Grünfeld, H. (1995). *Second International Workshop on Using the IMAGE 2 Model to Support Climate Negotiations*, Summary Report, Delft, Faculty of Systems

Engineering, Policy Analysis and Management, and National Institute of Public Health and Environment.

Van Daalen, E., & Grünfeld, H. (1996). *Third International Workshop on Using the IMAGE 2 Model to Support Climate Negotiations*, Summary Report, Delft, Faculty of Systems Engineering, Policy Analysis and Management, and National Institute of Public Health and Environment.

Van Daalen, C. E., Thissen, W. A. H., & Berk, M. M. (1998). The delft process: Experiences with a dialogue between policy makers and global modelers. In J. Alcamo, R. Leemans, & E. Kreileman (Eds.), *Global change scenarios of the 21st century, results from the IMAGE 2.1 model* (pp. 267–285). Oxford: Elsevier Science.

Van de Kerkhof, M. (2004). *Debating climate change: A study of stakeholder participation in an integrated assessment of long-term climate policy in the Netherlands*. Utrecht: Lemma.

Van de Kerkhof, M., & Leroy, P. (2000). Recent environmental research in the Netherlands: Towards post-normal science? *Futures, 32*(9–10), 899.

Van der Sluijs, J. P. (1997). *Anchoring amid uncertainty: On the management of uncertainties in risk assessment of anthropogenic climate change*. Ph.D. thesis Faculteit Scheikunde, Utrecht University, Utrecht.

Van der Sluijs, J. P., Craye, M., Funtowicz, S., Kloprogge, P., Ravetz, J., & Risbey, J. (2005). Combining quantitative and qualitative measures of uncertainty in model based foresight studies: The NUSAP system. *Risk Analysis, 25*(2), 481–492.

Van Ierland, E. C., de Groot, R. S., Kuikman, P. J., Martens, P., Amelung, B., Daan, N., Huynen, M., Kramer, K., Szonyi, J., Veraart, J., Verhagen, A., van Vliet, A., van Walsum, P. E. V., & Westein, E. (2001). *Integrated assessment of vulnerability to climate change and adaptation options in the Netherlands*. Bilthoven: RIVM.

Yearley, S., Forrester, J., & Bailey, P. (2001). Participation and expert knowledge: A case study analysis of scientific models and their publics. In M. Hisschemöller, R. Hoppe, W. Dunn, & J. Ravetz (Eds.), *Knowledge, power, and participation in environmental policy analysis* (pp. 47–72). New Jersey: Transaction.

Chapter 13
Linking Narrative Storylines and Quantitative Models to Combat Desertification in the Guadalentín Watershed (Spain)

Kasper Kok and Hedwig van Delden

13.1 Introduction

Desertification in Spain is largely a society-driven problem, which can be effectively managed only through a thorough understanding of the principal ecological, sociocultural, and economic driving forces (UNCCD 1994). This calls for a more active role of decision-makers and other stakeholders. We present a promising approach, involving stakeholders in the scenario-development process and linking these narrative storylines with an integrated quantitative model. Within the framework of a larger EC-financed project, dealing with desertification in the Mediterranean region, multi-scale scenarios were developed for Europe, the Northern Mediterranean, and four local areas. In the same project, a policy-support system (PSS) was developed. The main objective of this exercise was to establish a link between the qualitative scenarios and the PSS for the watershed of the Guadalentín river in Spain. From the results of two scenario workshops, three scenarios were selected, all linked to the same Mediterranean scenario. Our selection aimed to maximize both the variety in the narrative storylines and the expected output of the PSS. The scenarios were subsequently formalized, ensuring that the same information was present for all three scenarios; semi-quantified (translated) by linking them to the main entry points of the PSS; and quantified by parameterizing the model. The results indicate the potential of the constructed quantitative scenarios. This chapter illustrates the practical potential and pitfalls of linking qualitative storylines and quantitative models.

K. Kok
Environmental Sciences, Wageningen Universiteit,
P.O. Box 47, 6700 AA Wageningen, the Netherlands
e-mail: kasper.kok@wur.nl

H. van Delden (✉)
Research Institute for Knowledge Systems (RIKS),
463, Maastricht, P.O. Box 463, 6200 AL, The Netherlands
e-mail: hvdelden@riks.nl

M. Giaoutzi and B. Sapio (eds.), *Recent Developments in Foresight Methodologies*, Complex Networks and Dynamic Systems 1, DOI 10.1007/978-1-4614-5215-7_13,

© Springer Science+Business Media New York 2013

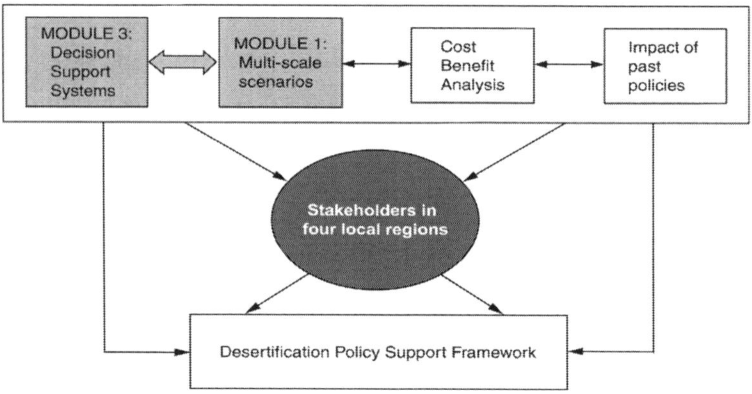

Fig. 13.1 Simplified flow chart of main activities within MedAction. Note: Grey shading indicates the components which are important in this chapter

Future research should, however, also focus on the more fundamental theoretical obstacles that are easily overlooked.

The integrated assessment approach used also calls for a much more active role of decision-makers and other local stakeholders during all phases of the process (Rotmans 1998). A particularly pressing issue is establishing the link between qualitative outputs from employing participatory methods and quantitative, data-demanding, spatially explicit models. To tackle the problem, various different methods are being developed, including, for example, agent-based models (Parker et al. 2002) that can be directly parameterized by stakeholders (Barreteau et al. 2001). Others have recently advocated the combination of qualitative and quantitative scenarios (Alcamo 2008). The approach presented here puts the framework presented by Alcamo to the test. The work was part of a larger European project, MedAction. Details on the methodology and results described here can be found in Kok and van Delden (2009).

13.1.1 MedAction

MedAction (see Appendix A; De Groot and Rotmans 2004) was an EC-financed project within which an information and decision-support base on land degradation was developed to assist decision-makers from the local to the European level in the formal and informal decision- and policy-making process to combat desertification in the Northern Mediterranean region.

The specific problems of desertification and mitigation measures were addressed at the European, Mediterranean, and local scale, with the ultimate goal to aid local decision-making with regard to policy formulation for sustainable land management at the local level.

Work was carried out in four local case studies: the Guadalentín (Spain), Val d'Agri (Italy), Alentejo (Portugal), and the island of Lesbos (Greece). A simplified flow chart of the main activities within MedAction is given in Fig. 13.1, highlighting components important in this chapter.

Module 1 of MedAction was coordinated at the International Centre for Integrative Studies (ICIS) in Maastricht, the Netherlands, and dealt with scenario development at the European, Mediterranean, and local scale (see Kok et al. 2006a, b). Local scenarios for the various local case studies were developed during a series of workshops with 20–25 local and regional stakeholders. Various scenario-development methods were tested that resulted in four main products: a story of the present characterizing the perception of the local stakeholders on the situation in their region, a story of the future in 2030 that was obtained during a forecasting (see also Kasemir et al. 2000) session, an extension of the present representing the situation in the near future based on an extrapolation of current trends, and a backcasting exercise (Dreborg 1996; Robinson 2003), reasoning back from a desirable end point in 2030 to short-term measurements that are necessary to realize this future (methods and results can be downloaded on http://www.icis.unimaas.nl/medaction/ download.html).

Module 3 dealt with, inter alia, the development of a policy-support system: a software instrument to support policy making at the regional level, developed by the Research Institute for Knowledge Systems (www.riks.nl). The MedAction PSS has been developed with the objective to address a number of policy themes concerning water resources, sustainable agriculture, desertification, and land degradation in Mediterranean regions. Problems, goals, policy options, and policy indicators have been collected and structured for each of these themes and translated into a conceptual framework. From this conceptual framework, a policy-support system was designed and developed incorporating socio-economic as well as physical models. The PSS supports policymakers in understanding the impacts of autonomous developments within a region, as well as the impacts of external influences on the region, such as economic and demographic growth or climate change. All impacts can be measured by means of a number of policy-relevant indicators (e.g. profits in the agricultural sector, forested area, suitability of the soil for agriculture or natural vegetation, water use and availability, and land use), which change dynamically during the run of a simulation (for a period up to 2030). The PSS was developed as a generic system and implemented with the GEONAMICA® application framework, which is specifically designed to build PSSs featuring complexly linked multi-scale models. The PSS was applied in particular to the Guadalentín river basin in Spain. Previous versions of this system have also been applied to the Marina Baixa region in Spain and the Argolidas region in Greece. The MedAction PSS is described in detail in Van Delden et al. (2007).

13.1.2 Objectives

The main objective of this chapter is to describe an approach to link qualitative scenarios and a PSS, as developed within the MedAction project. This chapter focuses on the practical application of this approach in one of the local case studies, the watershed of the Guadalentín in Southeastern Spain.

Table 13.1 Summary of the formalized scenarios used as input in the PSS, by the main factors, sectors, and actors

FAS	Scenario I	Scenario II	Scenario III
Factors			
Water availability	Increasingly limited due to drought	Limited, distribution favours agriculture	Strongly limited, no Ebro water
Migration	Rural–urban migration European Sunbelt	Fewer permanent tourists	Strong rural–urban migration, less immigrants from Morocco
Sectors			
Agriculture	Increasingly difficult position	Multifunctional, favoured	Lack of water, although still favoured
Tourism	Booming	Ecotourism, reduced in numbers	Lack of water stops expansion
Actors			
Businesses	Large-scale, mass tourism, smallholders disappear. Industry important	Small-scale favoured, industry under pressure	Lack of water limits developments

13.2 From Storyline to Model Input

13.2.1 Selecting the Qualitative Scenarios

In the Guadalentín, a series of three workshops were held in 2002–2003, attended by approximately 25 stakeholders (mainly farmers, local government officials, NGOs, and scientists). A variety of different methods yielded a wealth of scenarios, all based on three Mediterranean scenarios. The results varied from short-term (until 2010) trends to long-term projections and to long-term visions using a backcasting methodology. During the first workshop, three forecasting scenarios were developed, linked to the Mediterranean scenarios; during the backcasting workshop, four desirable futures were outlined, partly linked to the higher-level scenarios. For the purpose of this chapter, we selected one forecasting and one backcasting scenario, both linked with the European scenario *Convulsive Change*. This Mediterranean scenario can be described as follows. Climate change accelerates and many regions in Northern Europe are overwhelmed by floods, while the South experiences severe droughts, leading to permanent deserts. In the Mediterranean, an extensive network of new water pipelines eventually increases water availability. At the same time, there is a slow but fundamental change in general attitude towards a more sustainable way of living.

The forecasting scenario (Scenario I: *Likely future*, see Table 13.1) provides the most likely future under these Mediterranean developments; the backcasting scenario (Scenario II: *Desired future*) is based on the desirable future of a strong agricultural sector combined with a strong growth of ecotourism. A third scenario (Scenario III: *Water shortage*) was added, where one of the key assumptions in the first two scenarios – the construction of a canal from the Ebro River – was omitted, thus strongly limiting water availability with ensuing effects. This third scenario was thus not directly formulated by the stakeholders, although the possibility was discussed during the workshops.

Scenarios were chosen with the purpose to maximize the variety present in the narrative stories, the variety of the expected spatially explicit results, and the number of variables in the PSS that could be quantified.

13.2.2 Formalizing the Scenarios: Key Assumptions in the Narratives

These three scenarios were subsequently formalized, using what is called the factor–actor–sector framework that was also used in the development of the European and Mediterranean scenarios (see Table 13.1), thus maintaining the link with higher-level scenarios. The FAS framework includes a pre-selected number of key elements in which a *factor* represents an aspect of a social or natural system around which clusters a set of broad policy issues of particular interest, an *actor* represents an individual or organization of individuals with the capacity to effect and/or influence change, and a *sector* represents a subcomponent of a natural or social system (see Rotmans et al. 2000). In MedAction, there were four factors, actors, and sectors for scenarios at all levels. Table 13.1 shows the key assumptions for the most important sectors (tourism and agriculture) and related actors and sectors.

13.2.3 Translating the Scenarios

These formalized stories were then semi-quantified by linking them to the main entry points of the PSS. First, a selection was made of parameters in the PSS that have a link with the scenarios. For each of these parameters, it was then indicated what the expected change was in each of the three scenarios. Change was semi-quantitative, ranging from "+++" (very strong increase) to "–––" (very strong decrease). In general, we followed the linguistics of the storylines that were relatively straightforward for the most important variables. Particularly detailed was the information on migration flows, water availability, and developments of the tourist sector. Less informative was the information on other factors (notably economic stability), sectors (civic), and actors (governments). Where necessary, parameters were estimated by the authors. All scenarios make use of the same climate scenario (ECHAM, see Roeckner et al. 1996) with an annually imposed extra rainfall shortage in order to approximate the assumptions of the European *Convulsive Change* scenario. This methodology was also applied in earlier work by White et al. (2004). Below are the most important parameters in the PSS that were considered during translation; Fig. 13.2 then shows the basic structure of the PSS and the main relationships between modules:

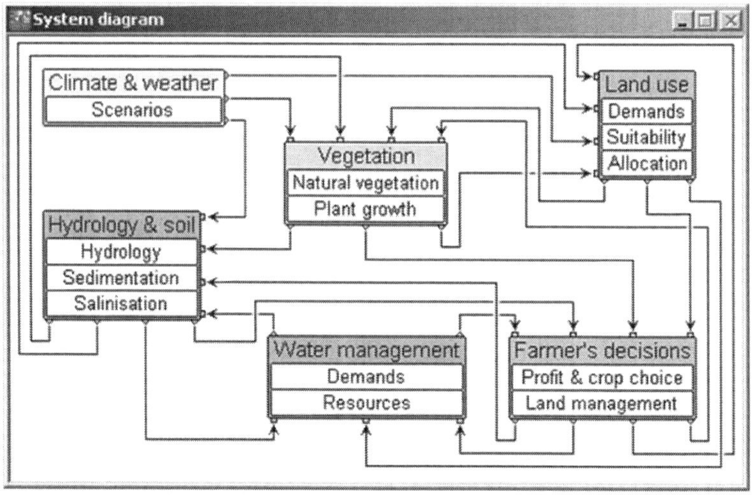

Source: Adapted from: Kok and Van Delden (2009).

Fig. 13.2 Simplified structure of the policy-support systems as developed in Module 3 of MedAction (Source: Adapted from Kok and van Delden (2009))

Land-use module: Total land demand for agriculture, rural residential, dense residential, industry and commercial areas, tourism, expats, forest reserves.

Climate and weather module: Scenario for future climate change, based on IPCC scenarios. Main factors are precipitation, temperature, and radiation.

Water module: Defined by resource (aquifers, reservoirs (including Tajo and Ebro water), desalinated sea water) and by function for the demands. Three main parameters were included: price (also per source), quantity (per source, per sector, or per person/hectare), and distribution (per sector).

Farmer's decision module: Parameters adapted based on the scenarios were market prices, subsidies, farmers' resistance to change, and the introduction of new crops which are more resistant to dry soils.

Irrigation module: Binary map showing where irrigation from each water source is possible: choice between drip and spray irrigation.

Other: Policy-relevant parameters include zoning maps for each function, construction of new roads, canals and maintenance of dams, dredging of the reservoirs, terracing, and ploughing.

13.2.4 Quantifying the Assumptions

The last step was the actual parameterizing of the PSS. A version of the PSS, which was calibrated using data of the present, was run until 2030 assuming constant trends, thus providing a baseline scenario. This baseline scenario was used for those parameters that had no relation with the narrative storylines. For example, there are detailed modules for hydrology, soil erosion, salinization, and plant growth, for which not much information could be extracted from the qualitative scenarios. Lack of space prevents a full analysis of the quantification of all variables. Good examples, however, are those variables that were related to either the total amount of a certain land-use type or the attraction between land-use types as defined by cellular automata rules. Table 13.2 presents some of the assumptions used. In the parameterizing process of the semi-quantified variables, we were as consistent as possible. For total area changes, "+++" usually translated into a 3% increase per year and "–" into a 3% decrease per year. However, given the time horizon of the scenarios (30 years), based on expert knowledge, we lowered some of the numbers. For example, a 3% yearly increase in area for tourism would mean a doubling within the scenario period, which was deemed to be not plausible. For the cellular automata settings, "++" translated into twice as strong as the baseline settings and "–" to a tendency to spread around. Yet, many small additional assumptions were necessary. For example, the total land area needs to add up to 100% at all times, a restriction that was not strictly complied with in the original stories, or with the semi-quantification. Scenario I, for example, would lead to demand for land that exceeds 120% of the total area.

13.3 Running the Model

Figure 13.3 shows some of the resulting maps of the PSS. Under Scenarios I and III, large areas of dryland agriculture are abandoned. Irrigated areas remain limited to the valley bottom, and dryland agriculture in the centre is maintained. The effects of strongly decreasing precipitation have a very large influence on land-use patterns. Under Scenario I, golf courses and residential areas strongly expand. Under Scenario III, on the other hand, the region is without prospects: The lack of water induces a strong outmigration and land abandonment and limits urban expansion. Interestingly, the results of Scenario II strongly resemble the land-use patterns in 2000. This lack of spatial change is due to the fact that the desired future perspective of the local stakeholders is mostly in market structures and sociocultural changes, while maintaining the current situation.

13.4 Discussion and Conclusions

13.4.1 Added Values of Linking Stories and Quantitative Models

Linking stories and models can generate added value in the following ways:

1. *Increase consistency of storylines*. The output of a quantitative model can be used to check the internal consistency of a storyline during a stakeholder

Table 13.2 Total area change and assumptions for cellular automata rule

	Baseline	Scenario I	Scenario II	Scenario III
Land use (total area)				
Agriculture	0	– –	0	– – –
Industry	+	++	+/++	–
Tourism	+	+++	+	0
Expatriates	+	+++	++	0
Urban	+	++	+	–
Cellular automata settings				
Agriculture	+	+	+	+
Tourism	+	++	–	+
Urban	+	++	++	++

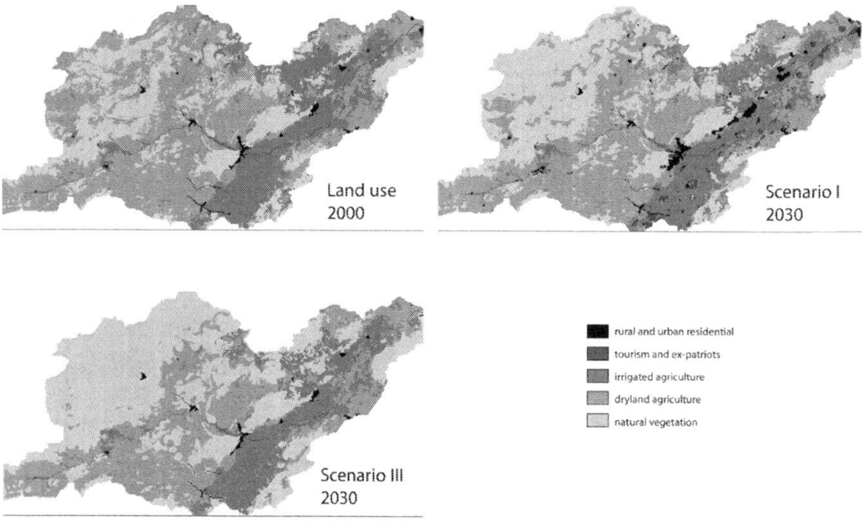

Source: Adapted from: Kok et al (2009).

Fig. 13.3 Land-use map of 2000 for the Guadalentín watershed and the output of the PSS for two scenarios (only five classes out of more than 30 are shown) (*Source*: Adapted from Kok and van Delden (2009))

workshop. We identified a number of impossibilities in the assumptions on, for example, land-use changes as present in the storylines.

2. *Visualization tool.* The possibilities to visualize the spatial consequences of storylines with a PSS are evidently large.
3. *Create integrated scenarios.* By using a storyline as input for a PSS, clusters of parameters can be changed simultaneously, thus providing guidance for policymakers beyond the current state of the art of using models as scenario generators.

4. *Involve stakeholders*. It goes without saying that combining the two tools can open the way for more successful management strategies and the combating of land degradation by directly involving stakeholders in the scenario-building process.

13.4.2 Potential Pitfalls

Despite the potentially large added value of linking narratives and models, a number of methodological issues remain. They are partly inherent to the methodology and therefore difficult to tackle, and partly practical and thus surmountable.

13.4.2.1 Methodological Issues

Key to the described methodology is the link between a product based on the perceptions of stakeholders and a product based on scientific expertise. These perceptions will differ more often than not, and consequently, the link will necessarily be incomplete. A good example is a storyline that envisions radical fundamental system changes, which cannot be adequately represented in a model that is built on current system characteristics. One of the scenarios used within MedAction – Big is Beautiful – entails such fundamental changes (see Kok et al. 2007). A careful selection of stakeholders might limit but not prevent this issue. A second fundamental issue is the large role of scientists in the actual quantification step. Particularly when local stakeholders are involved, it is close to impossible to involve them in the parameterization of the model. Yet models can be sensitive to the parameter settings, and consequently, scientists will continue to have a strong role in determining the model output. In general, the more complicated the model is, the larger the role of the scientists.

13.4.2.2 Practical Limitations

There are some additional practical limitations that could hamper the successful link between PSS and narratives. First of all, it cannot be stressed enough that the development of participatory narrative storylines and the construction of a PSS are expensive and, in the case of MedAction, took years to complete. Although methods for carrying out a similar process will become more standardized and resource demands will decrease, they will nevertheless remain high. Secondly, as in any participatory process, success is not guaranteed. Particularly in MedAction, the additional aim to stimulate an open discussion and to initiate longer-term participation was not always compatible with the aim of developing actual future scenarios. In fact, the series of stakeholder workshops was completed in only two out of four local case studies (Spain and Italy). And finally, only a partial link between stories and the PSS could be established. Important assumptions in the storylines were not represented in the model, while key variables in the model could not be parameterized with information from the storylines. Additionally, as Scenario II showed, a spatial representation might obscure part of the processes that are viewed as important by the stakeholders.

13.4.3 Conclusions

On the basis of the experiences in MedAction, we conclude that there are large advantages of linking narrative storylines and a spatial PSS. Developing storylines ensures the active participation of a large range of stakeholders, additionally offering the possibility to develop highly integrated scenarios. The PSS provides a spatially detailed and quantitative output, which can also be used to check the internal consistency of the qualitative scenarios. Linking stories and models can thus open the way for more successful management strategies to combat land degradation. However, combining these two methods is very resource-demanding, and success is not guaranteed.

There are several possibilities for improving the link between narrative storylines and spatial models, some of which are currently being explored in more recent scenario-development exercises. The most promising in the field of land use and land degradation is (participatory) group model building (see Vennix 1999). This approach is being embraced by both those developing agent-based models (see Parker et al. 2002) and those that develop models during participatory workshops. A good example from the latter approach is fuzzy cognitive maps that help in structuring stakeholder knowledge (Kok 2009; Van Vliet et al. 2010). Another possible improvement is the use of a highly iterative procedure. Alcamo's story-and-simulation approach (Alcamo 2008) is currently being tested in various projects that attempt to go through at least two iterations between stories and models, while quantifying a number of key model parameters during these stakeholder workshops (see Kämäri et al. 2008; Kok et al. 2011). Together, these initiatives will build a larger body of evidence on the possibilities of involving stakeholders and modellers in the process of scenario building and decision-making.

13.5 Appendix A

MedAction: Policies to combat desertification in the Northern Mediterranean region. Research project supported by the European Commission under the Fifth Framework Programme and contributing to the implementation of Key Action 2: Global Change, Climate and Biodiversity; Subaction 2.3.3 Fighting Land Degradation and Desertification. Research period: 2001–2004.

Website: www.icis.unimaas.nl/ medaction

References

Alcamo, J. (2008). The SAS approach: Combining qualitative and quantitative knowledge in environmental scenarios. In J. Alcamo (Ed.), *Environmental futures: The practice of environmental scenario analysis* (Developments in integrated environmental assessment, Vol. 2, pp. 123–150). Amsterdam: Elsevier.

Barreteau, O., Bousquet, F., & Attonaty, J. -M. (2001). Agent-based modelling, game theory and natural resource management issues, *Journal of Artificial Societies and Social Simulation, 4*(2). http://www.soc.surrey.ac.uk/JASSS/4/2/5.html.

De Groot, D., & Rotmans, J. (2004). MedAction, final report, submitted to the Research and Development Directorate, European Commission EVK2-2000-00085, Maastricht.

Dreborg, K. H. (1996). Essence of backcasting. *Futures, 28*, 813–828.

Kämäri, J., Alcamo, J., Bärlund, I., Duel, H., Farquharson, F., Flörke, M., Fry, M., Houghton-Carr, H., Kabat, P., Kaljonen, M., Kok, K., Meijer, K. S., Rekolainen, S., Sendzimir, S., Varjopuro, R., & Villars, N. (2008). Envisioning the future of water in Europe – the SCENES project, *E-Water Journal from EWA*. http://www.ewaonline.de/portale/ewa/ewa.nsf/home?reaform&objectid=37F6CB0C75D068CFC125748E004F6FDB.

Kasemir, B., Dahinden, U., Swartling, Å. G., Schüle, R., Tabara, D., & Jaeger, C. C. (2000). Citizens' perspectives on climate change and energy use. *Global Environmental Change, 10*, 169–184.

Kok, K. (2009). The potential of Fuzzy cognitive maps for semi-quantitative scenario development, with an example from Brazil. *Global Environmental Change, 19*, 122–133.

Kok, K., & Van Delden, H. (2009). Combining two approaches of integrated scenario development to combat desertification in the Guadalentín watershed, Spain. *Environment and Planning B, 36*, 49–66.

Kok, K., Rothman, D. S., & Patel, M. (2006a). Multi-scale narratives from an IA perspective: Part I. European and Mediterranean scenario development. *Futures, 38*(3), 261–284.

Kok, K., Patel, M., Rothman, D. S., & Quaranta, G. (2006b). Multi-scale narratives from an IA perspective: Part II. Participatory local scenario development. *Futures, 38*(3), 285–311.

Kok, K., Biggs, R., & Zurek, M. (2007). Multi-scale scenario development methodologies. Experiences from Southern Africa and the Mediterranean. *Ecology and Society, 12*(1), 8. http://www.ecologyandsociety.org/vol12/iss1/art8/.

Kok, K., van Vliet, M., Bärlund, I., Dubel, A., & Sendzimir, J. (2011). Combining participative backcasting and explorative scenario development: Experiences from the SCENES project. *Technological Forecasting and Social Change, 78*(5), 835–851.

Parker, D. C., Berger, T., Manson, S. M., & McConnell, W. J. (2002). Agent-based models of land-use and land-cover change. Proceedings of an International Workshop, 4–7 Oct 2001, Irvine. Report no. 6, LUCC report series, Focus 1 Office, Indiana University.

Robinson, J. (2003). Future subjunctive: Backcasting as social learning. *Futures, 35*, 839–856.

Roeckner, E., Arpe, K., Bengtsson, L., Christoph, M., Claussen, M., Dümenil, L., Esch, M., Giorgetta, M., Schlese, U., & Schulzweida, U. (1996). *The atmospheric general circulation model ECHAM-4: Model description and simulation of present-day climate*, Report no. 218, Max-Planck Institute for Meteorology, Hamburg.

Rotmans, J. (1998). Methods for IA: The challenges and opportunities ahead. *Environmental Modelling and Assessment, 3*, 155–179.

Rotmans, J., van Asselt, M. B. A., Anastasi, S. C. H., Greeuw, S. C. H., Mellors, J., Peters, S., Rothman, D. S., & Rijkens, N. (2000). Visions for a sustainable Europe. *Futures, 32*(9/10), 809–831.

UNCCD (1994). *United Nations convention to combat desertification.* Convention text as of September 1994. http://www.unccd.int/convention/menu.php.

Van Delden, H., Luja, P., & Engelen, G. (2007). Integration of multi-scale dynamic spatial models of socio-economic and physical processes for river basin management. *Environmental Modelling and Software, 22*, 223–238.

Van Vliet, M., Kok, K., & Veldkamp, T. (2010). Linking stakeholders and modellers in scenario studies: The use of fuzzy cognitive maps as a communication and learning tool. *Futures, 42*(1), 1–14.

Vennix, J. A. M. (1999). Group-model building: Tackling messy problems. *System Dynamics Review, 15*, 379–401.

White, R., Straatman, B., & Engelen, G. (2004). Planning scenario visualization and assessment: A cellular automata based integrated spatial decision support system. In M. F. Goodchild & D. Janelle (Eds.), *Spatially integrated social science* (pp. 420–442). New York: Oxford University Press.

Chapter 14
Scenario Planning as a Tool in Foresight Exercises: The LIPSOR Approach

Anastasia Stratigea and Maria Giaoutzi

14.1 Introduction

Foresight is a future-oriented activity that supports decision-making processes by focusing on the management of the complexity involved within a turbulent environment in a long-term planning context (Giaoutzi et al. 2012). Scenario planning, as a strategic and effective planning and learning tool, should constitute an integral part of any foresight exercise.

The regional level, in scenario planning, appears as the most appropriate level for foresight applications that deal with the emerging challenges introduced by increasing globalization, which is motivating both processes and changes (Ringland 1998, 2002; STRATA-ETAN 2002; Stimson et al. 2006; etc.).

Traditional tools used so far have serious drawbacks with respect to their potential for dealing with uncertainty and complexity, thus reflecting their inherent deterministic rationale (Giaoutzi et al. 2012). In recent research, there is an increasing emphasis on new tools to deal with uncertainty and complexity issues, such as mathematical tools that handle uncertainty and complexity; scenario-planning approaches; interactive participatory methods for involving a broader audience in the planning process; etc. (Giaoutzi and Stratigea 2010). Such developments indicate a shift from the view that 'the future is there to be predicted' to the view of the 'socially created future', where a systematic study of the future is nothing more than 'a tool for choosing and creating the most desirable future' (STRATA-ETAN 2002). The issue of participation is critical in this respect.

The focus of the present chapter is on the potential of participatory scenario planning for regional future studies. More precisely, Sect. 14.2 discusses the key concepts of scenario planning; Sect. 14.3 presents the scenario-planning participatory

---

A. Stratigea (✉) • M. Giaoutzi
Department of Geography and Regional Planning, National Technical University of Athens (NTUA), Heroon Polytechniou Str. 9, Zographou Campus, Athens 157 80, Greece
e-mail: stratige@central.ntua.gr; giaoutsi@central.ntua.gr

M. Giaoutzi and B. Sapio (eds.), *Recent Developments in Foresight Methodologies*, Complex Networks and Dynamic Systems 1, DOI 10.1007/978-1-4614-5215-7_14,
© Springer Science+Business Media New York 2013

framework LIPSOR, adopted in the present case study; Sect. 14.4 describes the study region (Heraklion–Crete); Sect. 14.5 elaborates on the results of the application of the LIPSOR participatory framework in the study region; and, finally, in Sect. 14.6 certain conclusions are drawn.

14.2 Scenario Planning as a Tool in Foresight Studies

The present section provides some background concepts of scenario planning in foresight studies.

Foresight studies refer to a medium- to long-term horizon, where the future appears multiple and uncertain. The longer the time horizon of a study, the greater the uncertainty and the potential discontinuities involved. Qualitative data in this context are elaborated on the basis of stochastic tools, where completely different future structures may emerge, due to, for example, a paradigm shift or non-linear changes, expressed by Lindgren and Bandhold (2003) as '*… the future is created*'.

Scenario planning is a strategic tool for medium- to long-term planning. It is also an effective learning tool, which helps planners to understand the potential paths of future developments in a study area. *Key attributes* of scenario planning are time horizon, purpose and focus, future thinking, system thinking and actor thinking.

The appropriate *time horizon* for scenario planning is a medium- to long-term perspective that is essential for the successful implementation of the approach. This should be both long enough to let changes happen but also short enough to allow for the building of possible scenarios.

The definition of the *purpose and focus* is quite essential in scenario planning. Different combinations of purpose and focus may lead to different orientations of the scenario-planning exercise, for example, *concept* or *strategy* development (see Lindgren and Bandhold 2003).

A critical issue in *future thinking* is the uncertainty involved. The scenario-planning approach may support decision-making processes in the following uncertainty contexts (Lindgren and Bandhold 2003):

- Dichotomous or discrete uncertainty, where different scenarios can be built based on the various outcomes of uncertainty dimensions.
- Uncertainty caused by speed of change, great complexity or both, the latter of which is the most common case for the scenario-planning approach applications. At moderate to high levels of complexity and speed, linear change dominates, and it is meaningful to talk about an uncertainty space. It is still possible to identify certain trends, to sort out the probable overall directions, and identify a limited number of key uncertainty dimensions, which could be used to reduce uncertainty to a manageable number of scenarios.

Systems thinking is considered the basis of the scenario-planning approach. It is about thinking in layers of change, dependencies and interdependencies. The basis

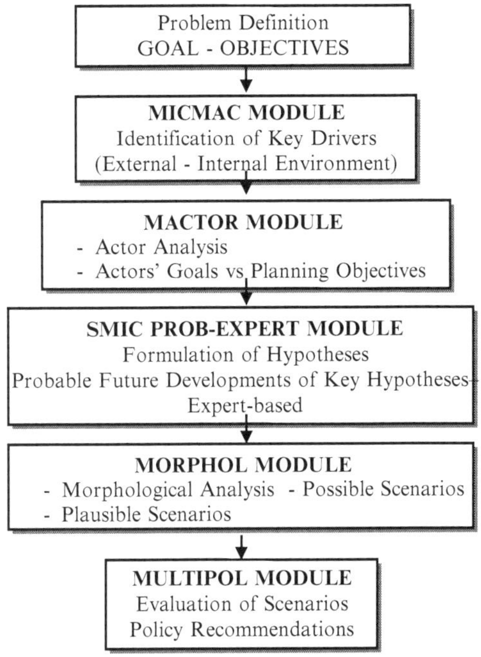

Fig. 14.1 The LIPSOR framework (*Source*: Godet et al. 2004)

for understanding a system is founded on the deep knowledge of its components (subsystems) and their interrelationships, which is crucial in order to reveal the key variables that drive a system's change (Boulding 1964).

Finally, *actor thinking* is also important, as the future is shaped by the decisions and actions of the various actors both inside and outside the study system. A deeper understanding of the future states may come from a thorough analysis of the actors' interests, power relationships, potential strategic moves, motives, attitudes, personal profiles, alliances, strengths and weaknesses, etc., which are among the issues deserving attention in the context of scenario planning (Lindgren and Bandhold 2003).

14.3 The LIPSOR Scenario-Planning Approach

The LIPSOR[1] model is an important participatory scenario-planning approach, developed by Godet (1999, 2001) in the 'Laboratory for the Investigation in Prospective and Strategy'. LIPSOR is used in the present study in order to design scenarios for the integrated development of the Heraklion–Crete region. It consists of the following five discrete parts (Fig. 14.1).

[1] LIPSOR: Lien Innovation, Prospective, Stratégie et Organisation.

The first part – *the MICMAC module* – explores the key aspects of the study system and formulates the basic questions relating to its future states. A 'structural analysis' is carried out, exploring the '*influence–dependence*' relationships among the selected key variables, corresponding to the attributes of the internal and external environment of the study system, which drive its future states. The selection of variables is conducted on the basis of their role as drivers of change and is carried out on the basis of a participatory process, involving groups of local stakeholders, experts and external advisors.

The *MACTOR module* identifies the role of the principal stakeholders in the study system. Various characteristics of these stakeholders are studied, such as their economic interests, potential strategic moves, projects in progress, motives, attitudes, personal profiles, strategic alliances, and strengths and weaknesses, as well as relationships among stakeholders. All these reveal the underlying power relationships in the study system. A very important attribute of this module is the capacity to reveal the attitudes (level of resistance) of the various actors/stakeholders against the objectives of the study. This may provide planners with valuable information for decision-making purposes.

The *SMIC PROB-EXPERT module* adopts an explorative approach based on qualitative information. It is an expert-based approach that provides information/opinions based on a certain number of hypotheses relating to the development of the study area. Single as well as conditional probabilities of these hypotheses are attached by the experts. The method attempts to foresee the future states of the study system as combinations of the above hypotheses. The results from the previous stages, namely, from MICMAC (key variables) and MACTOR (actors' behaviour), are quite useful in this stage.

Based on single and conditional probabilities of the above hypotheses, the module can provide:

- A set of probable scenarios as combinations of the above hypotheses, where certain probabilities have been attached. In this respect, a number of scenarios can be selected as being the most probable and compatible with both the internal and the external environments of the study area.
- Valuable information on the future developments of the above hypotheses, which can be used as input to the next step (MORPHOL module).

The *MORPHOL module* is used for a systematic exploration of possible future states of the study system, structured on the basis of all combinations of possible future states of the various key variables. The total number of combinations – possible scenarios forming the morphological space – can be further reduced into a useful subset of combinations on the basis of a set of selection (or exclusion) criteria, leading to the most 'plausible scenarios'. The module, at this stage, integrates all the information acquired in the previous modules.

The *MULTIPOL module* evaluates the scenarios structured in the previous module (by use of multiple-criteria analysis), attempting at the same time to define strategic directions (policies) and actions (policy measures) for the implementation of each scenario. The method does not conclude with the most preferred scenario but supports decision-makers by informing them which policies and measures could be more effective to pursue alternative future developments (scenarios). The *MULTIPOL*

Fig. 14.2 The Heraklion region (Source: www.ellada-diakopon.gr)

module incorporates two different types of evaluation, namely, the *actions/policies evaluation* that evaluates actions (measures) with respect to policies, indicating which actions best fit each single policy, and the *policies/scenarios evaluation* that indicates the policy which best fits each specific scenario.

The LIPSOR approach enables each module to function both independently and as a stepwise approach dealing with future studies in a coherent, systematic and analytical mode.

14.4 Description of the Study Region

In this section, the goals and objectives for the development of the Heraklion–Crete study region are presented, together with a description of the present situation in this region.

The *goal* of the present regional study is the structuring of an 'integrated agricultural development plan' for the study region, where the following objectives are pursued:

- Objective 1: environmental protection of both natural and cultural resources
- Objective 2: regional development
- Objective 3: economic efficiency, motivating local economic development
- Objective 4: social cohesion
- Objective 5: food safety and quality
- Objective 6: energy production (biomass and biofuel production)

The Heraklion region is located on the island of Crete in Greece (Fig. 14.2). The region is endowed with valuable natural and cultural resources. During the last decade, its population increased quite considerably by 8.7% (and by 50% from 1971 to 2001). In the total population of the study area, the share of urban population is 50%, while that of the agricultural population is up to 39% (2001). A large number of research and higher education centres are also located in the study region.

The share of the agricultural production of the region is 8.2% of the national agricultural production, and that of the service sector is 5.8% of the national level. The service sector accounts for 75% of the regional domestic product, with tourism as the most prevalent activity. The study area exhibits rather extensive renewable energy exploitation.

The *agri-sector* in the study region consists of small-scale agricultural firms, cultivating mainly traditional products. But during the last few decades, greenhouse and organic farming have also become important in local agricultural production.

The *secondary sector* in the area makes a thin contribution to job creation and share of the regional gross domestic product (GDP). The main sectors, based on size and yields, are the food sector, plastics, agricultural machinery, building materials and local art and clothing products.

The *service sector* is one of the most dynamic sectors, where tourism, trade, banking and health services are of high importance. This sector produces two third of the regional domestic product while employing more than 45% of the local population.

Transport infrastructure includes Heraklion *port*, a very important node for both national and international sea transportation in the Mediterranean basin; Heraklion *airport*, the largest tourist and trade 'gate' of the Crete region; the *northern road corridor*, part of the TEN-Transport network; the *southern road corridor*; and a rather well-developed *telecommunications infrastructure*.

14.5 Application of the LIPSOR Model

This section presents the application of the LIPSOR participatory approach for the study of the future perspectives of the Heraklion region.

14.5.1 The MICMAC Approach: Identification of Key Variables

The MICMAC module was applied in order to identify the key variables for the Heraklion study area. As a first step, a structural analysis was carried out, based on 51 variables, identified with the support of stakeholders, namely, experts, external advisors, etc. These refer to environmental aspects, agricultural population characteristics; the agricultural sector, the quality/safety of the agricultural products, interaction with the rest of the local economic sectors, energy, foreign investments, European funding in various sectors, inter/intraregional transport and telecommunications infrastructure.

The scope of the structural analysis is to focus on the '*influence–dependence*' relationships among the variables of both the *internal* and the *external* environments, which reveal the key drivers of change in the study region.

Based on the MICMAC module, four types of '*influence–dependence*' relationships between each specific set of variables were carried out, namely, '*direct*', '*indirect*', '*potential direct*' and '*potential indirect*'.

The analysis has revealed that certain key variables play a steadily highly influential role in the evolution of the study system, rating high in all four types of influence–dependence relationships. These are considered to be the key drivers of the study system, steering its future developments (Table 14.1).

14.5.2 The MACTOR Approach: Actors' Games

The focus of the MACTOR module is on the study of the actors' games that seeks to gauge the balance of power among actors and study their convergence with and divergence from a certain number of associated stakes and objectives (Godet et al. 2004). Explored in the study region are power relationships; goals and objectives; projects in progress; preferences; motivations; internal means of action (coherence); and past strategic behaviour, constraints, etc., of the actors involved.

The scope of the analysis is twofold: to obtain insights into the '*influence–dependence*' relationships between the various actors in the study region but also to study the actors' position with respect to the planning objectives pursued (convergence with or divergence from the objectives). The actors involved are selected on the basis of their influence on the key variables identified in the previous stage (see Table 14.2).

The '*influence–dependence*' relationships between actors and the attitude of the actors regarding the objectives set are described through two cross-impact matrices: an '*actor by actor*' and an '*actor by objective*' cross-impact matrix. The output of the module provides information on the actors' alliances and their positions relative to the objectives, which is quite important for planning and policymaking purposes.

The results of the MACTOR application show that the actors 'societal organizations' (Actor 3), 'scientific groups' (Actor 5), 'trade unions' (Actor 7) and 'marine transport companies' (Actor 10) are the most powerful, while 'local government agencies' (Actor 2) is the most dependent one. 'Air transport companies' (Actor 11) seem to be rather independent, while 'environmental agencies' (Actor 1), 'farmers' (Actor 4), 'manufacturing' (Actor 6), 'tourist firms' (Actor 8), 'road transport companies' (Actor 9) and 'companies for biofuel production' (Actor 12) are both influential and dependent (Fig. 14.3).

The position of actors vis-a-vis every single objective is also provided. In this way, the convergence and divergence aspects of the actors can be studied with respect to the research objectives, while policymakers may increase their knowledge on the study area by examining specific decisions and their impacts based on the convergence and divergence patterns appearing for each specific goal.

The results show that the objectives 'regional development', 'social cohesion', 'food safety and quality' and 'energy' are gaining full support by all the actors

Table 14.1 Key drivers of change in the Heraklion region

	Influence			
Variable	Direct influence	Indirect influence	Potential direct influence	Potential indirect influence
Water resources pollution – V3	+	+	+	+
Availability of water resources – V4	+	+		+
Agricultural population – V8	+	+	+	+
Structure of agricultural population (ageing) – V9	+	+		+
Educational level – V10	+	+	+	+
Intensification of agricultural production – V23	+	+	+	+
Products of designated origin – V25	+	+		+
Educational services – V33	+	+		+
Demand for qualitative products – V27	+	+		+
Technology – V34	+	+	+	+
Research and development – V35	+	+	+	+
Telecommunication network – V51	+	+	+	+

Table 14.2 Actors involved in the Heraklion case study

	Description
Actor	Description of the actor involved
Actor 1	Environmental agencies (agencies related to the study of environmental issues – support of environmental protection in the area of concern)
Actor 2	Local government agencies
Actor 3	Societal organizations
Actor 4	Farmers (actors employed in the agricultural sector)
Actor 5	Scientific groups (universities, research centres, etc.)
Actor 6	Manufacturing
Actor 7	Trade unions
Actor 8	Tourist firms
Actor 9	Road transport companies
Actor 10	Marine transport companies
Actor 11	Air transport companies
Actor 12	Companies for biofuel production

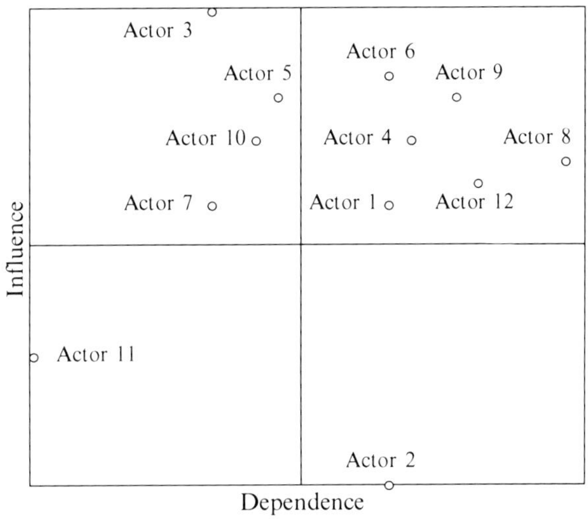

Fig. 14.3 Map of influence and dependence between actors

involved, while a certain divergence is exhibited by Actor 10 ('marine transport companies') and 11 ('air transport companies') for the objectives 'environmental protection' and 'economic efficiency' (see Fig. 14.4). On the other hand, Actor 1 ('environmental agencies'), Actor 2 ('local government agencies') and Actor 3 ('societal organizations') show a certain divergence away from the 'economic efficiency' objective. The above results are more clearly depicted in Fig. 14.5, where the mobilization of actors towards the planning objectives is depicted.

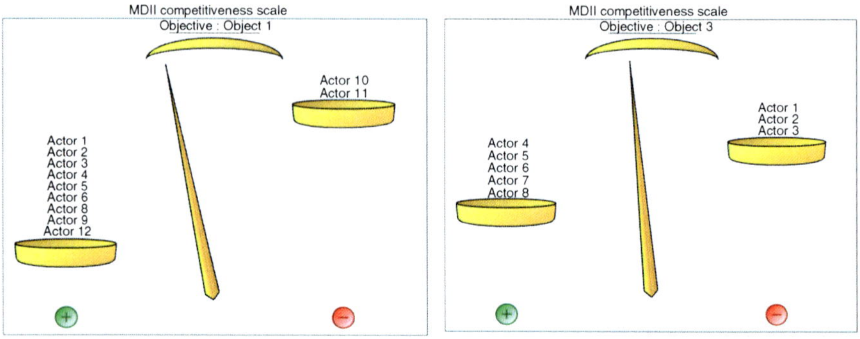

Fig. 14.4 Position of actors relative to the 'environmental protection' and 'economic efficiency' objectives, respectively

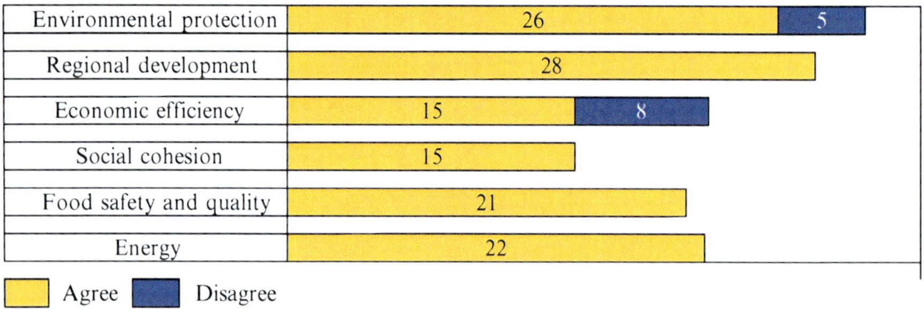

Fig. 14.5 Mobilization of actors relative to the objectives

14.5.3 *The SMIC PROB-EXPERT Approach: The Experts' Cross-Impact Analysis*

The third module of the LIPSOR model is the SMIC PROB-EXPERT module, which belongs to a greater group of explorative approaches. The SMIC PROB-EXPERT is an expert-based approach taking into account the opinions of a group of experts. The goal of this module is to explore 'probable' futures for the study region by constructing, as a first step, a set of hypotheses. Then expert-driven *single* and *conditional* probabilities are attached to these hypotheses, and finally a '*cross-impact* analysis' is used to describe their '*influence–dependence*' relationships, all of which leads to the construction of the 'probable' futures.

Based on the experience gained so far, the following set of six hypotheses was formulated (Stratigea et al. 2010):

– *H1: EU policy enhancing rural development* by attracting activities other than agriculture.
– *H2: The agricultural sector is becoming a high-technology-driven sector*, with increased productivity and competitiveness, due to technological advances.

- *H3: Ecotourism activities prevail in the region*, and a shift from mass to more environmentally friendly tourist activities is exhibited, preserving local assets.
- *H4: The food sector is developing rapidly* and transforming into a highly modernized, export-oriented sector.
- *H5: The strong exploitation of ICTs in the agricultural sector* favours networking and knowledge diffusion in all sectors.
- *H6: Agriculture is strongly involved in energy production*, reinforcing employment and income opportunities and share of renewable energy (RE).

These six hypotheses were presented to a number of experts with the request to attach both single and conditional probabilities. Based on this input, a cross-impact analysis of hypotheses leads to the construction of 'probable' scenarios. These are compared with the scenarios resulting from the MORPHOL module that follows, enabling the selection of a certain number of distinct futures for the study region.

14.5.4 The MORPHOL Approach: Scenario-Building Process

The MORPHOL module supports the scenario-building process by combining the results obtained in the previous steps, namely:

- Goal(s) and objectives
- Key drivers of change–output of the MICMAC module
- Information on the role of actors and their attitude as to the objectives–result of the MACTOR module
- The set of hypotheses and respective probabilities, identified in the SMIC PROB-EXPERT module

Based on the above, a number of key variables (drivers of change) are selected and respective hypotheses (on the future development of variables) are formulated (Table 14.3). These form the building blocks of all 'possible' scenarios in the MORPHOL module. For each hypothesis in Table 14.3, a specific probability has been assigned, based on the experience gained so far, but also the output provided by the previous modules and experts' opinion.

'Possible' scenarios are then identified as different combinations of future outcomes of key variables. The number of scenarios rises dramatically as the number of key variables increases. In order to limit this number, the MORPHOL module provides the option to exclude several non-feasible combinations of hypotheses and/or keep some combinations preferred by the planners and or decision centres.

The 'possible scenarios', which resulted from the MORPHOL module, were evaluated on the basis of their probability to appear. This resulted in a subset of highly probable scenarios, which were further evaluated in terms of their closeness, in order to select a few quite distinct and coherent scenarios. Finally, a combination of the existing knowledge of the region, together with the analytical results and stakeholders' views, resulted in three distinct scenarios (Fig. 14.6).

Table 14.3 Key variables and hypotheses in the Heraklion study

Variable	Hypotheses Set of hypotheses
Population growth patterns	H1: concentrated population pattern, rate of population increase 15%, educational level remaining at the same level as today
	H2: dispersed population pattern, rate of population increase 20%, educational level higher than today
	H3: other population development patterns
Economic structure	H1: continuation of the present development patterns – mass tourism and an agri-sector mainly export-oriented
	H2: continuation of the mass tourism development patterns and emphasis on the organic agricultural production and exports
	H3: alternative tourist development, a strong agri-sector specializing in organic production, and a high degree of R&D development and diffusion within all economic sectors
	H4: other
Technology and innovation in the agri-sector	H1: low diffusion of technology and innovation
	H2: medium diffusion of technology and innovation
	H3: high diffusion of technology and innovation
	H4: other diffusion patterns
ICTs	H1: medium diffusion – low exploitation from agri-sector
	H2: high diffusion – strong networking and knowledge diffusion
	H3: other ICTs diffusion patterns
Energy production patterns	H1: development of thermal plants and wind parks, same level of biomass energy production, and low share of RE
	H2: decreasing fossil fuels dependence with increasing RE production share
	H3: least dependence on fossil fuels, high share of RE production
	H4: other energy production patterns
Rural development	H1: agri-sector becomes less efficient, marked by a labour loss
	H2: agri-sector develops its multifunctional role (ecotourism and energy); emphasis on organic production and export trade
	H3: a very efficient technology/innovation-driven agri-sector; emphasis on multifunctionality (ecotourism and renewables)
	H4: other agri-sector development patterns

Source: Stratigea and Giaoutzi (2007) and Stratigea et al. (2010)

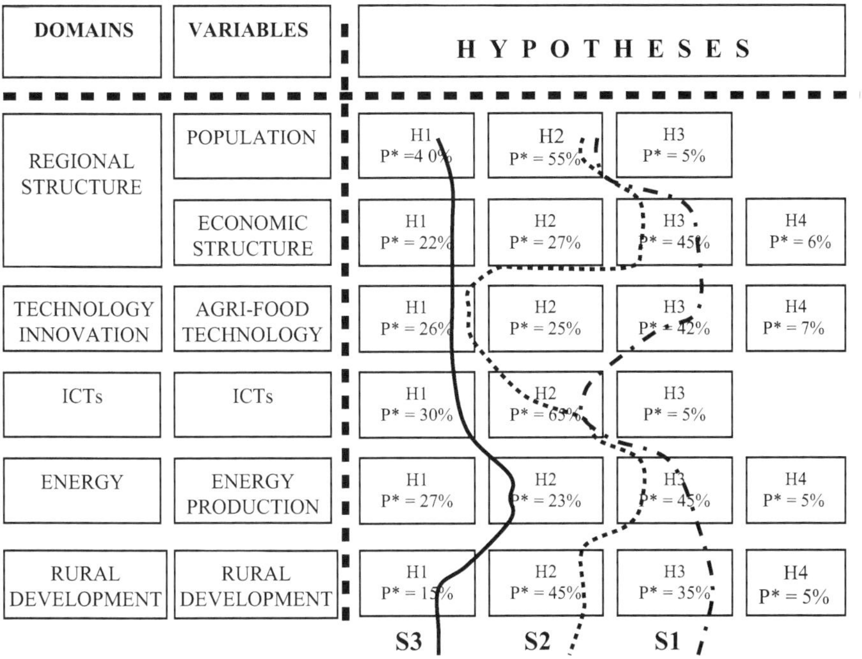

Fig. 14.6 The MORPHOL scenario-building process – Heraklion case study (*Source*: Stratigea and Giaoutzi 2007)

Key aspects in all three scenarios are *technology* and *innovation*.

The *first scenario* (S1) called the *'market first, scenario* (a combination of H2, H3, H3, H2, H3 and H3 for the respective variables – see Fig. 14.6) considers technology and innovation as means to increase efficiency. It refers to a future where population shows a rather dispersed spatial pattern, with highly skilled human resources. The economic structure of the region is based mainly on a technology-driven very efficient agri-food sector, placing emphasis on both cost and product differentiation (organic production). Mass tourist levels remain stable, while some kinds of alternative tourism are also developed. High rates of ICT diffusion contribute to strong networking among firms and increasing knowledge diffusion. In the energy sector, renewables have an increased share, thus contributing to a decreasing dependence on fossil fuels. The agri-sector takes advantage of its multifunctional role, placing emphasis on renewables and eco-activities. New promotion schemes, based on ICTs, enhance the potential of the tourist sector to increase its market share.

The *second scenario (S2) – the 'environment first' scenario* (a combination of H2, H3, H1, H2, H3 and H2 for the respective variables – see Fig. 14.6) – considers technology and innovation as means towards a more environmentally friendly future. The population exhibits a rather dispersed spatial pattern, with high skill levels. The local economic structure is characterized by an agri-food sector with a low technological profile and high environmental concerns, combined with traditional home-based agricultural development patterns, with an emphasis on organic farming and eco-activities based on small-scale farms. Mass tourism levels remain stable in the northern part of the region. ICTs are highly developed and used for networking and promotion purposes. A more environmentally responsible image of the region is promoted, attracting tourist flows and small-scale investments in eco-activities. Renewables attract high interest as a means to preserve local assets. Environmental culture prevails, while agriculture is the vehicle for preserving nature and developing small-scale eco-activities.

Finally, the *third scenario (S3)* is the *'worst-case scenario'* (a combination of H1, H1, H1, H1, H2 and H1 for the respective variables – see Fig. 14.6). It provides an image of the future, which is characterized by further concentration of population, the presence of low-skilled labour and a local economic structure based on mass tourism and trade of agri-products. There is a low/medium diffusion rate of knowledge and technology in the agri-food sector and a low/medium rate of ICT diffusion. Although renewables increase their share, energy production is still largely dependent on fossil fuels. The agricultural land is partly abandoned, due to the ageing of the rural population and the low level of competitiveness. The agricultural sector is gradually losing importance at the local level. The development of the region exhibits a polarized pattern of development (urban centres) and is lagging behind rural regions with respect to employment and income opportunities, skilled labour, accessibility to basic services, network infrastructure, ageing of population, etc.

14.5.5 The MULTIPOL Approach: Evaluation of Scenarios

The MULTIPOL module supports the final stage of the decision-making process. The evaluation of the previously constructed scenarios is based on multiple-criteria analysis and applies evaluation of both *actions vs. policies* and *policies vs. scenarios* (see Sect. 14.3). Both types of evaluation are based on a set of nine evaluation criteria, which, together with their priorities (weights), are based on local stakeholders' and experts' opinions (Table 14.4).

Main *inputs* to the MULTIPOL evaluation process are (Fig. 14.7):

- The *scenarios* as defined in the MORPHOL module
- A *pool of policies*
- *A pool of actions* (measures) in support of policies

The stakeholders and local representatives attached the following weights to three scenarios: the 'market first' scenario (S1) takes weight[2] 4; the 'environment first' scenario (S2), weight 5; and the 'worst-case' scenario (S3), weight 1.

[2] On a 5-point scale, with 1 the lowest and 5 the highest scenario priority.

Table 14.4 Evaluation criteria

a/a	Evaluation criteria		Weights[a]
1	Pesticide use – water pollution	K1	5
2	Biodiversity/landscape quality	K2	5
3	Accessibility	K3	5
4	Multifunctional role of agriculture	K4	4
5	Increasing competitiveness	K5	5
6	Social inclusion	K6	5
7	Renewable energy production	K7	3
8	Organic farming	K8	4
9	Product labelling	K9	3

Source: Stratigea et al. (2010)
[a]Measurement scale 1–5: with 1 the lowest and 5 the highest priority

The *'policies'* represent different policy directions, driving towards the fulfilment of the goal(s) and objectives. To each policy, a weight was attached by stakeholders and local representatives, reflecting the political priorities and values of the local society.

The *actions* are policy measures serving a specific policy.

Based on the above input, Table 14.5 below presents the performance of the selected actions (measures) with respect to policies. This is also presented schematically in Fig. 14.8 (actions/policies closeness map). From these results, it is evident that certain actions fit well with more than one policy, for example, Action A9 (agro-tourism – eco-activities) has a good performance in all policies considered (high value – low standard deviation); while other actions are policy-specific, for example, Actions A1 and A2, which perform best in Policy 1, while showing lower performance with respect to the rest of the policies considered.

From the information presented in Table 14.5 above, policy packages can be built for each specific policy, incorporating all actions (policy measures) that best serve this specific policy. For example, *Policy 1* is better served by the following set of actions: A1, A2, A5, A6, A7, A9, A10, A11 and A13; *Policy 2* by A2, A3, A5, A6, A7, A9, A10, A11 and A13; *Policy 3* by A4, A8, A12 and A13; and *Policy 4* by A5, A6, A9, A10 and A11 (see Fig. 14.8).

The results presented in Table 14.6 below map the performance of policies with respect to each scenario. All policies are serving all scenarios but to a different extent. Certain policies perform well in almost all scenarios, for example, Policy 3 ('Diffusion of ICT applications'). Other policies are more scenario-specific, as they fit better with a specific scenario, but not very well with the rest of the scenarios. Such an example is policy P1 ('R&D and innovation in the agri-sector'), which actually fits best in the 'market first' scenario (S1).

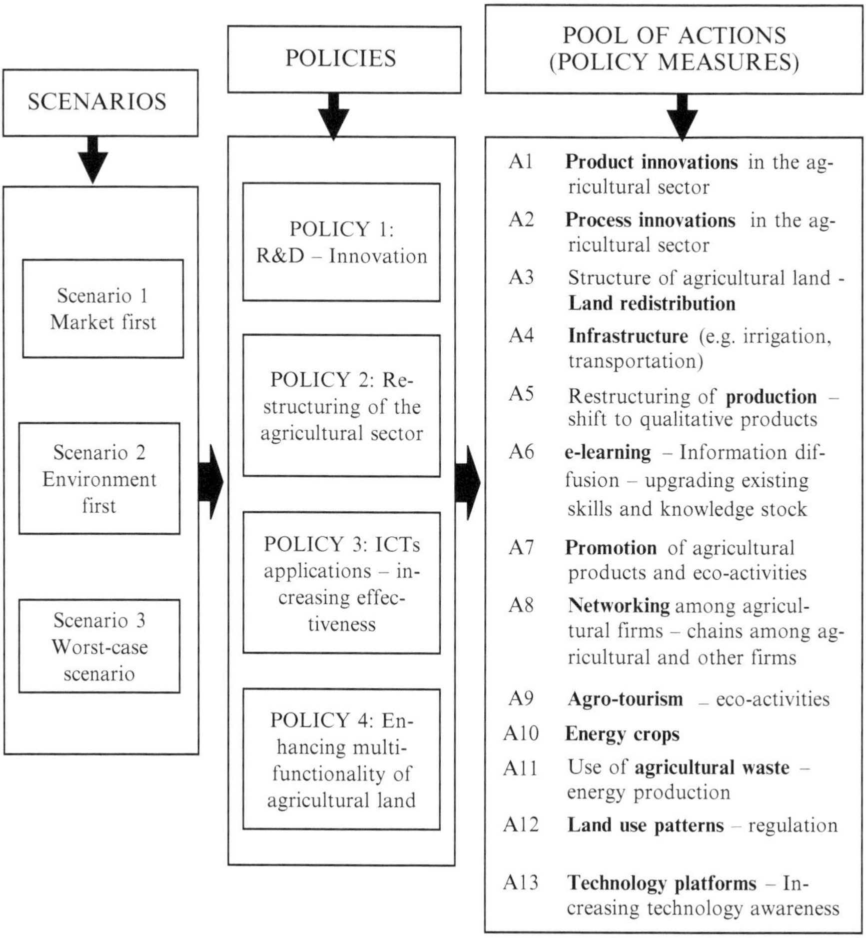

Fig. 14.7 Scenarios, policies and actions for the study region (*Source*: Stratigea et al. 2010)

On the basis of these results, it is possible from each specific scenario to draw conclusions on the policies relevant for its achievement. For example, the *'environment first'* scenario (S2) seems to be served by all four policy directions (P1, P2, P3 and P4), with P4 ('Enhancing multifunctionality of agricultural land') rated first, followed by P3 ('Diffusion of ICTs applications'), while P1 ('R&D – Innovation') and P2 ('Restructuring of the agricultural sector') are rated at a lower level as to their performance in this scenario (Fig. 14.9).

The *'market first'* scenario S1 is best served by P1 ('R&D and innovation'), P2 ('Restructuring of the agri-sector') and P3 ('ICT applications').

Finally, S3 – *the 'worst-case' scenario* – is better served by P2 ('Restructuring of the agri-sector'), P1 ('R&D and innovation') and P3 ('ICT applications') (in descending order), while last in priority is P4 ('Multifunctionality of agricultural land').

Table 14.5 Evaluation of actions with respect to policies (scores)

	Policy						
Action (measure)	P1	P2	P3	P4	Mean	St. dev.	Number
A1	9.4	7.3	6.6	4.9	7	1.6	2
A2	14.3	11.1	9.9	9	11	2	9
A3	10.1	9	9.1	7.9	9.1	0.8	4
A4	8.3	9.1	10.1	8.8	9.1	0.7	5
A5	14.4	12.5	11.1	11.2	12.2	1.4	11
A6	13.2	10.8	10.6	10.6	11.3	1.1	10
A7	8.2	6.8	6.7	5.7	6.8	0.9	1
A8	9.9	9.3	11.4	7.9	9.8	1.3	7
A9	12.9	13.2	12.4	13.1	12.9	0.3	13
A10	12.1	9	9.4	9.9	10.1	1.2	8
A11	10.7	8.9	9.2	8.9	9.4	0.7	6
A12	8.6	9.2	9.3	7.4	8.7	0.8	3
A13	14.5	12.5	13.4	10.6	12.8	1.5	12

Source: Stratigea et al. (2010)

Fig. 14.8 Actions/policies closeness map

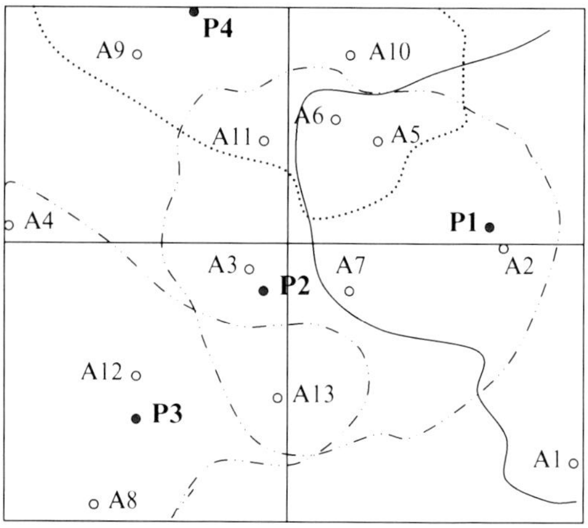

14.6 Conclusions

Regional foresight approaches, as future-focused and strategy-oriented tools, are quite promising for dealing with the future development perspectives of regions. They are capable of establishing a communication and learning platform at the regional level for gathering not only experts' intelligence but also common knowledge of the local population in order to deal with the complexity, risks and speed of changes at a global scale. Important in this respect is the level of involvement, commitment and sharing of a common vision through participatory approaches.

The LIPSOR methodological framework provides a future-focused, strategy-oriented and proactive tool, capable of dealing efficiently with uncertainty. It can be used as both a strategic planning and effective learning tool that aims to increase the performance of future-oriented studies.

LIPSOR is contributing to the decision-making process by managing the uncertainties involved in long-term planning and is dealing with the driving forces, key factors, key actors and their interaction, as well as interactions between internal and external environments in any study system. It also involves participatory aspects, enabling a broader commitment of the actors involved, by both increasing communication and common understanding on key issues in the study area and creating a shared vision by the local community, structured in a coherent and largely transparent way.

However, LIPSOR has certain *weaknesses*, mostly relating to the time-consuming processes involved in the design and running of the model; the discrepancies appearing in the points of view of the various actors involved, mostly as a result of their different backgrounds, interests, etc.; the subjective nature of the experts' judgements; the long list of the scenarios produced; etc.

Nevertheless, the weaknesses of the LIPSOR approach can be greatly overcome by consistent monitoring and planning of the process, a good knowledge of the problem, and high-skilled persons in charge of the exercise but also the involvement of a broader audience for lessening the subjectivity of judgements expressed during the process.

Table 14.6 Evaluation of policies with respect to scenarios (scores of policies with respect to scenarios)

	Scenario					
Policy direction	S1	S2	S3	Mean	St. dev.	Number
P1: 'R&D – Innovation'	14.9	10.8	11.3	12.5	2	4
P2: 'Restructuring of the agricultural sector'	12.2	10.9	11.9	11.5	0.6	3
P3: 'Diffusion of ICTs applications'	11.3	11.1	11.2	11.2	0.1	2
P4: 'Enhancing multifunctionality of agricultural land'	0.8	11.5	10.8	10.7	0.8	1

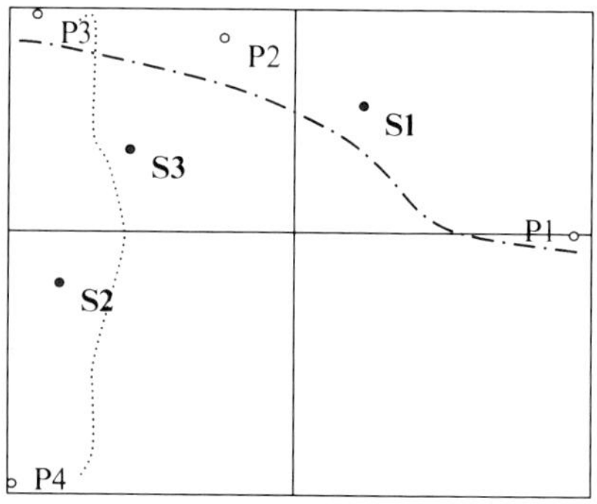

Fig. 14.9 Policies/scenarios closeness map

All in all, LIPSOR (Godet 1999) appears to be a rather promising tool for medium- to long-term regional future studies if certain improvements are carried out, increasing its user-friendliness and participatory potential.

The lessons learned from the application were that local participation in consecutive circles and discussion groups, at the various LIPSOR stages, has proved valuable for meeting the data requirements of the approach, also acting as a learning process for all sides involved. This has contributed to the establishment of a common ground for discussion and brainstorming on future developments of the region and, more specifically, of the agricultural sector. The analytical approach of scenario building for the region has been enriched by qualitative input from stakeholders' discussion, which has been used as a 'filter' for the quantitative output of the MORPHOL module, leading to three distinct scenario development paths for the study region. Moreover, stakeholders contributed to the evaluation stage by providing inputs on the weights of actions, policies and scenarios. The widening of the participatory context can provide more ideas for further elaboration, via which scenarios, policies and actions can be further elaborated.

References

Boulding, K. E. (1964). *The meaning of twentieth century*. New York: Prentice Hall.

Giaoutzi, M., & Stratigea, A. (2010). *General assessment of alternative policy scenarios, D6.1* (Final Report). AG2020 project, Foresight analysis for world agricultural markets (2020) and Europe. European Commission, Contract no.: 44280-AG2020, STREP, 2007–2009.

Giaoutzi, M., Stratigea, A., van Leeuwen, E., & Nijkamp, P. (2012). Scenario analysis as a foresight tool in agriculture. *International Journal of Foresight and Innovation Policy, 8*(2/3), 105–128.

Godet, M. (1999). Creating the future. *The Antidote, 22*, 11–16.

Godet, M. (2001). *Creating futures: Scenario planning as a strategic management tool*. London: Economica Ltd.

Godet, M., Monti, R., Meunier, F., & Roubelat, F. (2004). *Scenarios and strategies: A toolbox for problem solving*. Paris: Cahiers du LIPSOR, Laboratory for Investigation in Prospective and Strategy.

Lindgren, M., & Bandhold, H. (2003). *Scenario planning: The link between future and strategy*. New York: Palgrave Macmillan.

Ringland, G. (1998). *London in 2020*. London: Gresham College.

Ringland, G. (2002). *Scenarios in public policy*. West Sussex: Wiley.

Stimson, R., Stough, R., & Roberts, B. (2006). *Regional economic development: Analysis and planning strategy* (2nd ed.). Berlin: Springer.

STRATA-ETAN Expert Group. (2002). *The potential of regional foresight, mobilizing the regional foresight for an enlarged European Union – An essential contribution to strengthen the strategic basis of the European research area* (Final Report). EUR 20589.

Stratigea, A., & Giaoutzi, M. (2007, July 9–11). *Scenario planning as a tool in foresight exercises: A methodological framework.* Paper presented at the International COST A22 Conference "From Oracles to Dialogue: Exploring New Ways to Explore the Future", Athens.

Stratigea, A., Giaoutzi, M., & Papadopoulou, Ch.-A. (2010). *The AG2020 case studies.* Report on the Kastelli-Heraklion case study. AG2020 project, Foresight analysis for world agricultural markets (2020) and Europe. European Commission, Contract No.: 44280-AG2020, STREP.

Chapter 15
Foresights, Scenarios, and Sustainable Development: A Pluriformity Perspective

Eveline van Leeuwen, Peter Nijkamp, Aliye Ahu Akgün, and Masood Gheasi

15.1 Introduction

> If everybody in this world of ours were six feet tall and a foot and half wide and a foot thick (and that is making people a little bigger than they usually are), then the whole of the human race … could be packed into a box measuring half a mile in each direction…. H. W. van Loon (1932), p. 3

'*If* … *then* …' is a conditional proposition that describes precisely a logical causal statement about possible future events. Obtaining due insight into an uncertain future has been a permanent source of rational speculation in the history of mankind. In the Hellenistic period, the foundation for systematic foresight analysis was already laid by the Oracle of Delphi which – in contrast to popular wisdom – was not based on the incoherent utterances of an ancient intoxicated goddess but on evidence-based information collected by her through listening to the subordinates of any political figure who wanted to pick up a useful hint on how to face the future. The medieval and premodern literature was also full of seemingly rational attempts to predict uncertain future events, such as catastrophes or wars. The aim to acquire political power was often an inspiration for obtaining strategic future information on unknown territories, as is clearly reflected in the support of leading dynasties in European countries for the great voyages of discovery from the fifteenth to the eighteenth century.

E. van Leeuwen (✉) • P. Nijkamp • M. Gheasi
Department of Spatial Economics, VU University Amsterdam,
De Boelelaan 1105, Amsterdam, 1081HV, The Netherlands
e-mail: e.s.van.leeuwen@vu.nl; p.nijkamp@vu.nl; m.a.g.gheasi@vu.nl

A.A. Akgün
Department of Urban and Regional Planning, Istanbul Technical University,
Taskisla, Taksim, Istanbul, 34437, Turkey
e-mail: gulumser@itu.edu.tr

M. Giaoutzi and B. Sapio (eds.), *Recent Developments in Foresight Methodologies*, Complex Networks and Dynamic Systems 1, DOI 10.1007/978-1-4614-5215-7_15,
© Springer Science+Business Media New York 2013

Over the past few decades, the control of future circumstances that might adversely affect current or future economic or political developments has led to many scientific efforts to uncover the driving forces of potential drastic changes in the near or distant future. One of the first well-documented studies on future developments can be found in Kahn and Wiener (1967), who made a scientific analysis of the bandwidth within which the year 2000 could be rationally explored ('a framework for speculation'). Their investigation was inspired by control principles derived, inter alia, from cybernetics. The application of advanced modelling experiments was in particular advocated by Tinbergen (1956), who was able to construct system-wide models for economic policy and forecasting. Early attempts to provide national forecasts on the success conditions of economic systems were also made by Ayn Rand (1957). In subsequent decades similar attempts were made by, amongst others, Alvin Toffler (1970), whose foresight analysis was mainly based on collecting a wealth of (sometimes selective) trend information to map out the contours of likely future mega-trends. The scientific interest in future development has even led to a new orientation in the planning discipline, sometimes called 'futurology'. In our modern era, the exploration of possible futures has led to the great popularity of scenario analysis, for instance, in the energy, environmental, or transportation sector (see, e.g. Nijkamp et al. 1998; Giaoutzi et al. 2012).

In the present study, we adopt a future scenario approach to take account of the sustainability interests of local or regional development plans, in which local or regional stakeholders have an important say. The aim is thus to test the feasibility of scenario analysis for local or regional sustainable development strategies, using the viewpoints of relevant stakeholders as the main anchor points for our analysis.

This chapter is organized as follows. In Sect. 15.2, we elaborate on how thinking about sustainable futures evolved in our society. In Sect. 15.3, we outline various elements of a sustainable future. Then, Sects. 15.4 and 15.5 are devoted to the description of various future scenarios, to be applied in five case-study areas in Europe. A description of the methodology and of the main findings is provided in Sect. 15.6, followed by a concluding section.

15.2 Focus on Sustainable Futures

> Worldwide, there is only a market for five computers. Thomas J. Watson, President-Director IBM (1958)

The debate on a sustainable future has indeed prompted a renewed interest in the contours of – and the conditions for – environmentally friendly developments that might ensure the durable use of the earth's scarce resources. The concern about social, economic, and environmental sustainability dates back to the end of the 1980s, with the publication of the WCED report *Our Common Future* (1987) (often called the Brundtland Report). But its origins can already be found earlier in the post-war period, while very early attempts to address environmental and future issues in the social sciences can be found a few centuries back. A few illustrations will be given here to indicate the interest of earlier scholars in ecologically benign developments.

The trade-off between one's own interests and others' interest can be traced back to one of the great philosophers and the founding father of modern economics, Adam Smith, who in his *Theory of Moral Sentiments* (1759), introduced the concept of a '*man of humanity*' to illustrate the tensions between the present and the future. This man hears about an unprecedented earthquake in China and then reflects for a while on the transience of life and thinks also of the economic consequences for Europe and himself. But then, he returns to his normal business. However, if on that day he were to be told that he was to lose his little finger in the near future, he would be tormented at all times and would find no peace. Adam Smith then puts forward a moral dilemma and asks: '*If the injury to or loss of the finger is subjectively so great a catastrophe and the earthquake in China such a minor one, would this mean that the "man of humanity" would prefer the obliteration of millions to the rescue of this little finger if such a choice existed?*'

Self-interest has, over the past decades, been the foundation of economic behaviour. But it prompts questions on how to handle the effects of actions that influence someone else's well-being without being included in market or price transactions. A well-known example from the early days of the steam engine is the locomotive whose sparks may set on fire the crops cultivated by farmers on land adjacent to railways. Clearly, full compensation costs would have to be paid by the railway company, but, as a consequence, more farmers would grow crops near railways, as this might give them a guaranteed income in case of fire, irrespective of the probability of crops being destroyed by bad weather. This situation might lead to a misallocation of scarce resources, as normal entrepreneurial risk would not be included in these transaction costs. To take account of such externalities in the market system, Pigou (1930) introduced the notion of financial compensation through the principles of taxation, so that all (direct and indirect) costs would be incorporated in the '*measuring rod of money*'. This principle has played a prominent role in environmental policy, where it is nowadays known as the '*polluter pays principle*'. Would the market system then be able to ensure a sustainable future?

In the post-war thinking on environmental, resource, and climate issues, various stages can be distinguished as the:

- Intuitive phase, where mainly anecdotal evidence was presented on environmental decay. A clear example is Rachel Carson's '*Silent Spring*' (1962).
- Systemic phase, where population, resources, environment, and growth were analysed from a global systemic perspective. An illustrative contribution is the study of Meadows et al. (1972) entitled *The Limits to Growth.*
- Sustainability phase, where the long-run balance for the use of the earth' scarce resources was put in the perspective of both the future and the north-south interests. The seminal study of the WCED (1987) heralded a new epoch in environmental thinking. But also at a local and regional scale, various analytical studies on sustainable futures were undertaken (see, e.g. Giaoutzi and Nijkamp 1993).
- Climate change phase, in which as of the beginning of the twenty-first century, the focus has been directed towards long-range climatological changes on a worldwide scale.

These global developments have prompted many innovative concepts and policies for the balanced development of our planet. But, also at local levels, various initiatives have been launched, such as the sustainable city initiatives (see Nijkamp and Perrels 1994).

In many environmental economic policies at various spatial levels, price and market perspectives have played a dominant role. Market-oriented sustainability policies have largely adopted similar principles, for example, in resource policies and in emission rights policies. This strategy has meant a powerful policy contribution to the achievement of environmentally benign future developments and also constitutes a critical part in the implementation of the Kyoto Protocol and subsequent initiatives.

Nevertheless, there are many cases where a straightforward price and market principle concerning future sustainable developments may be problematic, especially at local or regional levels where areal development is based more on a commonly accepted, multi-stakeholder future perspective – leaving scope for much pluriformity in viewpoints on sustainable trajectories – than on a strict system of price or tax incentives. The main issue is then whether under such conditions a sustainable future can be mapped out and ensured.

15.3 In Search of a Sustainable Future

> Doubt is not a pleasant situation, but certainty is absurd. Voltaire (1694–1787)

A multiplicity of studies have been published on sustainable development. Many of them have a global or national orientation, while others are more instrumental and policy-oriented in nature. There are also many sectoral-oriented sustainability studies, while we observe an increasing number of spatial sustainability studies, for example, on sustainable cities or regions.

Sustainability is a hard-to-define concept, but it became popular and very much 'en vogue' after the publication of *Our Common Future*, also known as the Brundtland Report by the WCED in 1987. According to this report, sustainable development refers to a development of countries or regions that meets the needs of the present without compromising the ability of future generations to meet their own needs (WCED 1987). Sustainable development is not only limited to environmental aspects but also includes other developmental issues, such as socio-economic objectives.

Given the fact that sustainable development has a very complex nature from a policy and process perspective, a systemic approach may offer a practical frame of reference. In general, a systems approach aims at portraying the processes and relationships in a complex system that encompasses various components which are mutually connected by means of functional, technical, institutional, or behavioural linkages (Harvey 1969). Systems thinking advocates the scientific treatment of systems as interlinked units, composed of mutually related elements. According to

Hwang (2000), systems thinking enables us to see the overlapping and ever-expanding relationships among multifaceted systems in multiple dimensions, ranging from problem formulation to problem-solving in (organizational) practice. Moreover, Stewart and Ayres (2001) advocate a systems approach in policymaking by emphasizing the following points:

- A systems approach offers policymakers a fresh set of perspectives on the integrated fundamentals of policy analysis.
- Policy design is as much a matter of choosing structures and relationships as of choosing instruments.
- Understanding causation means acknowledging two-way influences and the role of complicated feedbacks.

In developing a coherent approach to spatial sustainable development, we have to think first of all about policy-related factors. To favour or enforce sustainable development, appropriate policies, regulations, and incentives are needed. However, governments and the focus of policies can change as a result of several (inter)national socio-economic shifts. But, apart from such developments, also the nature of policies and the scope of governance may be important. For example, the level of coherence and seamlessness of political decisions is important to stimulate sustainable development. Efficient incentives, the transparency of measures, sustainability goals in all policy fields, and regulatory tools can all be seen as factors necessary for success.

These factors can be even more important when the government has to look into the interests of private partners. Public-private partnerships can improve the understanding and trust on both sides and, therefore, the uptake of new (innovative) ideas in the business environment, as well as in the public sector. However, this requires interaction between different policy levels (local, regional, and national) and between public and private partners.

When thinking in business terms, trust and mutual understanding are very important to change entrepreneurial behaviour. This also requires insight into the behaviour of SMEs and into the structure of existing market networks. It should be noted that the economic structure including ownership issues, the presence of big international firms, and the influence of local firms significantly affect the manoeuvre space for sustainable development. In addition, not only the current economic structure but also the current and future economic business climate is relevant.

It should be added that innovation and change partly depend on social values, social acceptance, and the absorptive capacity of society. When entrepreneurs and households are aware of the importance of sustainable development and if they are open to new ideas, then this may induce new developments that are promising for a sustainable future. In addition, local leadership is also an important success factor. Another important aspect of sustainability is the operational and visible quality and quantity of ecological areas and the ecological processes taking place. Synergies between the protection of the natural/cultural heritage and economic development of tourism and land-based industries can also ensure sustainability success ranging from local to global scales. Moreover, insights into the current state and resilience of local/regional ecosystems, as well as insights into (economic) benefits of ecological values, can help decision-making and stimulate balanced sustainable development at various geographical scales.

In our empirical approach described next in Sect. 15.3 and subsequent sections, we follow a stepwise approach, involving:

- The design and definition of four sustainable development scenarios at a local or regional level on the basis of five case studies in Europe
- The identification and assessment of the impacts of these scenarios using the systematics of a multidimensional 'amoeba' diagram (a visualized 'impact matrix')
- The elicitation of the interest of the relevant stakeholders by developing a stakeholder's priority scheme ('weighting scheme')
- A performance analysis of each scenario with a view to the identification of the best-fit scenario for a sustainable future by means of a multi-criteria evaluation analysis

15.4 Design of Systemic Scenarios for Sustainable Development

> As far as the laws of mathematics refer to reality, they are not certain; and as far as they are certain, they do not refer to reality. Albert Einstein

The key issues in applying the concept of sustainable development to the five case studies under consideration are the competitive organization of production and consumption (economic and social), the aim of continuity (institutional and ecological), the consideration of capacity (ecological and economic), and the need of coherence (social and ecological). Clearly, the application of the sustainability concept will lead to different analytical issues and outcomes depending on local circumstances and interests. Therefore, we have deliberately introduced a great diversity in case studies from different countries in Europe. The aim of this approach is not to make a comparison of findings from different case studies but to test the robustness of our analytical apparatus by applying it to different experimental conditions. For this reason, we have consulted different types of stakeholders in the course of five distinct case studies: the forest sector in Finland, a Scottish national park, the Romanian energy sector, a Spanish region addressing sustainable progress, and the agricultural sector in an Italian region. These case studies are meant to provide a complementary view of different sustainability problems in various geographical, environmental, social, economic, developmental, and cultural contexts in order to test the potential value of our strategic scenarios for future sustainable development. Before presenting the definition of scenarios and the methodology used in this chapter, we briefly introduce the sectors targeted in each case study.

Finland: The focus of the sustainability case study in Finland is on the ecosystem and its utilization of forest by humans. This case study is a nationwide case study for the next 20 years. It focuses on the forest ecosystem, and therefore, the plan covers ecosystem quality and the ecosystem well-being of forest areas in Finland. The economic aspect of the case study is to analyse the possibilities to regenerate and sustain the ecosystem's quality.

Italy: The aim of this case study is to understand the complex interactions and metabolism of the agricultural sector in Italy. This study focuses on three spatial scales, viz. local scale (farm level, with three farms being selected), regional scale (Campania region), and national scale (Italy).

Romania: The focus of this case study is on the energy sector including the integrated social, economic, and environmental aspects. This case study also focuses on the transitional economy at a sectoral level on the basis of the metabolism of the system in terms of its flows of energy, materials, and money.

Scotland: This case study focuses on assessing the trends in the Cairngorms National Park (CNP) in Scotland and the implications for this park from the National Park Act via the Cairngorms National Park Plan. The CNP Plan is a strategic spatial planning document that is structured around three main themes: conserving and enhancing the park, living and working in the park, and enjoying and understanding the park. The plan has 22 strategic objectives to be achieved by the year 2030, as well as seven policy priorities for action to be achieved by 2012.

Spain: The target of this case study concerns the sustainable development of Catalonia. The aim is to test the possibility of carrying out the analysis of a metabolic pattern across geographic levels also using spatial analysis (data supported by GIS). The goal of this case study, to be realized by means of the quantitative results, is to obtain new insights into technical challenges, the possibility of gathering the required data, and the policy relevance of the results.

In the next section, we present the definition of our scenarios for the sustainable future of the areas concerned, and subsequently, we discuss the data used for this study.

15.5 Pluriform Scenarios for a Sustainable Future: Definition and Data Collection

15.5.1 Definition

This part of our analysis specifies the nature of the scenarios operationalized through the use of empirical stakeholders' questionnaires administered in each of the five cases. The different valuation of the various stakeholders suggests that there would be great diversity in sustainable futures. Consequently, we have developed four different scenarios with a high degree of pluriformity. These scenarios will now be briefly described.

- *Scenario 1*: *Competitiveness (Economic and Social)*. The main aim is to reach sustainable development by first improving the economic situation. This means that, apart from creating a satisfactory physical system (e.g. infrastructure), uncertainties in, for example, prices will be decreased and the economic diversity of sectors will be optimized. In this way, the economy will be less sensitive to economic crises, and income will be more equally distributed. As a result, the quality of social networks will increase, while budgets and technologies are available for the protection of the ecological environment. This will then result in stronger competitiveness and more sustainable economic development.
- *Scenario 2*: *Continuity (Institutional and Ecological)*. Here, the main aim is to protect the natural environment, as well as the diversity of ecosystems. Therefore, the scenario focuses on the increase in ecological quality and the choice of environmentally friendly sectors for economic development. In order to do so, there is a need for a high level of administrative and management involvement in terms of effectiveness of policies, sustainability inclusion, continuity, and the integration of institutional systems in the sector concerned and a respect for sustainability. Both continuity in the institutional system and the developments related to the ecological system are indispensable.
- *Scenario 3*: *Capacity (Ecological and Economic)*. In this scenario, the main aim is to obtain a high-quality natural environment, as well as a healthy economic and physical environment. By focusing on both economic and ecological development, the environmental capacity of an area will increase. Often these two aims can be integrated by using environmentally friendly ways of production. This means that producers should not only choose environmentally safer inputs for production but should also reduce unnecessary waste and transport movements.
- *Scenario 4*: Coherence (Social and Ecological). The main aim is to first develop the social environment in terms of the quality of social networks, for instance, to increase the awareness of sustainable development in general (e.g. through education), and of the importance of ecological and social networks more specifically. Therefore, investments in education levels and skills to use new technologies are very important to both the young and the elderly generation. As a result, the level of tolerance and openness of society to new developments and the level of involvement and understanding will be increased. This will then lead to the protection of the ecological system and a decrease of negative environmental impacts.

These four scenarios will now be applied to each of our five case studies. To that end, a comprehensive systematic database is necessary.

15.5.2 Database

We developed an online questionnaire to trace the different opinions of various groups of stakeholders. A digital questionnaire was sent out to carefully selected stakeholders of the five case studies. The questionnaire took about 10–15 min to complete and had

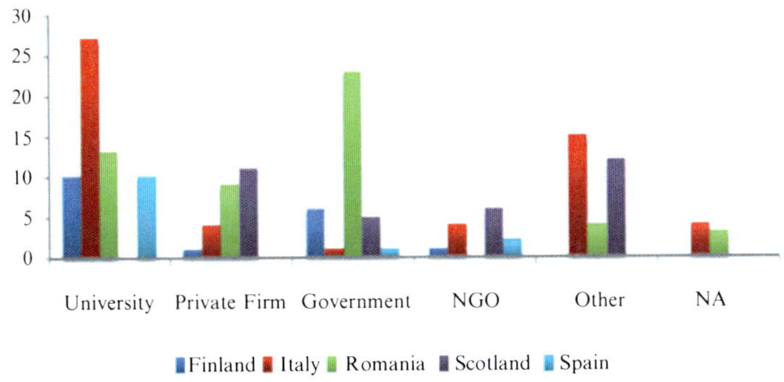

Fig. 15.1 Distribution of stakeholders by institution and country

the aim to collect a wide range of opinions and experiences regarding sustainable development from different stakeholders. The questionnaire consisted of four parts. First, we asked the respondents to express their views on the relative importance of five basic systems compared with each other (pairwise comparison). Secondly, we asked them to rate various aspects of these five systems. The third part had four scenarios that described how to reach sustainable development. We asked the respondents to allocate 10 points over these four scenarios in order to express their priorities for these scenarios. Finally, we also asked a few personal questions.

In total, 172 questionnaires were filled out: 18 from Finland, 55 from Italy, 52 from Romania, 34 from Scotland, and 13 from Spain. As Fig. 15.1 shows, the biggest group of respondents is associated with academia, especially in Italy. The second biggest group of respondents are those working for the government, in particular in Romania, in connection with the energy sector. Most small and large private firms that responded are from Scotland and Romania. Furthermore, in Italy has responded a relatively large number of small firms (farm owners in this case). The smallest group of stakeholders is the NGO group, which originates mainly from Italy and Scotland. We also have a group of 'others', containing those stakeholders who did not fill out information on their own institution but who did fill out the rest of the form. The category NA (not available), which consists of seven persons, did not provide any personal information (see Fig. 15.1).

The next step in our future sustainability analysis will be a systematic treatment of all data, for each of the five case studies, each of the four scenarios, and each of the relevant policy parameters under consideration. This will be done in the next section.

15.6 Methodology and Analysis

We have – as mentioned above – defined four distinct sustainable development scenarios in order to identify the best-fit sustainable development scenario and to see which stakeholders prefer which scenario alternative and how they rank these on the basis of their perceptions and preferences. To that end, we will use an 'amoeba diagram' to map out the various positions. Next, we will apply multi-criteria analysis (MCA) to the results. MCA comprises various classes of decision-making approaches (for a review, see Nijkamp et al. 1990). We now briefly describe these two steps.

15.6.1 'Amoeba' Diagrams

The 'amoeba' diagram is based on a multi-attribute visualization of a composite phenomenon (e.g. a good, a person, and a region). It takes for granted that, in a comparative sense, the most characteristic features of a phenomenon can be depicted in an amoeba-like diagram. The question of how many characteristics to include depends mainly on the aim of the research. In various policy studies (e.g. Capello et al. 2000), it has been demonstrated that in many cases five representative key factors can be distinguished that describe adequately the most critical attributes of a policy alternative. This is known in the literature as the 'pentagon' model. In our empirical study on sustainable development strategies of the five European regions under consideration, we are also able to distinguish five main drivers of sustainable development (for details, see Akgün et al. 2011). The five 'pentagon factors' identified in our comparative case study approach are as follows:

- Economic factors
- Ecological factors
- Physical factors
- Social factors
- Institutional factors

Clearly, we now have five pentagon factors, four scenarios, and five regions, as well as multiple stakeholders. This means that we have different ways to represent the information in an 'amoeba' diagram, viz. by combining the pentagon factors with either the five regions or with the four scenarios, while the priorities for each of the policy factors are obtained from the relevant stakeholders. All this is visualized in Figs. 15.2 and 15.3, respectively.

For the sake of illustration, we will present here some 'amoeba' diagrams for our comparative scenario experiment by mapping out the empirical features of both the various classes of stakeholders and the case studies on the basis of the five pentagon factors (see Figs. 15.4 and 15.5).

Information on these five pentagon factors (in either a cardinal or a ranking system) allows us to use an MCA, in order to identify the most acceptable future scenario (through the use of preferences expressed by stakeholders) or the highest performing regions (or cases) for each of the individual future scenarios. This will be further discussed in Sect. 15.6.2.

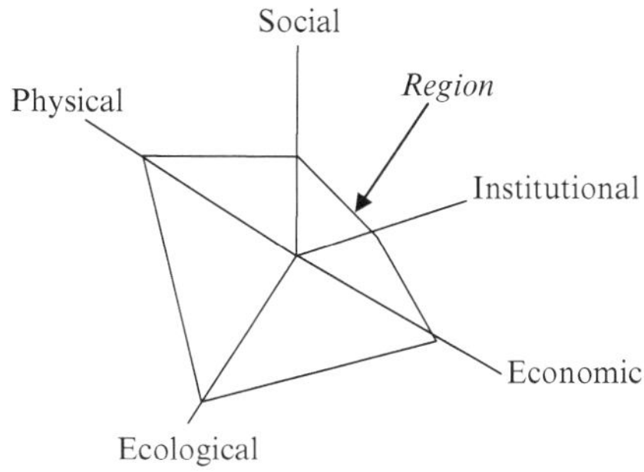

Fig. 15.2 An 'amoeba' diagram for five pentagon factors and a given region

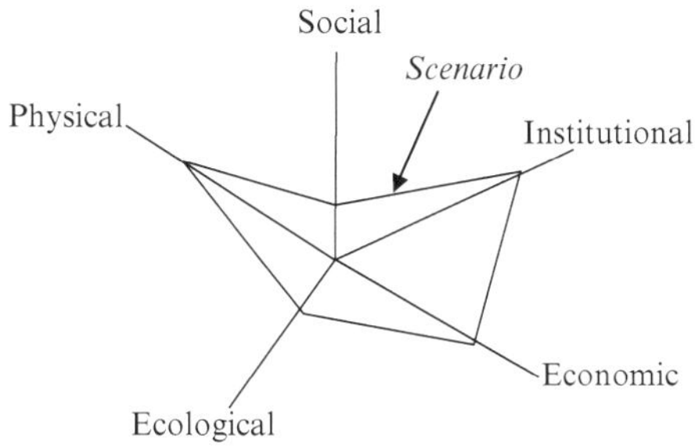

Fig. 15.3 An 'amoeba' diagram for five pentagon factors and a given scenario

15.6.2 A Multi-criteria Model

There are various MCA methods, but in this study, the regime method was applied (for details, see Nijkamp et al. 1990). The regime method presupposes a distinct set of a priori defined alternatives and a distinct set of a priori defined evaluation criteria. For all criteria together this then leads to what is called a 'regime matrix'. Then, by adding a weight vector for each criterion, the relative dominance of each alternative can be assessed in the form of a performance (or success) indicator.

The scenarios differ in the relative importance of each of the five pentagon factors, as is shown in the 'amoeba' diagram. This information was also provided to the respondents. Hence, this basically forms the well-known MCA impact matrix. For instance, the most important factor for the competitiveness scenario is formed by the economic system. In other words, in the first scenario, the main focus is to deal with the economic system, in order to obtain a sustainable future. All other scenarios are also scored by means of these five critical factors in the same way (for a full presentation in an ordinal format, see Fig. 15.6).

In order to prioritize the factors and understand their critical ranking, we have collected priority data from all stakeholders in the form of pairwise comparisons. This allows us to calculate the weights for each factor from the perspective of a variety of stakeholders. Weights calculated from the results of pairwise comparisons form basically the prioritization rank order of the pentagon factors, and in the literature on regime analysis, those rankings are referred to as weights.

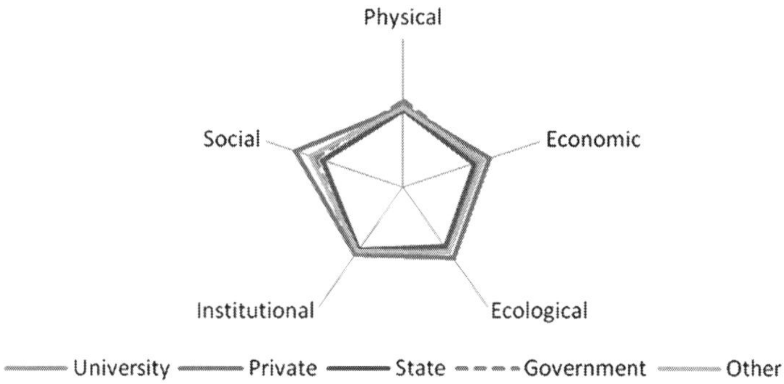

Fig. 15.4 'Amoeba' diagram of pentagon factors by stakeholder group

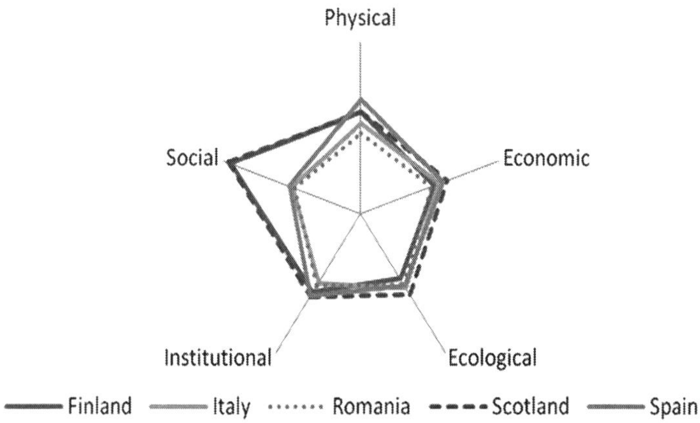

Fig. 15.5 'Amoeba' diagram of pentagon factors by case study

By using the weighted average of the pairwise comparison of different stakeholders, we can generate a prioritization table of pairwise comparisons, after which the ranking of the factors takes place. In our empirical work, this exercise allowed us to distinguish 11 sets of weights, that is, their priority rankings of the five pentagon factors (Table 15.1).[1] During the assessment of these groups, we used an 'equally important' ranking, when the stakeholders were not – or not entirely – consistent in their pairwise comparison. In other words, when a stakeholder evaluates social systems more important than economic systems, economic systems as more important than ecological systems, but ecological systems more important than social systems, this means that the ranking of the stakeholder is inconsistent (violation of the 'transitivity' rule); therefore, the relative weight of these factors is ambiguous, and consequently, all factors are equally weighted for this stakeholder. Apparently, he/she cannot decide which one is more important than the others. While constructing the weights distinction, we thus have three classes: 'agree', 'disagree', and 'unsure'. We now provide a brief interpretation of Table 15.1.

[1] The order of the groups has no meaning; they are in principle all equally important.

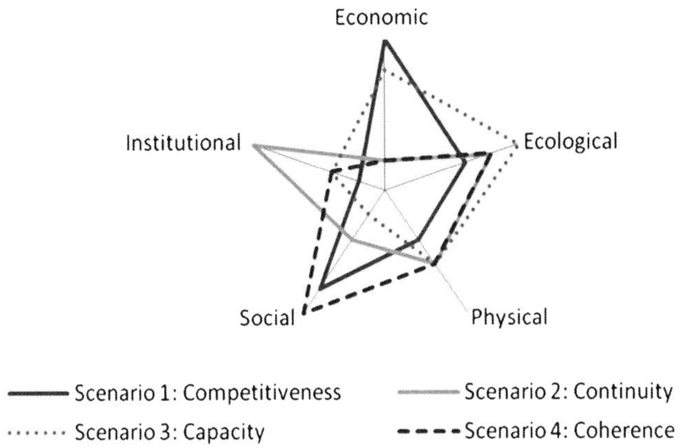

Fig. 15.6 'Amoeba' diagram of pentagon factors by scenarios

Table 15.1 Pairwise comparisons: set of weights

	Set of weights	Group of stakeholder
Group 1	Ecological = social = physical = institutional = economic	Education: bachelor and high school or less Occupation: students Institution: other Country: Italy and Romania
Group 2	Ecology > social > physical = institution = economy	*General* Gender: female and male Education: master Occupation: other
Group 3	Ecology > social > economy > physical > institution	Occupation: manager
Group 4	Ecology > social > institution > economy > physical	Education: doctorate
Group 5	Ecology > social > physical > institution > economy	Occupation: researcher
Group 6	Ecology = social = physical = economy > institution	Education: other Institution: university and civil service
Group 7	Social = physical = economy > ecology > institutional	Institution: NGO
Group 8	Physical > social > economy > institutional > ecology	Institution: private
Group 9	Ecology > physical > social > institutional > economy	Country: Finland
Group 10	Ecology = physical = institutional = economy > social	Country: Scotland
Group 11	Physical > ecology > economy = social = institutional	Country: Spain

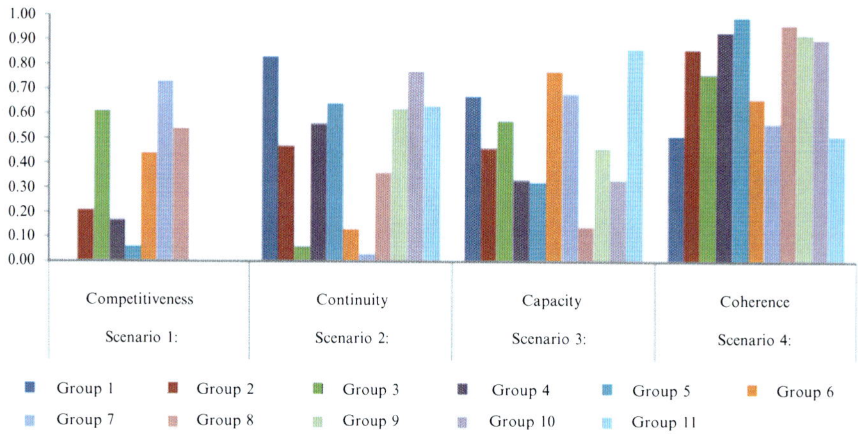

Fig. 15.7 Scenario choices of different stakeholders

The first group distinguished in Table 15.1 (i.e. Group 1) is the one where all factors are equally weighted. All factors are thus equally in favour of sustainable development, both among higher-educated groups (high school and bachelor degree graduates) and lower-educated groups. Next, in terms of gender, both female and male stakeholders believe that ecological systems should be given a high priority; these categories are then followed by social systems and physical systems, which are equally important as institutional and economic systems. Amongst the different stakeholder groups, we also included the general view by calculating the average mean of each stakeholder's valuation and labelled it 'general'. Thus, in general terms, without a differentiation of stakeholders groups, the same ranking as gender and master's graduates is given and can be grouped as Group 2. We see almost the same grouping with the criticality of sub-factors of the pentagon factors. For instance, university and state employees have similar preferences, even when other occupations think differently.

On the basis of the stakeholders' prioritization table (Table 15.1) and the impact scores on which the amoeba diagram in Fig. 15.6 is based, we are now able to run our regime analysis. We have run the regime analysis for each group of pentagon factors separately and organized the table of the performance indicators – in the MCA case – for each group, so that we are able to identify the most-preferred scenario from all stakeholders across all five case study areas (see Fig. 15.7). In order to understand the performance of the four scenarios, the choices of the 11 groups of stakeholders are visualized in a histogram.

The numerical interpretation of the bars in the histogram of Fig. 15.7 is as follows. The MCA software used in the regime method is able to calculate in a cardinal sense (on a scale from 0 to 1 the maximum performance) the relative performance rates of each of the alternatives (e.g. scenarios), based on an underlying ranking system for the stakeholders' preferences for the five pentagon factors. This offers a rather robust quantitative outcome for the comparative analysis of our four scenarios. From the histogram we can easily observe that Scenario 4 (coherence) is the most-preferred one and Scenario 3 (capacity) is the second most preferred. Scenario 2 (continuity) and Scenario 1 (competitiveness) next follow in rank order.

It is noteworthy that we can also observe some clear differences between the 11 groups. Scenario 1, for example, is in particular promising for groups 7 and 3, the NGOs and managers, but absolutely not for the first and the last three groups. In addition, it appears that stakeholders grouped by their geographical information do not prefer the first scenario, which stresses competitiveness as a future sustainable development. These stakeholder groups have a relatively strong interest in the continuity scenario.

According to the results of the regime analysis, the most-preferred sustainable future appears to be the coherence scenario (Scenario 4), in which ecological and social systems are attached more importance than other sustainability factors. While the third scenario, on capacity, follows the coherence scenario, here the most critical factors are economic and ecological factors. In addition, the scenario which focuses on the ecological and institutional systems ranks as the third one, while the competitiveness scenario, which does not pay much attention to ecological systems, is the least-preferred sustainable future image.

On the basis of our stakeholders' preferences, sustainable future development appears to find much support for a basic concern for 'the ecological system', which is next mainly followed by the social and economic systems. In other words, policies continuing or planning to focus on the ecological aspects of future sustainable environments are strongly supported.

15.7 Concluding Remarks

The previous foresight experiments on desirable future sustainability scenarios have demonstrated that a pluriformity perspective – with multiple stakeholders, multiple case studies, and multiple strategic policy factors – offers great potential for a systemic approach to ecological policy against the background of economic and social factors. The rather unambiguous preference for ecological quality – among different regions and among different stakeholders – is an interesting sign of societal consensus formation, provided that information on choices is given in a transparent and accessible manner.

Clearly, more solid research on such issues is still needed, in particular, a broad coverage of the relevant stakeholders. In addition, a broader set of empirical case studies also requires due attention in follow-up research. And finally, the underlying database – expressed in the five pentagon factors and underlying detailed case study data – might need more thorough attention.

References

Akgün, A., van Leeuwen, E., & Nijkamp, P. (2011). Analytical support tools for sustainable futures. *Romanian Journal of Regional Science, 5*(2), 1–24.

Capello, R., Nijkamp, P., & Pepping, G. (2000). *Sustainable cities and energy policies*. Berlin: Springer.

Carson, R. (1962). *Silent spring*. Boston: Houghton Mifflin.

Giaoutzi, M., & Nijkamp, P. (1993). *Decision support models for regional sustainable development*. Aldershot: Avebury.

Giaoutzi, M., Stratigea, A., van Leeuwen, E., & Nijkamp, P. (2012). Scenario analysis as a foresight tool in agriculture. *International Journal of Foresight and Innovation Policy, 8*(2/3), 105–128.

Harvey, D. (1969). *Explanation in geography*. London: Edward Arnold.

Hwang, A. (2000). Toward fostering systems learning in organizational contexts. *Systemic Practice and Action Research, 13*, 329–343.

Kahn, H., & Wiener, A. J. (1967). *The year 2000*. Stockholm: Beckmans.

Meadows, D. H., Meadows, D. L., Randers, J., & Behrens, W. H., III. (1972). *The limits to growth*. New York: Universe Books.

Nijkamp, P., & Perrels, A. (1994). *Sustainable cities in Europe*. London: Earthscan.

Nijkamp, P., Rietveld, P., & Voogd, H. (1990). *Multicriteria evaluation in physical planning*. Amsterdam: North-Holland.

Nijkamp, P., Rienstra, S. A., & Vleugel, J. M. (1998). *Transportation planning and the future*. New York: Wiley.

Pigou, A. (1930). *The economics of welfare*. London: Macmillan.

Rand, A. (1957). *Atlas shrugged*. New York: Random House.

Smith, A. (1759). *The theory of moral sentiments*. London: A. Miller.

Stewart, J., & Ayres, R. (2001). Systems theory and policy practice: An exploration. *Policy Science, 34*, 79–94.

Tinbergen, J. (1956). *Economic policy: Principles and design*. Amsterdam: North-Holland.

Toffler, A. (1970). *Future shock*. New York: Random House.

Van Loon, H. W. (1932). *Van Loon's geography, the story of the world we live in*. New York: Simon & Schuster. p. 3.

WCED. (1987). *Our common future*. Oxford: Oxford University Press.

Chapter 16
Methodological Challenges in Combining Quantitative and Qualitative Foresight Methods for Sustainable Energy Futures: The SEPIA Project

Erik Laes, Da Ruan, Fre Maes, and Aviel Verbruggen

16.1 Introduction

This chapter presents a reflection on the challenges of combining participatory fuzzy-set multi-criteria analysis (MCA) with narrative scenario building and energy modelling, in the context of the SEPIA project.[1] SEPIA aims to investigate participatory decision support systems for sustainable energy policymaking. More precisely, SEPIA elaborates on aspects of sustainability assessment (SA) in the energy policy context in order to reach consensus among the stakeholders involved. SEPIA provides the basis for an SA procedure adapted to the context of Belgian energy governance.

The authors wish to dedicate this chapter to the memory of our friend and colleague Da Ruan. We will remember Da as a motivated scientist with an indefatigable but also very congenial personality.

[1] The SEPIA project is being carried out by five partners: the University of Antwerp (UA, acting as the coordinator), the Free University of Brussels (VUB), the University of Liège (ULg), the Flemish Institute for Technological Research (VITO) and the Belgian Nuclear Research Centre (SCK•CEN). It is funded by the Belgian Office of Science Policy. Further details on this project can be found at the project's website www.ua.ac.be/sepia.

E. Laes (✉)
Unit 'Transition Energy and Environment' (TEM), VITO (Flemish Institute for Technological Research),
Boeretang 200, 2400 Mol, Belgium
e-mail: erik.laes@vito.be

F. Maes
Monitoring and climate policy cell, climate change service, Federal Public Service Public health, Safety of the food chain and Environment,
Victor Hortaplein, 40 bus 10, 1060 Brussels, Belgium
e-mail: fre.maes@health.fgov.be

A. Verbruggen
University of Antwerp (UA), Prinsstraat 13, 2000 Antwerpen, Belgium
e-mail: aviel.verbruggen@ua.ac.be

M. Giaoutzi and B. Sapio (eds.), *Recent Developments in Foresight Methodologies*, Complex Networks and Dynamic Systems 1, DOI 10.1007/978-1-4614-5215-7_16,
© Springer Science+Business Media New York 2013

This chapter addresses methodological challenges phased in SEPIA, as follows. First Sect. 16.2 presents the 'state of the art' in sustainability assessment, foresight methodologies and multi-criteria analysis. Sect. 16.3 discusses how these three domains were combined in the SEPIA project. The chapter ends in Sect. 16.4 with some preliminary conclusions and observations.

16.2 Methods for Strategic Decision Making on Sustainable Energy Development

16.2.1 Sustainability Assessment

The different approaches to integrated sustainability assessment can be illustrated when placed in the broader governance framework. Paredis et al. (2006) make a useful distinction between two ideal-typical governance 'styles', called, respectively, 'policy as calculus' and 'policy as discourse'. These 'styles' illustrate the two extremes of a spectrum of choices available to policymakers interested in setting up governance mechanisms for sustainability. They see sustainable development as a process of change engaging an entire network of actors, institutions, technical artefacts, etc. However, both perspectives differ in the way they approach the generation of strategic knowledge needed for steering this process of change towards a sustainable future. In brief, 'policy as calculus' represents a 'closed' process heavily predicated on expert input and agreement, whereas 'policy as discourse' 'opens up' to a wider range of actors, disciplines and concerns. Both perspectives are compared with a number of attributes in Table 16.1 and a SWOT analysis presented in Table 16.2 below.

'Policy as calculus' assumes that knowledge-based decision support – and the decision processes built on this support – can be conceptualised separately from its 'socio-technical object' (e.g. the energy system). For recommending how to steer socio-technical change in more sustainable directions, expert analysts should 'step outside' the system to objectify its workings.

Governance is characterised in terms of exogenous 'mechanistic' interventions. In all of this, an important role is attributed to 'expert input'. This does not exclude stakeholder involvement for providing 'inputs' to the assessment process.

But separate stakeholders are each assumed to hold a piece of the 'jigsaw puzzle' that experts collect and lay out to compose a picture of the 'socio-technical object'.

Stakeholders as such are nothing more than 'carriers' of policy alternatives, information and value judgements. It is assumed that all stakeholders observe 'the same' object, but each of them tends to focus on a limited set of aspects related to this object. Once the relevant pieces of the puzzle are collected (i.e. objectives are clearly defined and agreed upon, all necessary data are available, cause-effect

Table 16.1 Two different views on governance for sustainability (Based on Paredis et al. 2006; Smith and Stirling 2007)

	Policy as calculus	Policy as discourse
Role of sustainability assessment	Sustainability assessment as a tool for selecting the best alternatives in order to reduce negative sustainability impacts	Sustainability assessment as a framing process of deliberation on ends and means
What matters for political planning?	Uniform solutions based on technical and economic expertise	'Framings', deliberation and perspective-based testing of hypotheses involving a wide range of disciplines (including but not limited to economics and engineering)
Leading actors (networking)	Context dependent, with a focus on academics (with demonstrable expertise in the relevant scientific disciplines) and government actors	Context dependent, with a focus on experts (e.g. academics, professionals with experience in relevant fields), stakeholders (representative of the different 'problem framings') and government actors
Foresight methods	Mostly quantitative (i.e. modelling), explorative trend analysis (based on 'what if' reasoning) Government actors and/or stakeholders as 'clients'	Mostly qualitative (i.e. sociological) analysis (based on 'what is desirable' reasoning) with quantitative analysis as a support Government actors and/or stakeholders providing crucial inputs
Methods and tools (futuring, planning, networking)	'Standard' scientific methods, for example, mathematical models, cost-benefit analysis, cost-effectiveness analysis, checklists, matrices	Deliberative methods (e.g. scenario workshops, expert panels, focus groups) with 'standard' scientific methods as supportive
What is maximised?	Planning – that is, simple answers to complex problems and clear-cut recommendations about specific proposals	Networking – that is, interdisciplinary scientific knowledge, participation, deliberation, individual and societal learning effects
Procedurally effective if...	The optimal alternative has been identified Trade-offs are based on scientifically tested methodologies The proposal is of better quality (in the sense that negative impacts are avoided or mitigated) after the realisation of the assessment	Ideally, the deliberative process produces consensus by actually changing minds through reasoned argument A political community has been created around an issue Decision-making culture and practice have changed Sustainability assessment is iterative and fully integrated within the policy process, giving adequate and timely inputs to policy formation Transformative effect – acceptance of new goals and guiding principles for the energy transition

(continued)

Table 16.1 (continued)

	Policy as calculus	Policy as discourse
Procedurally efficient if…	A solution is found with minimum expenditure of available resources (time, money) and expertise (state-of-the-art knowledge) for the sustainability assessment	The sustainability assessment is carried out according to a clear and achievable timetable, giving enough time and resources for preparation of the process and stakeholder engagement
Procedurally fair if…	The recommended alternative(s) is justified by established expert authority, for example, accredited research institutes, peer review and lauded academics	No legitimate point of view is excluded a priori from the assessment Power differentials between social actors are neutralised

Source: Based on Paredis et al. (2006); Smith and Stirling (2007)

Table 16.2 SWOT of 'policy as calculus' and 'policy as discourse'

	Policy as calculus	Policy as discourse
Strengths	Practical instrument resulting in univocal recommendations from a 'narrow' framing perspective Part of the existing decision-making process in many countries	Sustainability raised as a collective concern Improved decision-making process
Opportunities	Political demand for this kind of exercises Use of existing knowledge and know-how Practical experience with similar exercises (environmental impact assessment, regulatory impact assessment)	Can build on existing participatory arrangements Scientific and political momentum in favour of sustainable development; acceleration of global change signals calls for ambitious action
Weaknesses	Attempt to include all aspects of sustainability in quantitative models faced with difficulties: unavailable data, uncertainties, etc. Environmental, governance and equity concerns are marginalised Acceptance of unlimited substitutability implies 'weak sustainability'	Representativeness of involved and missing stakeholders Potential to yield practical recommendations in due time Difficult to institutionalise Additional (and multidisciplinary) expertise, data, tools and time required compared with 'policy as calculus'
Threats	Technocracy and bureaucracy Reductionist perspectives are encouraged Risk of imbalance towards incremental approaches and consequent marginalisation of long-term sustainable development objectives	Lack of practical experience in conducting sustainability assessment exercises, leading to unrealistic expectations Manipulative interventions by some participants, eventually ending in demagogy Resistance against potentially transformative power of the sustainability assessment

relations are established), the 'solution' to the governance problem follows 'logically' from aggregating the different perspectives by using economic optimisation models, multi-attribute utility theory, etc.

The appraisal process 'closes down' on the single socio-technical object – that is, it is about '...finding the right questions, recruiting the appropriate actors (actors with "relevant" insights), highlighting the most likely outcomes and therefore also defining the best options...' (Smith and Stirling 2007: p. 6).

Once the appraisal procedure has aggregated all relevant information, the instruments for intervening in the dynamics of socio-technical objects follow mechanically (e.g. when economic evaluation finds nuclear power as the 'best option', policy instruments must clear the 'barriers' of a full nuclear deployment). Politically, this approach implies that 'relevant actors' bring their commitments in line with the recommendations from the appraisal. The alignment job is left to the political decision makers, in devising appropriate tools to persuade, entice or simply force actors to realise the path set out by 'the experts'.

'Policy as discourse' starts from the premise that there is no unique 'objectively rational' position from which a 'socio-technical object' (e.g. the energy system) can be observed. System boundaries, interrelations between system components, opinions on what causes change, etc., (in short: 'framings') vary according the actors' perspectives and may change during various stages of the appraisal. Because different 'framings' imply different methodologies for arriving at 'relevant' knowledge about the 'socio-technical object', input to the sustainability assessment cannot be 'imposed', but has to be negotiated. The same applies for the criteria guiding the sustainability assessment, which have to be checked for legitimacy and acceptance. Assessment does not identify the 'best possible' pathway for the evolution of the 'socio-technical object', but rather tests its evolution under the different 'framings' brought to the table by stakeholders. As a consequence, no unique set of ideal policy instruments can be identified; recommendations will always be much more 'conditional' (e.g. 'option x is the preferred option under framings a and b, but does not score well under framing c', 'option y scores rather well under all framings, and can therefore be considered as a robust option').

A word of caution is warranted here. The difference between 'policy as calculus' and 'policy as discourse' should not be conceived along the lines of a stark dichotomy between '...established, narrow, rigid, quantitative, opaque, exclusive, expert-based, analytic procedures tending to privilege economic considerations and incumbent interests...' and the '...new, relatively unconstrained, qualitative, sensitive, inclusive, transparent, deliberative, democratically legitimate, participatory processes promising greater emphasis on otherwise marginal issues and interests such as the environment, health, and fairness...' (Stirling 2008: p. 267). In other words, according to Stirling (2008), the detailed context and implementation of a particular governance approach are more important factors to understand what happens in practice. Instead of an illustration of the opposition between an 'expert-based' and a 'deliberative' governance approach, the difference between 'policy as calculus' and 'policy as discourse' should be seen as an illustration of how assessments and/or commitments can be 'closed down' (in the case of 'policy as calculus') or 'opened up' (in the case of 'policy as discourse') in an institutional environment which is structured and pervaded by power relationships. If appraisal is about 'closing down' the formation of commitments to policy instruments or technological options, then the aim of the assessment is to assist policymakers by providing a direct means to justify their choices. If, on the other hand, the assessment is aimed at 'opening up' a process of social choice, then the emphasis lies on revealing to the wider policy discourse any inherent indeterminacies, contingencies or capacities for action. Of course, expert-based analytic approaches such as cost-benefit or cost-effectiveness assessment are frequently practised as part of a 'policy as calculus' approach, but these techniques might equally lend themselves to an 'opening up' philosophy (Stagl 2009).

In order to define adequately the features SEPIA adopts, a thorough analysis of the existing energy policy context and the institutional landscape is necessary. In practice, the dominant approach in Belgium to decision support in energy policy has followed more or less the 'policy as calculus' philosophy. Therefore, we consider there is both in academic discussion and in policy practice some scope for a more symmetrical interest in processes for 'opening up' the debate on long-term sustainable energy strategies. SEPIA had to find an adequate balance between assessments of 'opening up' and 'closing down' and choose the appropriate methods accordingly. These methodological choices are explained further in Sect. 16.3.

16.2.2 Foresight Methodology

Sustainability assessment is necessarily predicated (to a greater or lesser extent) on 'foresight' abilities, that is, abilities of thinking, shaping or debating the future. In practice, foresight comes in many different shapes and forms (van Notten et al. 2003). A first distinction is between *predicting* and *exploring* the future. Earlier attempts at forecasting have proven to be largely unsuccessful (particularly in the case of long-term energy foresight) and are increasingly being abandoned by foresight practitioners – although expectations of correct prediction on the part of policymakers are still apparent. Next, there is the difference between *quantitative* (modelling) and qualitative (narrative) traditions, with the former prevailing in the field of energy. Hybrid approaches combine narrative scenario development with quantitative modelling. Also there are those futuring approaches distinguished as *descriptive* or *exploratory* which describe possible developments starting from what is known about current conditions and trends and from normative, *anticipatory* or backcasting approaches constructing scenario pathways to a desirable future.

Neither approach is 'value free', since both embody extra-scientific judgements, for example, about 'reasonable' assumptions. But the objectives of the scenario development exercise determine the choice between exploratory and anticipatory approaches. Exploratory (or 'what if') analysis articulates different plausible future outcomes and explores their consequences. By prioritising technological choices, technical and economic experts perform the analysis in a relatively closed process, with government actors mostly assuming the role of clients (they 'order' the analysis).

Anticipatory scenarios represent organised attempts to evaluate the feasibility and consequences of achieving certain desired outcomes or avoiding undesirable ones.

Finally, *trend* scenarios based on extrapolations of (perceived) dominant trends differ from *peripheral* scenarios focusing on unexpected developments and genuine 'surprising' events. Several choices on the most suitable foresight methodology need to be made. The SEPIA choices are elucidated later in Sect. 16.3.1.

16.2.3 Multi-criteria Decision Support

By accepting the view that energy systems are multidimensional in nature, it has also to be accepted that the evaluation of public plans or strategic decisions has to be based on procedures that explicitly require the integration of a broad set of (possibly conflicting) points of view. Consequently, multi-criteria (MC) evaluation is the appropriate decision framework to apply in principle (Kowalski et al. 2009).

A great variety of MC decision support tools exist and can be used in the context of sustainability assessment under both the 'policy as discourse' and the 'policy as calculus' philosophy. Therefore, the choice of a particular method must be guided by its fitness for the problem characteristics and the desired scope/features of analysis. Each analysis method is based on specific assumptions and supports only a certain type of analysis. A promising start for reflection is provided by Munda (2004) and Granat and Makowski (2006).

For complex decision-making problems such as deciding on long-term energy policy strategies, Munda has developed an MC decision support technique called 'social multi-criteria evaluation' (SMCE) and discusses its application to the problem of a wind farm location (Gamboa and Munda 2007). Granat and Makowski discuss the requirements of an MC decision analysis tool for an application very similar to ours – that is, a stakeholder evaluation of energy technologies and scenarios, albeit at the European level. According to these authors, MC decision support respecting the principles of the 'policy as discourse' philosophy has to show the following characteristics:

- The MC method is able to handle criterion scores of a different nature ('crisp' scores, stochastic scores, 'fuzzy' scores, etc.).
- In general, simplicity is a very desirable characteristic of the MC decision process – that is, the number of ad hoc parameters used should be limited (preferably only information on weights and scores should be used as exogenous inputs).
- Criterion weights should be seen as 'importance coefficients' (and not as numerical values allowing for full compensability between criteria or as indicators of a 'trade-off' between different criteria).
- Information on all possible rankings for each actor should be given (and not only on the 'optimal' one, since taking into account second-best or third-best options can reveal a space for compromise solutions compared with other actors' rankings).

- The MC appraisal should include a 'conflict analysis' (i.e. an analysis of the 'distance' between the different actor perspectives, revealing possible groupings into major 'world views'). As win-win situations are not always achievable, some trade-offs will have to be made. These trade-offs will then appear in the discussions on values stimulated by the use of the MC appraisal and will give normative input to the consequences of selecting one alternative over another. Mathematical models can then be of assistance in the selection of the most consensual alternative, the regrouping of alternatives according to the results of the conflict analysis, etc.

Section 16.3 below gives further details on the particular approach adopted in the SEPIA project.

16.3 Towards a Sustainable Energy Policy Integrated Assessment in Belgium

16.3.1 Foresight Methodology

Following SEPIA's 'opening up' logic, the foresight methodology explicitly acknowledges the possibility of different 'framings' of the energy system (the 'socio-technical object' under consideration) and of the factors that cause long-term changes in this system. Narrative scenario building is particularly well suited for 'opening up' the system description to, and for the exploration of, fundamental complexities and uncertainties (Bunn and Salo 1993). The construction of scenarios for exploring alternative future developments under a set of assumed 'driving forces' has a long tradition in strategic decision making, especially in the context of energy policy (Kowalski et al. 2009). Exploratory scenario building is criticised for its propensity to limit the space of the possible to only a few probable 'storylines' (Granger Morgan and Keith 2008). The backcasting approach is more suited for long-term and complex problems – such as sustainable development – requiring solutions which shift society away from business-as-usual trends. Backcasting is, however, often criticised for defining utopian futures with little value for decision makers in the 'real world'.

For combining the strengths of explorative and (traditional) backcasting methodologies, SEPIA developed a 'hybrid backcasting' approach. Following a hybrid backcasting approach, scenario building takes place starting from a systematic exploration of futures, by studying many combinations resulting from the breakdown of the energy system. The process of 'breaking down' the system implies the definition of a set of factors, which could each influence the development of the energy system in different directions. These possible developments are formulated as 'hypotheses' or 'possible configurations'. The total number of combinations represents a 'morphological space', which must then be reduced to a number of coherent sets by formulating transition conditions ('exclusions' and 'compromises') congruent with reaching sustainability visions defined by the stakeholder panel

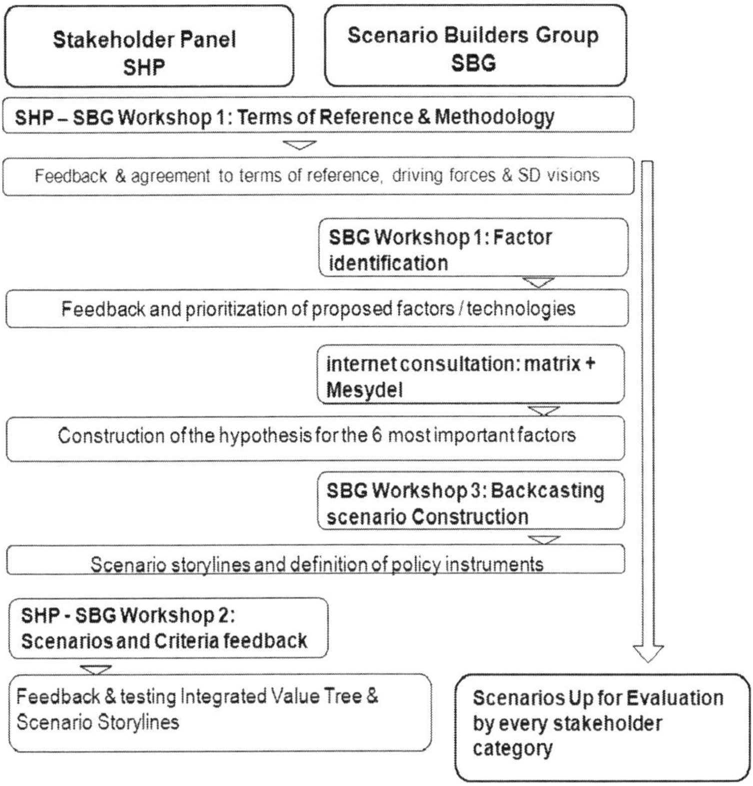

Fig. 16.1 Scenario-building steps in the context of the SEPIA project

(SHP). For this process, we proceeded in a number of separate steps (see Fig. 16.1). These steps are explained in Sects. 16.3.1.1, 16.3.1.2, 16.3.1.3, 16.3.1.4, 16.3.1.5, and 16.3.1.6. The scenario-building phase relied on qualitative in-depth deliberative workshops with the scenario builders group (SBG), where the SEPIA team acted as 'scientific secretariat', delivering input materials for the workshops (e.g. information sheets) and processing the outcomes. Scenarios were reviewed by the SHP.

Social mapping was used for composing the SBG and SHP groups, meeting the following criteria:

- *Scenario builders group (SBG)*: The SBG is responsible for developing the long-term energy scenarios describing the different possible visions of a sustainable energy future (horizon 2050) and the pathways (including policy instruments) needed to realise those visions. We expected each participant to contribute their expertise and personal experience to the discussions. The scenario builders were asked to participate in their personal capacity and not as a representative of the organisation in which they are active. The participants were generally willing to engage in an open, creative, non-judgemental foresight process. Members of the SBG are contacted by the SEPIA team and submitted for approval to the steering committee.

- *Stakeholder panel (SHP)*: The SHP is mainly responsible for evaluating the long-term energy scenarios developed by the SBG though they will also be given an important role in setting the general directions for these scenarios and providing feedback on scenario assumptions before the LEAP modelling will take place. This group aims to be representative of the 'stakes' in the Belgian energy sector. Therefore, it was important to ensure that all the potential social groups with a current or potential interest in the problem had the possibility of being included in the process. When deciding on the composition of groups taking part in participative processes, inclusiveness refers to ideas of *representativeness*, although *not in a statistical sense*. Rather, participants should be selected to represent constituencies that are known to have *diverse and, especially, opposing interests*. No stakeholder group should be composed of a preponderance of representatives who are known to have a similar position or who have already formed an alliance for common purpose. In the case of experts – who are presumed not to have constituencies but ideas – they should be chosen to represent whatever *differing theories or paradigms* may exist with regard to a particular task.

16.3.1.1 SHP-SBG-W1: Terms of Reference and Methodology

It is clear that, before starting to formulate sustainable energy strategies, policy makers and/or relevant stakeholder groups will already have some general ideas about the possible alternative solutions. Before entering the multi-criteria assessment phase (in which a decision about the significance of the possible impacts of the alternatives in terms of furthering the sustainable development agenda has to be made), these general ideas will already have to be worked out to a greater level of detail.

It is only as a result of the detailed 'scoping' of the sustainability assessment that the decision alternatives will take on their definitive shape – that is, the 'scoping' provides the necessary consensual ground rules for deciding what counts as a 'reasonable' alternative, the range of alternatives to be taken into account, the level of detail needed to explore each alternative, etc. Scoping is therefore an essential part of the sustainability assessment and should form the basis of a negotiated 'contract' between the project team, stakeholders, experts and steering committee involved in the project. This 'contract' is called the 'terms of reference' (TOR). The SEPIA terms of reference were thoroughly discussed in a full-day workshop.[2] Since the (hybrid) backcasting approach adopted in the project essentially relies on normative inputs for the development of desirable end points, the first workshop was for a large part devoted to finding a consensus on sustainability principles. An *integrated value tree* was developed which discusses the sustainability goals specific to the development of energy systems in more detail. A value tree identifies and organises the values of an individual or group with respect to possible decision options. It

[2] The final version of the SEPIA TOR can be downloaded from the project website (www.ua.ac.be/sepia).

structures values, criteria and corresponding attributes in a hierarchy, with general values and concerns at the top and specific attributes at the bottom. The integrated value tree integrates *fundamental sustainable development (SD) objectives* (to be reached in 2050), *SD (sub-) dimensions* (a further specification of the objectives) and *SD indicators* (quantitative or qualitative scores indicating how well a particular scenario contributes to reaching the objectives).

16.3.1.2 SBG-W1: Factor Identification

For the first SBG workshop, the SEPIA project team developed brief explanations and 'fact sheets' for about 50 major factors (trends, tendencies)/technological developments which were expected to have an impact on long-term Belgian energy system development. A 'factor' was defined as anything that could influence energy system development in the long run. This workshop was meant to explore the possible factors of change without making an opinion on the desirability of certain evolutions. Only in the later process steps were possible factor evolutions connected with desirable visions on the long-term energy future. During the workshop comments, suggestions and remarks on the current state, predictability, possible states (hypotheses) and the time horizon of change (slow evolution vs. sudden change) of different factors were elicited.

The afternoon session of the workshop continued with the identification and selection of about 20 most important factors rated according to their impact on reaching sustainable development objectives in 2050.

The output of the individual point allocation (green and red dot stickers), as well as the bailout points (blue dot stickers), had as a result the definition of the guiding factors for the SEPIA exercise. The participants agreed on selecting 22 factors instead of 20 as to avoid wasting valuable time in discussions. The final list of 22 factors was accepted after the question 'Do we all agree on this?' (see Table 16.3).

16.3.1.3 Internet Consultation: Matrix Exercise

The list of 22 factors with a likely influence on energy system development was consequently submitted to the SBG in an Internet consultation in order to perform a cross-impact analysis of interdependencies between factors. The cross-impact analysis was performed by asking the members of the SBG to fill in a 22×22 matrix with the 22 factors represented in the rows and columns of the matrix. Each cell of the matrix represented the impact of the factor in the row on the evolution of the factor in the column (score between 0 and 3:0 = no impact; 3 = high influence). By adding together the scores of all members of the SBG, factors could be classified into the following groups:

- *Determinants*: Factors with a high influence on the development of other factors, without being influenced much in return. In other words, these factors act as 'motors' or 'restraints' for the development of energy systems.

Table 16.3 List of 22 factors selected during SBG-W1

T8 Advances in energy storage technologies
P2 EU internal energy market policy
T1 Competitiveness of energy conservation technologies for stationary end uses
Ex3 Structural changes to the Belgian economy in a globalised environment
Ex13 Location
P1 EU energy vulnerability strategy
P3 EU energy RD&D strategy
P4 Price instruments to internalise externalities
T13 The 'hydrogen economy'
T6 Advances in renewable energy technologies
T14 The 'electric economy'
Ex 11 Ecological and health constraints
T10 ICT technology innovations
B5 Active public involvement in environmental issues
Ex 12 Market environment
Ex 9 Energy price dynamics
P9 Land-use policies
B6 Risk perception and evaluation
B8 Shifts in demands for housing and living space/comfort
P8 Stranded assets and lock-in
P7 Importance of social policy
T2 Energy efficiency of various transport modes: technological progress

Note: RD&D = research development and deployment

- *Strategic variables*: Factors with both a high influence and dependence on other factors. These factors are likely candidates for the development of broad strategic actions plans, provided they can be 'steered' by political interventions.
- *Regulatory variables*: Factors with both a mid- to low influence and dependence on other factors. These factors can be taken into consideration when designing specific policy instruments, provided they can be 'steered' by political interventions.
- *Dependent variables*: Factors which are highly dependent on the evolution of other factors. These factors can be likely candidates for monitoring efforts.
- *Autonomous variables*: Factors which evolve largely independently of other factors.

Based on this matrix exercise, six factors were selected (three determinants and three strategic variables) that would serve as the 'backbone' for the scenario storylines (developed in SBG-W3):

- Ecological and health constraints
- Energy price dynamics
- Market environment
- Use of price instruments to internalise externalities
- EU energy research development and deployment (RD&D) strategy
- EU energy vulnerability strategy

16.3.1.4 Internet Consultation: Mesydel

At the start of the second phase of the Internet consultation, the project team developed two to three hypotheses with regard to the long-term evolution for each of the six most influential factors. These hypotheses were submitted to deliberative feedback by members of the SBG with the aid of the 'Mesydel' tool.[3] With Mesydel, questions are encoded on a central computer and access to the software is given to each expert. At any time, they could come back to the software and amend or augment their answers. The mediator, for his part, has access to a series of answers classification tools and the ability to mark the answer's relevance, to note if he will or will not work later on the question, to comment on the answers (these comments are for his exclusive use) and – the most interesting feature – to give 'tags' (keywords) to answers. These tags could then be classified according to topics selected by the mediator. These classification tools allow the mediator to have huge flexibility in his work and help to optimise his results by allowing him to find all relevant messages on a given topic very quickly. The 'Mesydel' round thus resulted in amended versions of the hypotheses developed for each of the factors.

16.3.1.5 SBG-W3: Construction of Scenario 'Skeletons' as Combinations of 'Favourable' Factor Projections for Different Sustainability World Views

Starting from the processed results of the Internet consultation (priority factors and a short description of possible alternative hypotheses for their evolution), the members of the SBG developed three scenario 'skeletons' composed of factor hypotheses and technological developments congruent with the logic of reaching the fundamental sustainability objectives. This can be done by a formal consistency check; however – in view of the highly resource-intensive mathematical character of this procedure (and the need for supporting software) – we chose a more *intuitive* manner of proceeding. Starting from a certain factor, a hypothesis was selected and then connected to other hypotheses (for the other factors) that were deemed to be consistent with the initial hypothesis. This combination of hypotheses could then be regarded as an alternative 'solution' to the problem of moving towards the attainment of the sustainability objectives in 2050. These combinations were then taken as a basis for the construction of a scenario, and the procedure was repeated until the SBG felt that they had covered the range of possibilities with their scenarios.

For each of the scenario skeletons (which both enable and constrain certain developments), the SBG group had to explore in which other factors (taken from the original list resulting from SBG-W1; see Sect. 16.3.1.1) – that is, technologies, behavioural changes, broad policy choices, etc., – 'critical' changes had to be achieved (compared with now) in order to achieve a certain vision of a Belgian energy system in 2050 which is supportive of the sustainability objectives. They

---

[3] For more information, see http://www.mesydel.com/mesydel.php.

also had to indicate an approximate timing of the changes needed in the 'critical' factors. Finally, in order to complete the pathways, the SBG group had to *backcast* the necessary policy interventions needed on the Belgian level for reaching the sustainability objectives, given a certain combination of a vision and pathway elements as the policy context. The backcast had to give an answer to the question: '*What is needed at the Belgian* (i.e. *federal and regional) level in order to realize the changes in the factors within the time frame indicated by a particular pathway*?' Although the workshop discussions led to many interesting suggestions, we did not succeed in constructing pathways in sufficient detail to serve as an input to the LEAP energy system model. A detailed backcast also proved to be too demanding a task, mainly due to the rather low attendance. A lot of decisions still had to be made. As a consequence, the project team decided to change the format of the final workshop to some extent, dedicating it also to the further construction of scenarios storylines.

16.3.1.6 SHP-SBG Workshop 2: Deliberative Feedback on Scenario Storylines and Proposed Value Tree Before Evaluation

The last workshop, which combined inputs from the SHP and SBG, served a dual purpose: deliberation and feedback on a draft value tree as proposed by the project team (with 'fact sheets' unequivocally explaining each indicator, potential data sources and possible measurements (e.g. quantitative/qualitative), taking into account uncertainties) and feedback and further development of the 'scenario skeletons' developed by the SBG in the previous workshops. The value tree was modified according to the feedback received.[4] Deliberative feedback on the scenario skeletons resulted in more detailed specifications on the scenarios to serve as an input into the LEAP modelling exercise; however, a lot of 'room for interpretation' was still left for the project team. At the time of writing this chapter, the SEPIA scenarios were still under development. Therefore, for the time being, we can only give a qualitative description of the three scenario storylines serving as an input for further modelling.

A first storyline called 'global consensus' starts from the assumption that climate change concerns dominate energy system development, in the sense that early and drastic emission cuts are called for (e.g. an EU target of −30% in 2020 compared with 1990). Energy RD&D spending on the EU level is increased substantially and is geared towards realising a common European vision – a low-carbon energy system with maximum penetration of renewable and distributed energy sources. RD&D focuses on technological 'breakthroughs' for the achievement of the common energy system vision (e.g. advances in ICT, large offshore wind parks, smart grids, energy storage technologies, nanotechnology). Those solutions mostly require big investments in new supply technology and/or new infrastructures (cf. the 'SuperSmart Grid').[5] Technologies that are labelled as 'risky' encounter strong public and politi-

[4] The final version can be downloaded from the project website (www.ua.ac.be/sepia).

[5] More information on the 'SuperSmart Grid' concept can be downloaded from www.supersmartgrid.net.

cal opposition. A combination of low public acceptance and unresolved waste, safety and proliferation issues leads to a rejection of the nuclear option: without public backing, investments in new nuclear power plants simply become too risky for private investors. Existing plants are shut down as they reach the end of their projected lifetime, and lifetime extensions are not considered. Public support for carbon capture and storage (CCS) is also reluctant. By 2050, energy supply is largely based on renewable energy sources.

In the 'oil shock(s)' storyline, the oil (and possibly also the gas) market goes through a series of crises in the period 2010–2030, caused by physical (peak production or refinery capacities are surpassed) or political factors (e.g. crisis in the Middle East), resulting in sudden and unpredictable price increments. Leading powers try to control the remaining resources by engaging in strategic alliances, as energy policy is to a large extent dictated by foreign policy and security considerations. Energy security is the main concern over the short to midterm, leading to a focus on energy efficiency (on the demand side) and on available technologies that alleviate the dependence on imported oil and gas (on the supply side): renewables (mainly wind energy and biofuels), coal (later equipped with carbon capture and storage) and prolonging the lifetime of existing nuclear power plants. Thanks to these measures, energy security concerns are alleviated over the period 2030–2050, allowing the climate change agenda to take over as a priority issue.

Finally, the 'confidence in RD&D' storyline stands for a scenario where a combination of high oil (and gas) prices, climate policy and competitive energy markets decisively influences the pace of transition to a low-carbon energy future in the OECD countries. In the EU, the Lisbon agenda (and possible successors) carries high priority. The EU protects and expands its previous economic achievements, including the internal energy markets. However, governments are still heavily involved in securing their external energy supplies (this goes for 'government', as well as on the EU and on the national level in Europe), albeit in a more subtle and indirect way than in the 'oil shock(s)' scenario. In general, market forces determine the investments choices made by the energy industry between renewables, 'clean coal' and nuclear power, but public and/or political perceptions sometimes lead to targeted interventions. The use of the nuclear option is especially closely associated to national preferences. Independently from the developments in the fields of nuclear energy, Europe is on its way to a smooth and accelerated transition towards renewable energy. The process is quite similar to the one described in 'global consensus', although the share of renewable energy sources is smaller. Large offshore wind farms are the most important renewable source for electricity production, and biomass is playing a major role in heating or cogeneration. In addition, because of the higher demand, highly efficient gas- and coal-fired power plants with CCS are needed in this scenario. Decentralised power generation is a growing trend in the coming 50 years. The increase in energy efficiency is also determined by market forces as new energy end-use technologies emerge in electricity use, space heating, 'smart' decentralised energy systems and transportation.

16.3.2 *Fuzzy-Set Multi-criteria Decision Support*

As mentioned in Sects. 16.1 and 16.2, the scenarios developed for the SEPIA project have not yet been evaluated with the aid of the multi-criteria decision support tool at the time of writing this chapter. To clarify the motivation for the use of fuzzy-set multi-criteria analysis, we briefly illustrate here the features of the fuzzy-logic multi-criteria group decision support tool called DECIDER used in SEPIA (Ruan et al. 2010).

16.3.2.1 Application of Fuzzy-Logic to Multi-criteria Analysis

Multi-criteria analysis (MCA) with linguistic variables, commonly known as fuzzy-set multi-criteria decision support, has been one of the fastest growing areas in decision making and operations research during the last three decades. The motivation for such a development is the large number of criteria that decision makers are expected to incorporate in their actions and the difficulty in practice of expressing decision makers' opinions by means of precise values. Group decision making takes into account how people work together in reaching a decision. Uncertain factors often appear in a group decision process: namely, with regard to decision makers' roles (weights), preferences (scores) for alternatives (scenarios) and judgements (weights) for criteria (indicators) (Lu et al. 2007). Moreover, MCA aims at supporting decision makers who are faced with making numerous and conflicting evaluations. It highlights these conflicts and derives a way to come to a compromise or to illustrate irreducible value conflicts in a transparent process. Firstly, as decision-aiding tools, such methods do not replace decision makers with a pure mathematical model, but support them to construct their solution by describing and evaluating their options. Secondly, instead of using a unique criterion capturing all aspects of the problem, in the multi-criteria decision-aid methods, one seeks to build multiple criteria, representing several points of view. In particular, fuzzy-set multi-criteria decision support respects the principles of the 'policy as discourse' approach as set out in Sect. 16.2.1. This will be illustrated next in Sect. 16.3.2.2 by illustrating one feature of the DECIDER software – that is, the possibility of opinion clustering.

16.3.2.2 Clustering of Opinions

Suppose we have 10 people (P1–P10) who have scored scenarios on the different (sub-) criteria and have given weights to these (sub-) criteria (the example here is taken from an earlier application of the DECIDER model). Mathematical functions allow us to calculate the 'distance' between the revealed preferences of the different people. This is represented graphically in Fig. 16.2, which gives us the following information:

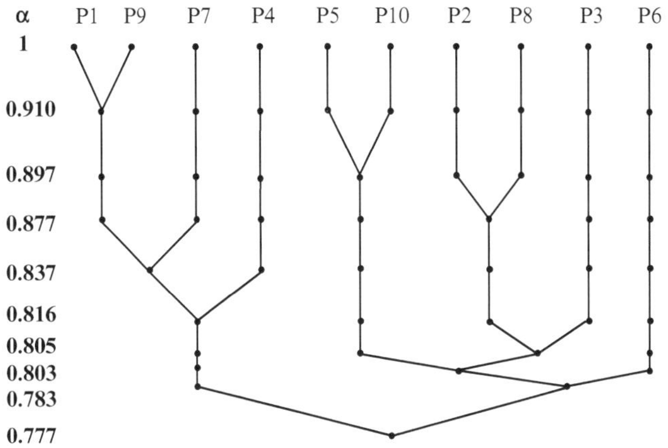

Fig. 16.2 Dendrogram of the cluster formation process

- The opinions of P1 and P9 are closest to each other (as expressed by the value of the parameter **α**). Therefore, they are the most likely candidates for a 'coalition'. Therefore, if we want to simplify the decision process and work with just 9 opinions instead of the original 10, P1 and P9 are the most likely candidates to be taken together without major conflicts (i.e. represented by an 'average opinion').
- The opinions of P5 and P10 are the second closest to each other. Therefore, if we want to simplify the decision process and work with just 8 opinions instead of the original 10, P5 and P10 can probably be taken together without major conflicts (i.e. represented by an 'average opinion'), next to P1 and P9.
- The list below indicates which 'coalitions' have to be considered if we want to represent just one opinion (i.e. the average for the whole group) concerning the ranking of scenarios, 10 different opinions (the original individual scores and weights) or any possible number of 'coalitions' between 1 and 10.

1. {P1, P2, P3, P4, P5, P6, P7, P8, P9, P10}
2. {P1, P4, P7, P9}, {P2, P3, P5, P6, P8, P10}
3. {P1, P4, P7, P9}, {P2, P3, P5, P8, P10}, {P6}
4. {P1, P4, P7, P9}, {P2, P3, P8}, {P5, P10}, {P6}
5. {P1, P4, P7, P9}, {P2, P8}, {P3}, {P5, P10}, {P6}
6. {P1, P7, P9}, {P2, P8}, {P3}, {P4}, {P5, P10}, {P6}
7. {P1, P9}, {P2, P8}, {P3}, {P4}, {P5, P10}, {P6}, {P7}
8. {P1, P9}, {P2}, {P3}, {P4}, {P5, P10}, {P6}, {P7}, {P8}
9. {P1, P9}, {P2}, {P3}, {P4}, {P5}, {P6}, {P7}, {P8}, {P10}
10. {P1}, {P2}, {P3}, {P4}, {P5}, {P6}, {P7}, {P8}, {P9}, {P10}

This 'clustering process' can be an important tool for policymakers. Instead of just relying on the average result for the whole group (which hides important value conflicts) or individual opinions (which give no information on a collectively preferred scenario), clustering can be used to investigate different possible rankings of scenarios based on different decision principles, such as:

- What happens if we give different weights to the different individuals or coalitions (i.e. policymakers might attach more importance to the opinion of some people over others)?
- What happens if we respect the majority principle?
- What happens if we give veto power to minority opinions (e.g. they can veto the scenario they prefer least)?
- Which scenarios provoke the strongest conflicts of opinion?

16.3.2.3 Further Development of the DECIDER Model in the Context of SEPIA

Owing to the potential difficulties of evaluating the quantitative and qualitative information (or data) obtained by different experts, the MCA in the above-mentioned DECIDER tool for decision support was further developed to suit the purpose of the SEPIA project.

Such quantitative and qualitative information (or data) by experts is of a very different nature; it may be heuristic or incomplete or data that is either of unknown origin, or may be out of date or imprecise, or not fully reliable, or conflicting, and even irrelevant. In order to allow an adequate interpretation of the information from the experts' evaluation and to reach a conclusion, there was a need to update the DECIDER tool so that it is able to deal with various uncertainties that result in various data formats in practice.

It was considered advantageous to have a sound and reliable mathematical framework available that provides a basis for synthesis across multidimensional information of varying quality, especially to deal with information that is not quantifiable due to its nature and that is too complex and ill defined, for which the traditional quantitative approach (e.g. the statistical approach) does not give an adequate answer.

Within the SEPIA project, DECIDER was further developed to deal with the following issues:

I. Information (data) presentation with different formats

Type A. *Numerical value* – It is the most common way of indicating information scale. Any information α takes values in a [0, C] interval, where 0 is the lowest and predetermined C value is the highest level of possible judgements. C = 1 and C = 100 cases are the most frequently used ones.

Type B. *Interval value* – Any interval of [0, C] may give sufficient information.

Type C. *Linguistic value* – It is sometimes more appropriate to indicate information with linguistic terms (fuzzy sets) instead of numerical values. In type C, α takes values from a predetermined linguistic terms set. Let S = {Si}, i = {0,…, m} be a finite and totally ordered term set. Any label, si, represents a possible value for a linguistic variable. The semantics of the finite term set S is given by fuzzy numbers defined in the [0, 1] interval, which are described by their membership functions. For instance, S = {Si}, i = {0,..., 6}, in which the following meanings are assigned to the terms: S0: none, S1: very low, S2: low, S3: medium, S4: high, S5: very high and S6: excellent.

Type D. *2-tuple (continuous linguistic value)* – When it is hard to make information with discrete linguistic terms, then one can indicate some information between S2 and S3 below.

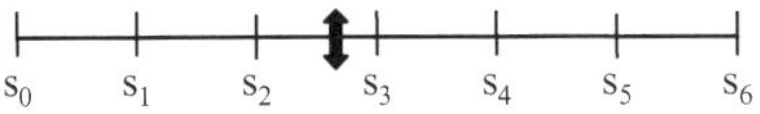

Type E. *Distribution over linguistic values*

A belief structure could be used, for instance, to represent general belief of the information with a given situation, such that to evaluate a performance of, say, scenarios vs. criteria, an expert may state that he is 20% sure it (the relationship between scenario x and criterion y) is S1, 50% sure it is S2 and 30% sure it is S3. In this statement, S1, S2 and S3 are linguistic evaluation grades, and percentage values of 20%, 50% and 30% are referred to as the belief degrees, which indicate the extents to which the corresponding grades are assessed.

II. Information aggregation with various certain and uncertain theories

After having obtained all formats of information, one can transfer all information from the types A, B, C and D to the type E. Thus all well-known theories such as set theory, probability theory, possibility theory, fuzzy-set theory and evidence theory can be selected and applied depending on the nature of uncertainty of the information. Different aggregation techniques can be also applied for different needs of the decision analysis support.

III. Final decision support scenarios

By using the type E-based approach in (I), one can deal with efficient uncertain information, especially when missing information appears during the decision analysis within the project. Typically, missing information could be as follows: (a) experts do not know/understand the information; (b) experts do not have any information; and (c) experts think the information is irrelevant. Most traditional approaches would have some difficulty in dealing with such missing information.

16.4 Concluding Observations

Sustainability assessment of energy policy strategies is performed at the interface between scientific theory building and political practice. Therefore, any practical implementation of a sustainability assessment will be judged by criteria related to scientific soundness, political legitimacy as well as practicability (in a real political setting). In this chapter, we have offered a reflection on how such criteria can be met, based on experiences gained in the SEPIA project. Indeed, the SEPIA project is predicated on the presumption that the issue of deciding on an appropriate (i.e. sustainable) long-term energy strategy is at least a suitable 'test case' for a more deliberative (discursive) governance arrangement, *ergo* that it is not a priori better handled by alternatives such as (a combination) of free market competition, lobbying and/or direct government regulation (top-down 'government' as opposed to bottom-up 'governance'). Further in-built presuppositions include that some particular composition of actors is thought to be capable of making decisions according to (voluntarily accepted and consensually deliberated) rules that will resolve conflicts to the maximum extent possible and (ideally) provide the resources necessary for dealing with the issue concerned and, moreover, that – the next presupposition – these decisions once implemented will be accepted as legitimate by those who did not participate and who have suffered or enjoyed their consequences. Apart from these considerations, one needs to keep in mind that unlike normal science, foresight knowledge is non-verifiable in nature, since it does not give a representation of an empirical reality. All in all, substantiating the quality of the SEPIA approach is certainly challenging, in theory as well as in practice, as demonstrated by the following observations.

Essentially, the SEPIA methodology is in line with a large body of theory building in the field of ecological economics, decision analysis, and science and technology studies, which all argue in favour of combining analytical and participatory research methods in the field of 'science for sustainability'. This view is motivated by the fact that sustainability problems are multidimensional (thus limiting the use of strictly monetary cost-benefit analysis) and of a long-term nature (thus involving significant uncertainties) and apply to complex socio-economic and biophysical systems (thus limiting the use of mono-disciplinary approaches). In principle, the advantages of combining a (hybrid backcasting) scenario approach with a (fuzzy-logic) multi-criteria decision-aiding tool are clear. Scenario exploration allows the (socio-economic and biophysical) complexities of energy system development to be taken into account so that uncertainties in the long term can be explored. Multi-criteria methods, and especially those based on fuzzy-set theory, are very useful in their ability to address problems that are characterised by conflicting assessments and have to deal with imprecise information, uncertainty and incommensurable values. Both methods are supported by a large body of scientific literature, ensuring that an effective check of 'scientific soundness' can be made through the peer review process. However, the application of these methods, and especially their participatory nature, poses significant challenges in practice. For instance, the combination of narrative scenario building and quantitative modelling in theory necessitates the need for a deliberative consensus on all parameters used in the model, which in practice turns out to be impossible to organise (the LEAP model requires hundreds of inputs). In any case, the scenario development phase had already turned out to be time intensive for stakeholder participants. We struggled with nonparticipation and dropouts of stakeholders; without proper investigation, for the time being, we cannot explain why participation fluctuated as it did. However, at least part of the explanation can probably be found in the general impression that the potential players in the Belgian energy system transition landscape – whatever their number may be – are rather scattered. In Belgium (as in many other countries), energy problems cross a varied set of policy domains and agendas, such as ensuring the correct functioning of liberalised energy markets, promoting renewables, environmental protection and climate policy. These are dealt with by different bureaucratic 'silos' and analysed by separate groups of experts and policymakers. As a result of this fragmentation, a lot of the key players struggle with overloaded agendas, organisation-specific expectations and performance criteria and hence find no time for explicit reflective/exchange moments in the context of a scientific project not directly connected to any actual decision-making process. There may be many contacts when events occur and by means of communication, but there is not a structured exchange of experiences, knowledge and mutual feedback ('structured' in the sense of embedded in a culture of working methods). This impression of fragmentation sharply contrasts with the high priority assigned to institutionalised networks and collaboration in the context of 'transition management'. Perhaps the best way to sum up the findings so far is as follows: assessing scenarios in the form of transition pathways towards a sustainable energy future with the aid of a participatory fuzzy-logic multi-criteria decision-aiding tool certainly has the potential to support a more robust and democratic decision-making process, which is able to address socio-technical complexities, and acknowledges multiple legitimate perspectives. However, these methods are time and resource intensive and require the support of adequate institutional settings for a proper functioning in real political settings. Participation in integrated energy policy assessment should therefore not be taken for granted. We hope that the experience gained so far in the context of the SEPIA project will allow future initiators of similar participatory projects to coordinate the project objectives, the participants' expectations and the political backing with each other, a prerequisite for successful participation in foresight exercises.

References

Bunn, D., & Salo, A. (1993). Forecasting with scenarios. *European Journal of Operational Research, 68*(3), 291–303.

Gamboa, G., & Munda, G. (2007). The problem of windfarm location: A social multi-criteria evaluation framework. *Energy Policy, 35*(3), 1564–1583.

Granat, J., & Makowski, M. (2006). Multicriteria methodology for the NEEDS project, NEEDS project deliverable T9.2 – RS2b. International Institute for Applied System Analysis (IIASA), Laxenburg.

Granger Morgan, M., & Keith, D. (2008). Improving the way we think about projecting future energy use and emissions of carbon dioxide. *Climatic Change, 90*(3), 189–215.

Kowalski, K., Stagl, S., Madlener, R., & Omann, I. (2009). Sustainable energy futures: Methodological challenges in combining scenarios and participatory multi-criteria analysis. *European Journal of Operational Research, 197*(3), 1063–1074.

Lu, J., Zhang, G., Ruan, D., & Wu, F. (2007). *Multi-objective group decision making – Methods, software and applications with fuzzy set technology*. London: Imperial College Press.

Munda, G. (2004). Social multi-criteria evaluation: Methodological foundations and operational consequences. *European Journal of Operational Research, 158*(3), 662–667.

Paredis, E., Heyerick, A., Doom, R., Bauler, T., Zaccaï, E., Waktare, M., Bonifazi, A., Boulanger, P.-M., Lussis, B., Varone, F., Lavrysen, L., Thomaes, P., & Risse, N. (2006). *Methodology and feasibility of sustainability impact assessment. Case: Federal policy making processes* (Final Report of a BELSPO Project). Belgian Science Policy, Brussels.

Ruan, D., Lu, J., Laes, E., Zhang, G., Ma, J., & Meskens, G. (2010). Multi-criteria group decision support with linguistic variables in long-term scenarios for Belgian energy policy. *Journal of Universal Computer Sciences, 15*(1), 103–120.

Smith, A., & Stirling, A. (2007). Moving inside or outside? Objectivation and reflexivity in the governance of socio-technical systems. *Journal of Environmental Policy and Planning, 8*(3–4), 1–23.

Stagl, S. (2009). *Value articulating institutions and changing social preferences* (REFGOV Working Paper Series GPS-11). Centre for Philosophy of Law, Université Catholique de Louvain, Louvain-La-Neuve.

Stirling, A. (2008). Opening up' and 'closing down. Power, participation, and pluralism in the social appraisal of technology. *Science, Technology & Human Values, 33*(2), 262–294.

van Notten, P., Rotmans, J., van Asselt, M., & Rothman, D. (2003). An updated scenario typology. *Futures, 35*, 423–443.

Chapter 17
Building Strategic Policy Scenarios for EU Agriculture: AG2020

Maria Giaoutzi and Anastasia Stratigea

17.1 Introduction

The agricultural sector is one of the most important production sectors of the global economy, as it largely determines not only the population's survival and quality of life but also the development potential of a significant part of the European territory – the rural regions. Agriculture is largely associated with the economic prosperity, tradition, production systems, culture, etc., of European farmers. Given the multifunctional role of the agricultural sector, which largely affects the environmental, social and economic development of rural regions, it has become an imperative for future policies in agriculture to focus on sustainability targets, incorporating at the same time the quality–safety dimension in agricultural production. To reach the goal of sustainable agricultural development, a process of exploring the future is required, determining trends, key drivers and uncertainties, which may form the basis for strategic decisions in the field.

The *focus* of this chapter is on the presentation of the strategic policy scenarios approach developed in the AG2020[1] project for EU agriculture, with emphasis on the description of the policy framework. Section 17.2 presents the building process of the AG2020 strategic policy scenarios framework for sustainable agriculture in Europe, and Sect. 17.3 presents the policy framework developed in the AG2020 context. Finally, Sect. 17.4 draws certain conclusions relating to policy and methodological aspects.

[1] AG2020 Project: Foresight Analysis for World Agricultural Markets (2020) and Europe, Contract No.: 44280-AG2020, STREP, January 2007–December 2009.

M. Giaoutzi (✉) • A. Stratigea
Department of Geography and Regional Planning,
National Technical University of Athens (NTUA), Heroon Polytechniou
Str. 9, Zographou Campus, Athens, 157 80, Greece
e-mail: giaoutsi@central.ntua.gr; stratige@central.ntua.gr

M. Giaoutzi and B. Sapio (eds.), *Recent Developments in Foresight Methodologies*, Complex Networks and Dynamic Systems 1, DOI 10.1007/978-1-4614-5215-7_17,

© Springer Science+Business Media New York 2013

17.2 The AG2020 Backcasting Policy Scenarios Framework

This section presents the *foresight methodology* developed in the AG2020 project for building strategic policy scenarios for the assessment of the impacts of probable and desirable future directions of the EU agriculture.

17.2.1 The AG2020 Strategic Policy Scenarios Approach

The AG2020 project has developed an innovative backcasting methodology for structuring policy scenarios at the European level for the period 2007–2020. The structuring of these strategic policy instruments was based on the following elements:

- Objectives and targets in AG2020
- Baseline scenario
- The *Images of the Future*
- The policy framework

In the first part of the AG2020 process, *objectives and targets* were set for orienting the future of the EU Common Agricultural Policy towards the desired ends. On this basis, the *baseline scenario* was structured to identify the scale of changes needed to pursue the selected targets. These, together with the contextual and strategic elements (see Fig. 17.1), were used for building the AG2020 *Images of the Future.*

In the final part, the alternative policy options were selected: these relate to the *policy measures*, *packages* and *paths*; the *scale* of required changes; and the *principles* for their implementation, based on the acceptability, lead times, dynamic effects and adaptability criteria, in the AG2020 context.

These are briefly described in the following subsections:

17.2.2 Objectives and Targets in AG2020

The selected objectives in AG2020 are as follows:

- Environmental protection
- Economic efficiency
- Regional development
- Social cohesion
- Food safety and quality
- Energy

More precisely, sustainability, in the AG2020 context, encompasses (Giaoutzi and Stratigea 2007):

- *Environmental aspects*: for preserving the ecological balance of physical and biological systems, for the present and future generations

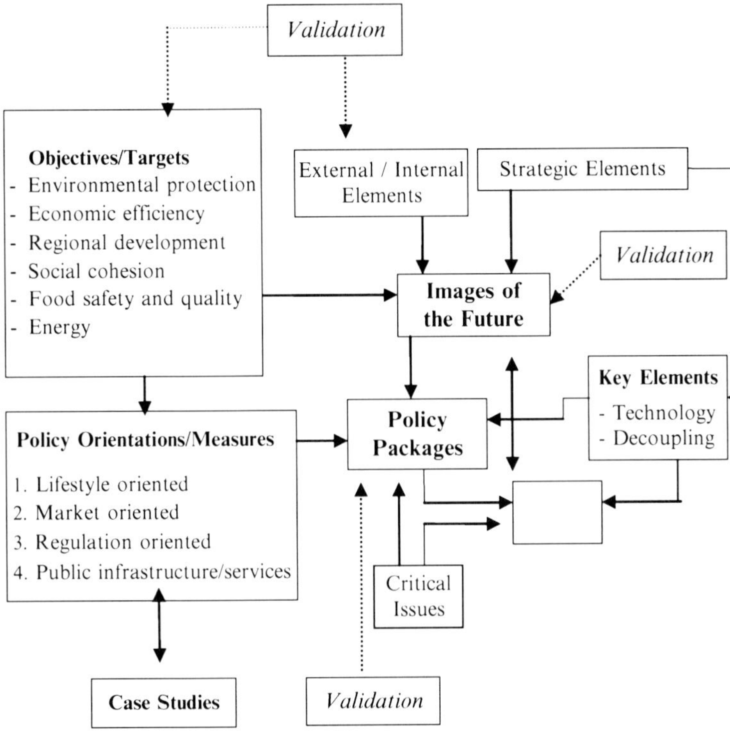

Fig. 17.1 Strategic policy scenarios – the AG2020 framework (Source: Giaoutzi and Stratigea 2009, 2010) (After POSSUM 1998)

- *Economic efficiency aspects*: based on the concept of '… attaining the maximum flow of income that can be created, while at least maintaining the renewable stocks or assets that yield these benefits' (Stimson et al. 2006: p. 40)
- *Regional development aspects*: which aim at the reduction of disparities in rural areas and the equal access to employment, services, etc.
- *Social cohesion aspects:* which aim at maintaining stability in social and cultural systems, by pursuing a healthy and productive life, in harmony with the environment
- *Food quality and safety aspects*: promoting food safety and trust in agricultural qualitative products for consumers
- *Energy production aspects*: which contribute to renewable energy production, according to the EU energy policy

In line with the above objectives, the selected targets, presented in Table 17.1, were the outcome of consortium discussions, literature review, stakeholders' consultation, etc. (Giaoutzi et al. 2008c; Stratigea et al. 2012).

Table 17.1 AG2020 targets

Targets	EU target year 2020	AG2020 targets	Source
GHG emissions (in CO_2 equiv.)	20 % decrease of GHG emissions by 2020 compared with 1990 emissions	N_2O, CH_4 and in CO_2 equivalents	EU
Biodiversity	Halt loss of biodiversity (set in 2001 for 2010)	Halting the loss of biodiversity by 2020 – high rate of halting	EU Council of the European Union, 2004, EURURALIS
Competitiveness/ efficiency	Economically viable regions	Strong competitiveness/ efficiency in the agri-food sector	EU – Lisbon Agenda
Multifunctionality	Increase the level of multifunctionality of agricultural regions	Multifunctionality of rural regions – high level	EU
Food and feed traceability	Traceability of feed and food	Food and feed traceability – high rate	EU
Bio-based economy	Blending targets:in transportation fuel 10 % (2020)	Blending targets in: Transportation fuel 10 % (2020) Electricity 7 % (2020) Chemicals 10 % (2020)	EU

Source: Giaoutzi and Stratigea (2007) and Giaoutzi et al. (2008b)

17.2.3 Baseline Scenarios

At this stage, it is necessary to construct a baseline scenario in order to measure the scale of changes needed when pursuing targets in the *Images of the Future*. Data from today is projected into the future 2020 in a business-as-usual scenario, where no trends have been broken (see Banse et al. 2008). This should indicate developments in the EU agricultural system, assuming that current policies in the field of agriculture remain unchanged until 2020. It is used to identify which sectors of the agri-food system will contribute most to the targets set and which will show the highest growth rates. This will give an indication of on which parts of the agri-food sector to focus but also highlight where dramatic trend breaking is necessary, in order to fulfil the pursued targets.

17.2.4 The AG2020 'Images of the Future'

The design of the *Images of the Future* is a central element in the backcasting process. The *images* should be clearly different from each other, plausible and discrete, and reflect a range of possible futures.

The AG2020 *Images of the Future* outline the characteristics of the economy, society, agriculture, etc., in the target year (2020) and are based on different combinations of *strategic* and *contextual elements*. The strategic elements are considered to be *technology* and *decoupling*, while the selected contextual elements are the three following alternatives:

- *Bilateral and EU cooperation* ('top-down' approach), where cooperation among global players is important
- *Local-multilateral cooperation* (combined approach), which promotes a kind of harmony between 'bottom-up' and 'top-down' politics
- *Local, regional, national and EU cooperation* ('bottom-up' politics), where policies are mainly driven by local and regional initiatives (see Giaoutzi et al. 2012)

There were nine possible alternative *Images of the Future*, which had a great degree of overlapping. Three of them were selected as the most prevailing *images* that represent, as distinctly as possible, settings and futures of the agricultural developments in Europe for the year 2020.

In the process of building blocks of *images*, the involvement of stakeholders and experts groups in the validation workshops was necessary, in order to reach an agreed output (see Giaoutzi et al. 2012).

The prevailing *Images of the Future* for AG2020 are the following:

More precisely:

In *image I – 'High-tech Europe: Global Cooperation for Sustainable Agriculture'* – Europe is politically and economically the strongest block in the world, playing a leading role in climate change policy. Its wealth is mainly based on its leading role in the high-tech sector.

GDP growth is high and large investments in science and research activities are made, with an emphasis on clean energy.

Lifestyles are consumption-oriented, exhibiting a high degree of trust in technology. There is a preference for convenience, functional, ethnic and fast food, while out-of-home consumption prevails.

ICTs dominate in international relations, contributing to strong networking and cooperation among EU businesses and citizens.

The agri-food sector is an intensive user of high tech, with a strong market orientation. Food quality is important, meeting the increasing demand of customers for qualitative and nutritive products. The EU agri-policy focuses on the support of investments in new technologies. There is a high degree of integration of agri-food systems in the international markets, where trade liberalization is based on bilateral agreements.

The energy sector is setting high blending targets for biofuels, placing emphasis on the second generation of biofuels.

Green issues are not greatly supported, while public participation is led by centralized initiatives (national and EU). There is a certain degree of green consciousness and an acceptance of policy measures, intended to mitigate the environmental problems.

In *image II – 'In Search of Balance: Accord on Sustainability'* – a balance is reached between 'bottom-up' and 'top-down' politics, based on the strong public involvement in local and regional affairs; development of focused technological innovations, supporting the transition of rural regions towards bio-based economies;

green values are widespread, while emphasis is also placed on the adoption/use of ICTs and their applications in both the private and business domain. Food patterns are mixed, driving technological cost-saving progress, traceability and monitored labelling developments in the agri-food sector.

In *image III – 'Active Regions and Reflexive Lifestyles'* – policies are mainly driven by local and regional initiatives. Local lifestyles and green values are widespread among the general public, further supported by ICT networks and their applications; people increasingly take responsibility for the common goods, while attitudes towards collective actions are positive, especially at the local and regional level; reflexive slow lifestyles, slow food, slow travelling, etc., are established, while agri-food production is more locally oriented, serving the demand of well-informed and aware consumers. There is a strong focus on quality of life, health, well-being, recreation, safety, etc.

The selected *images* were structured, completed and refined together with stakeholder and expert groups in an iterative process aiming to reach a converging output (see Giaoutzi et al. 2008a, c, 2012; Giaoutzi and Stratigea 2009, 2010).

Box 17.1 Images of the Future for AG2020

Image I: *'High-tech Europe: Global Cooperation for Sustainable Agriculture'*, where science and technology is of the utmost importance, with a focus on 'top-down' initiatives
Image II: *'In Search of Balance: Accord on Sustainability'* – a 'combined approach' where the focus is on economy and energy
Image III: *'Active Regions and Reflexive Lifestyles'*, where the emphasis is on behavioural changes and involves strong public participation ('bottom-up approach')

17.2.5 Policy Measures, Packages and Paths

This subsection presents the process of identifying *policy measures*, *packages*[2] and *paths.*[3] For the selection of policy measures pursuing AG2020 targets, the following five stages are used (see Fig. 17.2):

In the first stage:

- *Key elements* (areas of change) are defined that are considered of importance for driving changes towards the targets.
- Also, *critical issues* are selected that are considered of importance in the design of the AG2020 *Images of the Future*.
- Finally, the impact of *technology* and *decoupling* on each of the above key elements is assessed.

[2] *Policy packages*: proper combinations of sets of policy measures that are likely to work well together (create synergies).

[3] *Policy Paths*: combinations of policy packages and measures for achieving the targets set in the images.

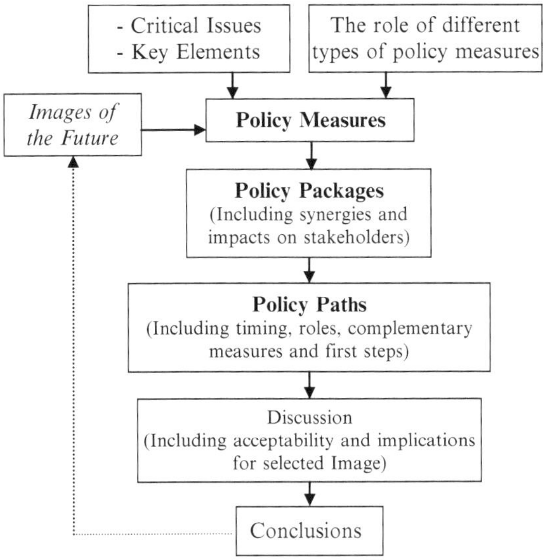

Fig. 17.2 The process of identifying policy measures, packages and paths in AG2020 (Source: POSSUM 1998; Giaoutzi and Stratigea 2009, 2010)

In the second stage:

- Existing *policy measures* are reviewed.
- Measures are selected based on their potential contribution to the targets and their influence on the key elements and critical issues.
- Finally, a list is made of the *prevailing* policy measures.

In the third stage:

- The above policy measures are properly combined to form *policy packages*.

In the fourth stage:

- The *policy paths* are formed from proper combinations of policy packages. Each policy path can drive developments towards a certain *Image of the Future*.

In the fifth stage:

- Issues of acceptability are considered but also implications for the selected image.

The policy framework is presented in more detail below.

17.3 The Policy Framework

The selection of policy measures and the building of policy packages and paths in AG2020 go through a number of stages described in Sect. 17.2.5 above (Fig. 17.2).

Sections 17.3.1 and 17.3.2, respectively, present in detail a number of *critical issues* addressed in the design of the *Images of the Future* and the *key elements* which drive changes in the AG2020 context.

17.3.1 Critical Issues (Hot Spots)

The critical issues considered in AG2020 are as follows (Giaoutzi and Stratigea 2009):

- Global environmental issues: as agricultural activities generate the largest share (63 %) of the world's anthropogenic non-CO_2 emissions (Prentice et al. 2001).
- Balance between agriculture and biodiversity is at stake due to the agricultural production patterns.
- *Food quality and safety* is gaining ground in consumer preferences, as a result of a 'food scare' but also the increasing consciousness of the role of quality food on human health.
- *Integration into global agri-food markets*: due to policy interventions at the EU level, which play a critical role in assisting the EU agri-food industry to adapt to global changes.
- *Regulated agricultural factor markets* that have an impact on the well functioning of the agricultural factor markets (land, water, etc.).
- *Land-use conflicts*: emerging in rural regions, both within the agricultural sector (land for competitive crops) and between agriculture and other economic sectors, e.g. biofuel production.
- *Rural development* that supports a sustainable exploitation of rural resources.

17.3.2 Key Elements

For the identification of the type and range of changes required, it is rather important to distinguish the *key elements*, which motivate policy orientations, such as *technological improvements, bio-based economy, regulated factor markets, rural development, global agri-food markets, energy production, food quality and safety*.

These represent important issues that should be dealt with by policy measures within each of the three AG2020 *Images of the Future* but also serve as a checklist for the identification of policy measures (Giaoutzi and Stratigea 2009, 2010).

17.3.3 Policy Measures

After carrying out the previous steps, a comprehensive list of policy measures is constructed. For a systematic presentation of the role of these measures, four distinct *policy orientations*[4] are considered (Giaoutzi and Stratigea 2009):

[4] By *policy orientation* is meant the generic rationale, which can be found behind different policy measures.

- *Lifestyle-oriented policy measures* support ongoing lifestyle changes that may strongly affect agri-food consumption, way of life, understanding the role of agriculture as a nature safeguard, etc.
- *Market-oriented policy measures* which promote best environmental practices and linkages between environmental sustainability and economic profitability.
- *Regulation-oriented policy measures* which rely on technical standards and norms (e.g. upper pesticide limits, traceability, GMO allowance), innovative planning methodologies (e.g. participatory planning, land-use planning) and government reforms.
- *Public infrastructure/services-oriented policy measures* which relate to the provision of infrastructure (roads, rail, telecommunications, irrigation infrastructure, etc.) and services (training, access to information, R&D services, technology systems and successful agri-practices).

The comprehensive list of the collected policy measures forms the basis for building *policy packages* and *paths* that drive future developments. The *three-step* selection process of policy measures used in the AG2020 project is presented below.

17.3.3.1 First Step: Identification of Policy Measures

In this step, a set of policy measures is presented that may have an impact on the key elements (see Fig. 17.2 above). For this purpose, a number of strategies are defined for each key element, followed by a set of indicative policy measures serving each strategy, together with their policy orientation and the level and time/scale of impact on the key element concerned.

As an *example* of this step, the stages followed for the identification of policy measures that influence the *key element 'bio-based economy'* are presented briefly.

First, the focus is on the operationalization of the bio-based economy concept, which is considered as being based on the adoption of policy measures supporting the four pillars of the bio-based economy (O'Brien 2009), namely, food–feed production and processing, value-added food processing, agri-environmental products and services, and energy and bio-processing.

Second, the *strategies* that may serve the transition towards a bio-based economy are considered to be:

- The *economic efficiency strategy* that seeks an increasing economic output from the agricultural sector
- The *environmental stewardship strategy* that promotes the delivery of a wide range of important and socially valuable agri-environmental products and services
- A *less resource-consuming production strategy* that promotes less resource-intensive agri-food production systems, preserves natural resources, reduces production costs (e.g. energy costs) and increases the competitiveness of the agri-food systems

- The *bioenergy and biomaterials production strategy* that refers to the production of bioenergy and other materials (e.g. chemicals) by use of agri-raw material, waste and forest biomass
- The *spatial planning strategy* that places emphasis on the protection of valuable ecosystems and agricultural land, the structure of a more effective transport network system, etc.

Third, the *policy orientations* with respect to the key element 'bio-based economy' are defined as follows (Giaoutzi and Stratigea 2009):

- *Lifestyle policy measures*: environmental respect is the most important issue in lifestyle policy measures in a bio-based economy context. Information plays a key role in raising environmental awareness. ICTs and the flow of information are very well established, and the citizens are seeking environmentally friendly products. Product 'identity' is a key issue in shaping preferences, being a compulsory dimension of any labelling system. Consumers' behaviour is driven by the principle of 'reduce, recycle and reuse', steering a less resource-consuming attitude in rural societies. A more local orientation of preferences is prevalent. Energy is a key issue and considerable effort is placed on rationalizing energy consumption.
- *Market-based policy measures* incorporate all policy measures (incentives) motivating market behaviour towards the production of environmentally friendly and bio-based products. Efficiency and environmentally responsible behaviour are key issues in agri-food production and processing. Measures aim at either promoting R&D investments/innovations in order to steer businesses towards knowledge-based competitive business paths and better exploitation of existing resources or penalizing them for irrational or excessive, non-environmentally friendly resource use. The emphasis is on energy production.
- *Regulation-based policy measures* are setting rules that enforce certain behaviour, the production/processing pattern, standards of product, etc. Policy measures may relate to setting standards for food/feed production and placing quality at the forefront (traceability and labelling), rationalizing the use of resources, setting obligatory blending shares of biofuels for increasing their share in the market, regulating land use, etc.
- *Public infrastructure/services policy measures* may focus on the support of agri-food businesses in rural areas towards the bio-based economy, unimpeded flow of all kinds of information and knowledge in rural regions/businesses (infrastructure), upgrading of human resources and provision of public services in a more effective way for both citizens and businesses (e-government).

On the basis of the above, a number of agri-sector policy measures are then selected that can support developments towards a bio-based economy.

Examples of these measures are presented in Table 17.2 (for more details, see Giaoutzi and Stratigea 2009).

Table 17.2 Potential agri-sector policy measures affecting bio-based economy

Policy measure	Strategy	Impact on AG2020 targets timescale
Lifestyle-oriented		
Eco-labelling of agri-products	Environmental stewardship	++ Medium term
Reduce, recycle and reuse waste	Environmental stewardship	+++
	Economic efficiency	Long term
	Bioenergy production	
Market-oriented		
Taxation on agri-products with high ecological footprint	Environmental stewardship	++ Long term
Promotion of renewable energy in agricultural regions	Environmental stewardship	+++
	Economic efficiency	Long term
Regulation-oriented		
Emission standards from agriculture	Environmental stewardship	+++ Long term
Land-use planning – protected zones in rural areas	Spatial planning	++
	Environmental stewardship	Medium term
Public infrastructure/services-oriented		
Promotion of GMO to increase yields and plant resistance to certain external factors, e.g. drought	Economic efficiency	+++
	Less resource-consuming production pattern	Long term
Infrastructure for waste management	Bioenergy production	+++
	Environmental stewardship	Medium term

Source: Giaoutzi and Stratigea (2009)
Notes: *Scale of impact*: '+' for small effect, '++' for medium effect and '+++' for strong effect. *Timescale*: 'short term', 'medium term' and 'long term'

17.3.3.2 Second Step: Identification of Technology and Decoupling-Oriented Policy Measures

This step explores the likely *contribution of strategic elements,* namely, *technology* and *decoupling*, to each of the key elements considered above.

Similarly, for each policy measure that relates to technology or decoupling, the strategy served by this measure is presented, together with the policy orientation, and the level and time/scale of impact on the key elements.

Following the *example* of the previous step, we now present briefly the stages followed for the identification of *technology policy measures* relating to the key element 'bio-based economy' (see Table 17.3).

First, a good level of knowledge on the potential role of technology in agriculture should be achieved.

Table 17.3 Technology-oriented policy measures for the transition to the bio-based economy

Policy measure	Strategy/sub-strategy	Impact on AG2020 targets/timescale
Lifestyle-oriented		
Technologies for labelling	Seize market opportunities Reduction of environmental load Raise awareness	+++ Medium term
Technologies supporting reduce, recycle and reuse of waste	Reduction of environmental load Renewable energy Raise awareness	+++ Medium term
Market-oriented		
Green biotechnology (agriculture)	Reduction of environmental load Seize market opportunities	+++ Long term
Promotion of eco-businesses	Reduction of environmental load	+++ Long term
Technologies for bioenergy production	Reduction of environmental load Renewable energy production	+++ Long term
Regulation-oriented.		
Promotion of GMO	Reduction of environmental load Seize market opportunities	+++ Medium term
Enforcement of technologies reducing greenhouse gases	Reduction of environmental load Green or low input agri-production	+++ Long term
Public infrastructure/services-oriented		
R&D on agri-environmental technology	Reduction of environmental load Green or low input agri-production	+++ Long term
R&D on environmentally sound management practices	Reduction of environmental load Seize market opportunities	+++ Long term

Source: Giaoutzi and Stratigea (2009)
Notes: Scale of impact: '+' small effect, '++' medium effect and '+++' strong effect. Timescale: 'short term', 'medium term' and 'long term'

Next, the *strategies* are defined that need to be served by technology-oriented policy measures in rural areas, in order to develop as bio-based economies. These are as follows (Giaoutzi and Stratigea 2009):

- *Seize market opportunities*, supporting the market orientation and flourishing of rural areas on the basis of either cost-effectiveness or quality of their production, e.g. biotechnology and green production
- *Reduction of environmental load*, based on innovations in the production process or processing of products, e.g. water-saving technologies, agri-environmental technologies and precise farming[5]
- *Production of renewable energy and energy efficiency*, e.g. bio-refinery technologies and energy-saving technologies
- *Raising awareness and knowledge of consumers* on the 'content' of a specific product as a tool for influencing consumers' preferences and behaviour

[5] *Precise farming* generally defined as a technology and information-intensive farm management system, used to identify, analyse, and manage field variability, for reaching both profitability and environmental objectives.

Based on the above information, Table 17.3 above presents an indicative list of technology-oriented policy measures.

The type of *decoupling* pursued is of key importance for the identification of policy measures, which affect the bio-based economy. In this example, the following types of decoupling are considered (Giaoutzi and Stratigea 2009):

- Decoupling of agricultural production from environmental impact
- Decoupling of rural development from agri-food production, as other important sectors are developed in bio-based economies as well, taking advantage of the local natural resources
- Decoupling of agricultural production from the production of biofuels

The *strategies* that need to be served by decoupling-oriented policy measures are as follows:

- *Seize market opportunities* by supporting the market orientation and flourishing of rural areas on the basis of the development of other sectors as well, e.g. tourism and energy production
- *Environmental stewardship*: reduction of environmental load based on the adoption of environmentally friendly agricultural practices
- *Renewable energy production*: by processing available biomass resources, through cost-efficient technologies
- *Accessibility* of rural population (both citizens and businesses) by means of ICTs and transport infrastructure

Table 17.4 below presents an indicative list of decoupling-oriented policy measures.

For all key elements considered in AG2020, lists are compiled of policy measures of the form presented in Tables 17.2, 17.3 and 17.4 that serve the strategies defined in each specific key element.

17.3.3.3 Third Step: Assessing the Impact of Policy Measures

Finally, the *third step* explores the impact of the above list of policy measures on the agricultural sector, the critical issues and the AG2020 targets.

The final outcome of this step is a *pool of 257 policy measures*, used for building policy packages and paths. These measures were also assessed by stakeholders and experts, in AG2020 workshops, as to their impact on the proposed AG2020 targets (level and timescale of impact), policy orientation, impact on the agricultural sector, impact on the critical issues and their relationship to technology and/or decoupling.

17.3.4 Policy Packages

The AG2020 policy packages are constructed by properly combining sets of policy measures that are likely to create synergies. Each of the 15 policy packages presented below is designed to serve a specific dimension of the *Images of the Future* (Giaoutzi and Stratigea 2009).

Table 17.4 Decoupling-oriented policy measures for the transition towards the bio-based economy

Policy measure	Strategy/sub-strategy	Impact on AG2020 targets/timescale
Lifestyle-oriented		
Eco-labelling of regional green products	Environmental stewardship	+++ Medium term
Waste management in household and agri-food businesses	Environmental stewardship	++
	Energy production	Medium term
Market-oriented		
Promotion of ecological tax reform	Environmental stewardship	++ Long term
Promotion of organic and low-input farming	Environmental stewardship	++ Medium term
Regulation-oriented		
Land-use planning – protection of valuable nature	Environmental stewardship	+++ Medium term
Promotion of regional biomass plans	Environmental stewardship	++
	Energy production	Long term
Public infrastructure/services-oriented		
Establishment of infrastructure for waste management	Environmental stewardship	++
	Energy production	Medium term
	Seize market opportunities	
Setting up farm management and forest advisory services – forest management	Environmental stewardship	++
	Energy production	Long term
	Seize market opportunities	

Source: Giaoutzi and Stratigea (2009)
Notes: *Scale of impact*: '+' small effect, '++' medium effect and '+++' strong effect. *Timescale*: 'short term', 'medium term' and 'long term'

17.3.4.1 Description of Policy Packages

The 15 policy packages constructed in AG2020 are as follows:

- *PP 1: Knowledge-driven diversified rural economies*
- This package aims to increase the knowledge stock of rural economies, based on a demand-driven development of knowledge and innovations, capable of serving the needs of a diversified pattern of rural businesses and population.
- *PP 2: Green entrepreneurship*
- The package is concerned with the environmental integrity of businesses and aims to merge environmental protection, economic efficiency, and business innovation objectives.

- *PP 3: Environmental and resource stewardship*
- The package focuses on the role of the agri-food and forest sectors as safeguards of environmental resources and pursues the environmental and resource stewardship, based on the sustainable use of natural resources.
- *PP 4: Reduce, reuse and recycle (the 3Rs)*
- This package focuses on an effective and multiple use of resources, based on the 3Rs, namely, reduce, reuse and recycle.
- *PP 5: Ecological tax reform*
- This package focuses on the idea behind an ecological tax reform which is based on the view that externalities of resource use and environmentally harmful activities are taxed too lightly, while labour is taxed too heavily.
- *PP 6: Log in the information economy*
- The focus of the sixth policy package is on the diffusion and applications of ICTs in rural regions in support of the interaction and knowledge exchange at both the intra- and the interregional level.
- *PP 7: Culture of 'regionality'*
- The focus in this package is on a locally driven perspective, based on the endogenous potential of rural regions, and realized by behavioural patterns that place emphasis on quality and authenticity (quality of food, quality of life, quality of social interaction, quality of the environment, etc.).
- *PP 8: Social responsibility*
- This package aims at strengthening the commitment of both citizens and businesses in rural regions to environmental and social progress.
- *PP 9: Administrative innovations – e-government*
- This policy package builds upon the need for upgrading administrative services in rural areas by the timely and cost-effective provision of services (e-government).
- *PP 10: R&D – bio-innovations*
- As rural regions are facing great challenges during the transition towards bio-based economies, the focus is on the development of knowledge and innovations that may support this transition.
- *PP 11: Public participation*
- This policy package focuses on the public participation in the decision-making process that may potentially enable a broader consensus on the future development of rural regions.
- *PP 12: Knowledge-intensive farm management*
- The aim of this policy package is on the promotion of a knowledge- and technology-intensive farm management approach, used to support profitability and environmental integrity in high-productivity systems.
- *PP 13: Development of human resources*
- This package focuses on the development of the knowledge base and skills of rural regions, in both the agri-food and other sectors.
- *PP 14: Accessibility to transport – ICTs infrastructure*
- This package aims at increasing the accessibility of rural regions, by transport and ICTs, in order to support the mobility of people and goods, but also information exchange both at the intra- and interregional level.

- *PP 15: Spatial planning*
- This policy package focuses on the development of spatial patterns in rural regions. Spatial planning is of the utmost importance in this respect, for both reaching a balanced distribution of land-use patterns and refraining from competition between land uses.

Provided for each policy package are the construction logic, the policy orientations, the main policy measures and the impact on stakeholders.

17.3.4.2 Construction Logic of Policy Packages

An example of the construction logic of 'PP 1: knowledge-driven diversified rural economies' is presented below (Giaoutzi and Stratigea 2009).

In PP 1 great emphasis is placed on the development of rural regions as *hubs of knowledge and innovation,* in the agri-food but also in other sectors as well. These hubs drive innovation and the competitiveness of local economies. Rural economies are characterized by strong links to R&D and knowledge sources, while high-tech industries in the agri-food sector are creating clusters, which fuel the other sectors of rural economies.

17.3.4.3 Policy Orientations

The *ranking* of policy orientations, serving PP 1, is as follows (in descending order):

- *Public infrastructure/services orientation*, supporting R&D, links to universities and research institutions, establishment of broadband connection in rural areas, lifelong training to rural population, provision of specialized services, etc.
- *Market orientation*, supporting export initiatives of rural businesses and networking.
- *Regulation orientation,* elaborating on the rules for cultivation, harvesting, processing and packaging of agri-food products (e.g. biotechnology and GMO).

17.3.4.4 Main Policy Measures

Key policy measures in PP 1 are as follows:

- Increased accessibility of rural areas to knowledge infrastructure and R&D (universities, research institutions, etc.)
- Provision of broadband connection
- Skilled and ICT-experienced labour force – lifelong training for the rural population
- Openness of rural regions to the external world – strong trade orientation
- Strong networking among businesses, especially in the agri-food sector (agri-food chains)

- Diversification of local economic structure (agri-food, trade, services supporting the agri-food sector, other services)
- R&D support – development of new knowledge and technologies in the agri-food sector
- Promotion of entrepreneurship in rural regions

17.3.4.5 Impact on Stakeholders of PP1

The implementation of PP 1 will be to the benefit of local businesses in rural regions as they gain access to knowledge and information (re)sources that will enhance their potential to innovate and grasp new opportunities (e.g. new products, new production processes, increase of market share) and increase competitiveness. Agricultural businesses can also benefit from the increasing knowledge stock on agri-input, agri-food processes, plants, etc., and accessibility to R&D through ICTs that will ensure the continuous updating of relevant information, which drive competitiveness and market penetration.

On the other hand, increasing accessibility to knowledge (re)sources may have a positive impact on the quality of the local labour force. More precisely, access to lifelong training may enable the upgrading of skills and capabilities, thus contributing both to the improvement of opportunities for the rural population and to the increase of productivity in local firms.

All the above, combined with the increasing knowledge stock and accessibility to R&D through ICTs networks, makes the rural regions more attractive destinations for the location of knowledge-intensive firms, which reinforces the economic base and diversity of high value-added agri-food activities in rural regions.

17.3.5 Policy Paths

This subsection presents the construction of target-driven *policy paths* to the AG2020 *Images of the Future*. This is based on a proper combination of policy packages, serving the AG2020 objectives in each of the *images* (*key states are considered* that describe the pace of changes needed) (Giaoutzi and Stratigea 2009, 2010). The output of this process are three policy paths (Table 17.5), serving each of the specific AG2020 *Images of the Future*.

The description of each policy path incorporates the following:

- General context
- Main policy packages – areas of change
- Priorities in policy orientations involved
- Linkages and synergies among policy packages
- An indicative list of policy measures serving the specific path

Table 17.5 Policy paths in AG2020

Policy orientation	Image		
	Image I (top-down)	Image II (combined)	Image III (ottom-up)
Lifestyle orientation Market orientation Regulation orientation Public infrastructure/service orientation	Path 1	Path 2	Path 3

Source: Giaoutzi and Stratigea (2009, 2010)

The following is a brief description of the three policy paths constructed in AG2020 (Giaoutzi and Stratigea 2009, 2010).

17.3.5.1 Policy Path to Image I: Path 1

In *image I, High-tech Europe: Global Cooperation for Sustainable Agriculture*, science and technology is of the utmost importance, together with a focus on 'top-down' initiatives. The prevailing policy orientation of this path is the 'public infrastructure/services orientation', followed by the market orientation (Giaoutzi and Stratigea 2009).

At the core of this path is *policy package 6* (log in the information society), provided that a proper network infrastructure is already deployed in rural regions (*PP 14* – access to ICTs and transport infrastructure). An important aspect is the creation of a communication platform to increase interaction between stakeholders in rural regions and the establishment of links with R&D institutions, research centres, universities, etc. The adoption and use of ICT applications in rural regions may support the development of an e-culture, which lies at the core of personal and business development (Stratigea 2011), and is beneficial for rural areas.

Additionally, *PP6* is closely linked to *PP1* (knowledge-driven diversified rural communities) and *PP12* (knowledge-intensive farm management), thus establishing a 'bridge' for the diffusion of knowledge and information to the various stakeholders. *PP6* also forms the backbone for the implementation of *PP13* (development of human resources).

At the same time, the fulfilment of *PP13* facilitates the pursuit of *PP1* (knowledge-driven diversified rural economies) and *PP12* (knowledge-intensive farm management), as it prepares the 'ground' for both the adoption and the use of technology and innovation.

Furthermore, *PP6* is very important for the implementation of *PP9* (administrative innovations – e-government). Moreover, the adoption and use of ICTs and their applications engenders trust in technology and skill acquisition, which in turn facilitates the development and use of e-government applications.

Another important policy package, in this respect, is *PP14* (accessibility to transport – ICTs infrastructure), which promotes the deployment of the necessary transport and telecommunications infrastructure that may facilitate the smooth flow of persons, goods and information from rural regions to and from the outer world. This supports a better exploitation of local resources and the flourishing of a variety of economic activities, which contributes to the diversification of rural economies (Giaoutzi and Stratigea 2009).

Finally, *PP15* (spatial planning) sets the rules for a balanced land-use distribution in rural regions, serving as a tool for land management, protection of valuable natural resources, control of competition among land uses, etc., contributing to, among other things, the improvement of the social and economic cohesion of the rural population.

17.3.5.2 Policy Path to Image II: Path 2

In image II – *In Search of Balance: Accord on Sustainability* – the focus is on the economy and energy. A certain balance of different policy orientations is necessary for the implementation of this path, although the public infrastructure/services orientation seems to be ranked higher in importance than the other three policy orientations (Giaoutzi and Stratigea 2009).

For reaching image II, it is necessary to combine the *technology* and *decoupling* elements. This requests certain improvements in the rates of adoption/use of technology, where an emphasis is placed on the support of R&D, and on both knowledge-driven innovations (*PP1* – knowledge-driven diversified economies) and knowledge-intensive farm management innovations *(PP12)*. In addition, *PP3* on 'environmental and resource stewardship', dealing with the use of technological innovations for a more environmentally friendly management of the agri-forest resources, is of great importance. Moreover, *PP10* (R&D innovations – bio-innovations) places emphasis on all kinds of innovations relating to the bio-based economy. The application of the above PPs creates *synergies* that motivate rural communities towards acquiring a broader knowledge base and available technologies for effective rural resource management. Furthermore, the upgrading of human resources (*PP13* – development of human resources), combined with the adoption/use of ICTs and their applications (*PP6* – Log-in the information society), creates the appropriate conditions for the unimpeded flow of information and knowledge among rural stakeholders, which is expected to have a positive impact on entrepreneurship, job creation, and the support of local income. Finally, good access of rural regions to infrastructure (*PP14* – access to transport and ICTs infrastructure) may support interaction between and trade in agri-food and forest products (Giaoutzi and Stratigea 2009).

As to the *decoupling element*, in image II, a more conscious attitude towards preserving natural and environmental resources is adopted, with stakeholders taking active roles for the protection and preservation of these resources. *PP2* (green entrepreneurship), *PP3* (environmental and resource stewardship) and *PP4* (reduce, reuse and recycle) are promoting a more conscious attitude towards consumption patterns of natural resources, which implies the development of a new culture of resource saving and multiple use of natural resources in rural regions. These also introduce a new spirit of resource management at both the household and the business level that may also have a positive impact on job creation and local income generation. Moreover, it forms the basis for the creation of agri-based and non-agri-based activities, which supports the levels of multifunctionality in rural regions (Giaoutzi and Stratigea 2009).

Finally, spatial planning *(PP15)* sets the framework for a more effective spatial organization of rural regions that aims to serve the needs of both population and activities, preserving at the same time local valuable ecosystems.

17.3.5.3 Policy Path to Image III: Path 3

The focus of image III – *Active Regions and Reflexive Lifestyles* – is on behavioural changes and involves strong public participation. The key policy orientation in this path is lifestyle orientation, for widening the already-existing commitment to green values, environmental responsibility, regionality, respect for local resources, culture and also quality of life and the local environment, food, etc. This is followed by the market orientation, which adopts market policy measures motivating business to follow 'green' behaviour and develop/adopt innovations accordingly (Giaoutzi and Stratigea 2009).

In path 3, where lifestyle orientation is important, participatory decision making (PP11 – public participation) forms the platform for interaction among local stakeholders and the creation of a vision for the future development of rural regions, based on respect for social, cultural and natural resources. This is facilitated by the adoption and use of ICTs and the creation of an ICT platform, supported by *PP9* (administrative innovations – e-government). Of importance is also the dedication to values, authenticity, cultural heritage, social interaction and demand for quality, expressed via *PP7* (culture of 'regionality') that drives behavioural patterns in rural regions, but also both population and businesses take responsibility with respect to the management of environmental, cultural and social resources, expressed by *PP8* (social responsibility). The above actors mobilize environmental and resource stewardship, based on resource preservation and management *(PP3)*, as well as a reduction and multiple use of resources and recycling *(PP4)*. Moreover, the spirit of e-culture, which is well established among the local population, is a key driver for access to information and knowledge that supports decisions on sustainable resource management (Giaoutzi and Stratigea 2009).

Of additional importance is the provision of incentives to business for gaining respect and competitiveness in the rural context. More precisely, *PP2* (Green entrepreneurship) promotes policy measures that motivate business innovations capable of supporting the integration of an environmentally friendly business behaviour and competitiveness objectives; *PP3* (environmental and resource stewardship) promotes policy measures that motivate agri-food and forest business to adopt environmentally friendly product and process practices that may support the decoupling of environmental quality from these activities; *PP8* (social responsibility) motivates business behaviour that takes into account local environmental and social resources; *PP12* (knowledge-intensive farm management) motivates business to adopt knowledge-intensive farm management practices that may support the decoupling of agricultural production from environmental quality; lastly, *PP10* (R&D innovations – bio-innovations) motivates business to develop/adopt innovations that may support biomaterial production. Of importance is the contribution of *PP6* (log in the information society), which creates the potential for both intra-business and/or inter-business innovations that may support firms' competitiveness, efficiency, sharing of resources and risks, etc. (Giaoutzi and Stratigea 2009; Stratigea 2011).

Also of importance are investments in public infrastructure and services that may strengthen the economic base of rural regions, by improving the accessibility of rural regions (*PP14* – access to ICTs and transport infrastructure), their knowledge base (*PP6* – log-in the information society), but also quality of local human resources (*PP13* – development of human resources).

Finally, of importance is a certain regulatory framework for the spatial organization of rural settlements (*PP15* – spatial planning) that may form the basis for strengthening the linkages between activities and an upgraded network infrastructure (*PP14* – access to transport and ICTs infrastructure). The protection of valuable natural systems (*PP3* – environmental and resource stewardship) is also of great importance. The above framework is setting the directions for the sustainable exploitation of resources (*PP4* – reduce, reuse and recycle) and a set of financial measures as disincentives for developing resource-intensive activities (*PP5* – ecological tax reform) (Giaoutzi and Stratigea 2009).

17.4 Conclusions

The focus of the present chapter has been on the presentation of the policy framework developed in the strategic policy scenarios in the AG2020 project. It provides a set of policy options (policy measures, policy packages and paths) and related impacts that might be of value for policy makers in preparing policy decisions for EU agriculture in 2020.

The *advantages* of the approach are considered to be:

- The creative approach developed, based on a number of iterative steps that enable a deeper insight into the agricultural sector per se and also into its interactions with the rest of the economic activities, the environment and the society.
- The detailed set of procedures followed, which may ensure *consistency* and *comprehensiveness* in policy formulation at the EU level.
- The addressing of important issues for policy implementation, such as the *time horizon* of the impact of policy measures and *scale of impact*, as important components for building policy packages and paths. It then becomes possible to assess the role that each policy measure may have in achieving these targets.
- The *participatory context* adopted that serves both as a *communication tool* for increasing awareness of the prospects and risks involved in the development of

the agricultural sector and as a *consensus-building tool* for the more effective implementation of policy decisions.

- The *validation* of the approach, the *Images of the Future*, and the policy framework by a range of experts and stakeholders, thus enriching both the process and the outcome of the whole effort.

Among the *difficulties* of the approach can be mentioned:

- The *communication barriers* involved in a multidisciplinary nature of such an effort
- The *time and cost constraints* involved in running participatory approaches

References

Banse, M., Jansson, T., Eickhout, B., Sulser, T., & Ringler, C. (2008). *Analysis of major challenges and constraints: Trade and market measures and policies, D3.1.* AG2020 project, Foresight analysis for world agricultural markets (2020) and Europe. Contract no.: 44280-AG2020, STREP, 2007–2009.

Giaoutzi, M., & Stratigea, A. (2007, May 7–9). *Defining objectives and targets in AG2020.* Paper prepared for the AG2020 Workshop, Copenhagen.

Giaoutzi, M., & Stratigea, A. (2009). *Assessment of alternative policy scenarios on a global level, D5.5*, AG2020 project, Foresight analysis for world agricultural markets (2020) and Europe. Contract no.: 44280-AG2020, STREP, 2007–2009.

Giaoutzi, M., & Stratigea, A. (2010). *General assessment of alternative policy scenarios, D5.6* (Final Report). AG2020 project, Foresight analysis for world agricultural markets (2020) and Europe. Contract no.: 44280-AG2020, STREP, 2007–2009.

Giaoutzi, M., Banse, M., Borch, K., Eichhout, B., Petzoldt, M., & Stratigea, A. (2008a). *Building images of the future in AG2020*, Working Paper No 12, AG2020 Project.

Giaoutzi, M., Stratigea, A., Nijkamp, P., van Leeuven, E., & Banse, M. (2008b). *Definition of policy scenarios, D5.3.* Foresight analysis for world agricultural markets (2020) and Europe. Contract no.: 44280-AG2020, STREP, 2007–2009.

Giaoutzi, M., Stratigea, A., & Grammatikogiannis, E. (2008c). *Stakeholder validation of policy scenarios, D5.4.* AG2020 project, Foresight analysis for world agricultural markets (2020) and Europe. Contract no.: 44280-AG2020, STREP, 2007–2009.

Giaoutzi, M., Stratigea, A., van Leeuwen, E., & Nijkamp, P. (2012). Scenario analysis as a foresight tool in agriculture. *International Journal of Foresight and Innovation Policy, 8*(2/3), 105–128.

O'Brien, L. (2009). Towards 2030: Teagasc's role in transforming Ireland's agri-food sector and the wider bioeconomy. *Rivista Di Economia Agraria, a. LXIV*(n. 1–2), Giugno.

POSSUM Project. (1998). *Final report.* European Commission, Transport RTD Programme, 4th Framework Programme.

Prentice, I. C., Farquhar, G. A., & Fasham, M. J. R. (2001). The carbon cycle and atmospheric carbon dioxide. In: *Intergovernmental Panel on Climate Change, Working Group I, 'Climate change 2001: The scientific basis'.* Cambridge: Cambridge University Press.

Stimson, R., Stough, R., & Roberts, B. (2006). *Regional economic development: Analysis and planning strategy* (2nd ed.). Berlin: Springer.

Stratigea, A. (2011), ICTs for rural development: potential applications and barriers involved, NETCOM, Vol. 25, No 3-4, pp. 179–204.

Stratigea, A., Grammatikogiannis, E., & Giaoutzi, M. (2012). How to approach narratives in foresight studies: Qualitative data analysis. *International Journal of Foresight and Innovation Policy, 8*(2/3), 105–128.

Chapter 18
Opportunities for Combining Quantitative and Qualitative Approaches in Scenario Building: The Experience of the 'Estonia 2010' Project

Erik Terk

18.1 Introduction

Linking the qualitative and quantitative approaches or, figuratively speaking, narratives and numbers is one of the most challenging problems in the development of the methodology of foresight/futures studies. This problem frequently emerges in the framework of implementing user-oriented scenarios. The multiple-scenario method is not just one of the futures studies methods out of many but is rather a broader methodological construction, within which other particular methods can be applied (Bell 1997: p. 239). The aforementioned problem can be presented in the following form: how could the qualitative and quantitative methods be combined in scenario building so as to ensure, on the one hand, the adequacy of the reality of the created constructs and, on the other hand, their convenient usability for the decision makers. It should be pointed out here that this task of combining two approaches can have quite different forms. The first variant emerges in the case of scenarios built on some quantitative model (let us not concentrate here on the fact that the model itself and its usability in a specific situation are based on a certain qualitative notion, which may not be easy to convey), which must be made 'palatable' for practical users unable to comprehend the model and the results of its use, that is, the numerical results should be complemented by a qualitative text to explain them. The second option, which is quite frequent, emerges in cases when the qualitative scenario description is written first and later made more concrete by including in the text some calculation results or expert opinions. Of course, a combination of the above two variants is also possible – the author does write (at least initially) a qualitative description of the scenario, but while doing that, he considers, either

---

E. Terk (✉)
Tallinn University, Lai Str. 34, Tallinn, 10133, Estonia
e-mail: erik.terk@lu.ee

M. Giaoutzi and B. Sapio (eds.), *Recent Developments in Foresight Methodologies*, Complex Networks and Dynamic Systems 1, DOI 10.1007/978-1-4614-5215-7_18,
© Springer Science+Business Media New York 2013

consciously or subconsciously, certain calculation results, qualitative indicators taken from similar development analogies, trend extrapolations, etc.

This chapter considers an even more complicated variant, one, where the emphasis on numbers or narratives alternated several times during the various stages of building scenarios and in their use.

18.2 The 'Estonia 2010' Project

The analysed case is the project 'Estonia 2010', which took place in 1995–1997. The 'Estonia 2010' package of scenarios was worked out by an interdisciplinary group of researchers from various institutions, some of them having experience of politics or administration in the early 1990s. The main authors of the work were Garri Raagmaa, Erik Terk, Marju Lauristin, Krista Loogma, Raivo Vilu, Boris Tamm and Alari Purju. The objective of building the scenario package was to 'stage' Estonia's development alternatives in a perspective of 15 years and to derive conclusions from it for government departments, politicians, managers of infrastructure enterprises and specialists concentrating on inter-Estonian regional development, as well as for the public. The project was financed by various ministries of Estonia and major infrastructure enterprises. Although the project was initially launched for performing one task, laying the foundation for drafting Estonia's spatial development guidelines, it appeared that the package would also be useful for solving a number of other related problems. It was used during a longer period by various user contingents, and the scenario texts were repeatedly modified depending on the needs of the specific sets of actors, further analytical work being carried out in between or when new information became available.

18.2.1 Preliminary Stage: Starting from Calculations

The 'Estonia 2010' project did not start from a blank page: it was possible to use some earlier works of the Estonian Institute for Futures Studies. Within the institute's research programme in 1993–1994, it was attempted to establish the probable growth rate and other macroeconomic parameters of Estonia's economy, in the case of two potential sets of conditions. In the first set, Estonia improves its access to the EU markets while retaining its normal opportunities for trading in the Russian and other CIS countries' markets. In the second set, Estonia's economic integration with the EU results in a sudden decline of economic ties with Russia. Deriving from hypotheses on market access and based on the foreign investments statistics of various countries, the probable inflow of investments was established for both scenarios by using expert estimates. Further, the correlation coefficients between investments (volume and structure) and economic growth of a number of countries in the latest period were used. Depending on the economic growth rate and earlier forecasts of Estonia's macroeconomic stabilization, some other macroeconomic indicators were derived, as well as probable shifts in the economic structure. A brief description was provided for either option (movement towards Western markets exclusively or combining the Western markets with those of the CIS), which built on the logic of economic development and the forecast indicators.[1]

Although that stage was not strictly speaking modelling and a number of operations were based on expert estimates (the selection of background countries, the connection between access to markets and the volume and structure of investments), one could claim that the building of scenarios at that stage of the work was certainly calculation based. The central axis was the statistical relationship between investments and economic growth (ICOR, Incremental Capital-Output Ratio), which was transferred from the practice of the countries selected for comparison with the practice of Estonia in the following period. The text of the scenarios, which was written at that stage of the work, did not have an independent meaning, but complemented the calculation results.

Since this work was actually completed before the start of the 'Estonia 2010' project but was used as the input of the project, we could also describe it as a 'zero stage' of the project.

18.2.2 The First Stage of the 'Estonia 2010' Project

This first stage involved the analysis of international trends, interviews and round table discussions. At this stage, the gathering of material and analysis were combined with the generating of ideas and attempts at synthesis. Since it was intended to treat Estonia's development in a significantly broader way than merely by markets and economic growth, the discussion involved a number of key factors which were presumed to have influence on Estonia's further developments. These factors were geopolitics; changes in lifestyle; development of technologies; changes in environment, environmental thinking and environmental policies; European integration; and changes in transport and logistics. Concerning all these subjects, texts in relatively free form were compiled after the discussion seminars, which attempted to define the global trends of these spheres, their impact, the life of the people and

[1] As W. Bell points out (Bell 1997: p. 246), one of the most complicated problems of futures studies is the selection of the leading indicators, on whose actions the forecast or scenarios should be built. It is usually not possible to prove definitely the choice of the particular leading indicators selected. This applies to our case, when access to various foreign markets was selected as the basis of the scenarios. The decision was based on the conviction gained from the previous works of the Estonian Institute for Futures Studies that Estonia's economy has, after the transition 'shock', either stabilized or is certainly stabilizing, and further progress of privatization is more or less predictable. A new, predominantly private capital-based economy had developed, and its growth in the following periods, due to Estonia's small size, depended primarily on foreign markets. The following progress of events validated the adequacy of this hypothesis.

the problems emerging in economic conditions, strengthening policies in the international arena, and any possible influence (threats, opportunities, challenges) on Estonia.

Based on these texts, the book *The world and Estonia. Future trends* (Maailm ja Eesti 1996) was later published. After the completion of the texts, additional interviews were carried out with some influential Estonian thinkers, and two round table discussions were held, which attempted to develop synthetic, 'big picture' speculations on Estonia's integrated development during the next 10–20 years and on the related goals and opportunities.

18.2.3 The Second Stage

This second stage involves choosing the focus of the foresight task and determining the main axes of the scenarios. After the gathering and analysis of material in the previous stage, the array of factors influencing Estonia's development and potentially significant relations was considerably expanded. However, in order to begin the staging of Estonia's probable development variants, it was imperative to determine those impact factors, which would play the central role in the following analysis, in order to define for the analysis the *main variables*, which should be observed in multiple variants; the *background factors*, whose changes should be presented as the central value forecasts; and the *dependent variables*, which could be defined as the result of the action of the main variables and background factors.

In effect, at this stage of the project, a hypothetical qualitative model of a post-socialist country's development for the post-stabilization period was built. No direct theoretical basis for such a model was available, because the theoretical bases of transition and transformation for countries escaping from state socialism, which were just being developed at that period, initially only reflected the realities of the first period of transition (macroeconomic stabilization, privatization, etc.), but not what would take place in the following, catching-up period.

The more significant intermediate decisions made in that stage of work were as follows:

(a) To continue considering the criterion of market access, which had already become important in the preliminary stage of the work, as one of the central parameters, but to treat it in a somewhat broader context: first, to link it more closely to Estonia's success in the official accession to the EU and, secondly, to differentiate regarding the term 'Western markets' between the neighbouring markets for Estonia (the Nordic countries, Finland, Sweden, etc.) and the more remote markets (the EU countries, as well as the USA)

(b) To consider the technological modernization of the economy, the second central axis in the development of Estonia's economy, especially in the second half of the period under observation. In turn, to consider the most important aspect of technological modernization, the implementation of the new opportunities available in the revolutionary development of ICT

(c) To use the premise that Estonia will join both the EU and NATO during the period under observation, while leaving the precise accession times open in the building of the scenarios, that is, the time and order of accession (whether the EU or NATO would come first) would depend on the particular scenario
(d) To treat the developments related to ecology and social changes as scenario based, that is, to presume that the emergence of trends like the aggravation of ecological problems, the strengthening of ecology-related policies, unemployment, the emergence of potential social and/or ethnic (within Estonia) tension has a significantly different probability in the different scenarios

Based on the above, the following 'grid' for the building of scenarios was developed (see Table 18.1).

The axes of the above matrix cannot be observed in isolation from each other, but have mutual influence, while the impact of the horizontal axis events on those of the vertical axis is stronger than vice versa, although this ratio can change in the second half of the period. As it is, if (especially in the first period of development) significant economic growth can be achieved by combining the different geographical markets, the state's, as well as the companies', opportunities for the technological modernization of the economy will increase in the event of successful EU integration, thanks to the availability of the EU structural funds for that purpose. (These connections were described in greater detail in Terk (1999)). It was presumed, however, that economic growth based exclusively on market access cannot last very long. An economy approaching normal price proportions (the low cost of infrastructure services inherited from the socialist period cannot last, because the infrastructure will require modernization), especially in a situation where the state intends to integrate in the EU economic space, will inevitably lose its competitiveness as an exporter of simpler and cheaper products, while continued economic growth would be possible only on the basis of a modernized and technologically more advanced economic structure.

The above logic is neither unequivocally externally deterministic nor voluntaristic. Achieving success along the horizontal axis will depend to a greater degree on external factors (geopolitical situation, East-West relations), but to some degree also on the state's foreign policy and the success of its policies aimed at expanding its export geography.

The developments along the vertical axis depend not only on the state's policies (innovation policy, cooperation between the public and private sectors, etc.) but also on how attractive a position Estonia could adopt in the geopolitical geo-economic arena for international organizations and private investors interested in more sophisticated types of production (Wright and Cairns 2011).

18.2.4 The Third Stage

This third stage involves the 'naming' of the scenarios and the writing of the scenario texts which is closely related to metaphors, narratives and storylines. The stage started with the specification and 'naming' of the states to be observed

Table 18.1 The 'grid' for building scenarios

	No chances of success	Partial success possible	Complete success possible
Leader			
Drag-along			
Failure			

Estonia's success in making use of the opportunities of technological progress
Geopolitical configuration in the Baltic Sea region as a basis for gaining access to markets and investments (potential and making use of it)

along either axis. Concerning the horizontal axis, it was decided to operate with two probable states, or, in other words, with two possible geopolitical positions of Estonia between the West and the East (Russia and the CIS). In the case of the first one, we used the 'Bastion' metaphor (dominated by geopolitical opposition, and contraction of economic relations between East and West), while the other was designated as 'gateway'. The latter presumed positive tendencies in East-West relations with the EU not only expanding to the Baltic states, but parallel to that the EU-Russian economic integration increasing. In the ideal case, the Baltic States could become a base for the EU's economic (positive) expansion to Russia and the CIS.

The third theoretical position, the failure of Estonia's Western integration and its return to being under Russian control was considered to have too low a probability for further analysis in the context of this package of scenarios.

As a complementary premise it was set that, in case of the 'Bastion' role, active economic relations (if any) could be expected with Estonia's closest neighbours in the EU, while, in the case of the 'gateway' position, the activity of more remote EU great powers (Germany, the UK, France), as well as investments from the great global economies (the USA, Japan), could be expected. An exception would be US investments potentially motivated by military or geopolitical considerations.

Along the vertical axis, three states were initially set: 'leader', 'drag-along' and 'failure'. The third state was discarded in further analysis, since Estonia's ICT development had taken off quite fast at that time, and failure was actually improbable.

The scenario texts were drafted to cover 5–6 pages (each), starting with a general explanation of the dynamics, followed by a description of the developing situation regarding the given spheres: economic growth and the development of the economic structure, prevailing values and the type of social affairs, ecological development, inter-Estonian regional development, etc. The text closed with a brief definition of the situation developed by the end of the period.

As early as during the drafting of the scenarios, the issue emerged of defining the development typical of each scenario in a brief and concise manner, that is, how to sum up the description in a short and memorable name for the scenario.[2] It was

[2] It can be argued that an effective name, capable of conveying a synthetic image, can be found after the main outlines of the scenario description have been drafted.

decided after long debates to designate the most positive scenario (multidirectional geographic cooperation combined with technological modernization) as '*interface*' and the scenario of limited geographic integration and 'non-modernization' as '*periphery of Scandinavia*'. The variant, where Estonia would remain in the 'Bastion' situation but managed to carry out successful technological modernization, was unanimously considered of very low probability (within the logic of this scenario, it would be difficult to imagine the source of the cash flow necessary for the country's technological modernization).

As a theoretical construction, one could only imagine geopolitically and militarily motivated US and NATO aid to the 'Bastion' which they deemed significant, combined with Estonia's considerable success in joining and making use of military technological development programmes. Thus, the scenario was designated as '*military info-oasis*'.

The last scenario is based on the combination of the 'gateway' position, which would not be accompanied by ambitions and success in technological modernization and increasing the industry's value added. This is different from the '*periphery of Scandinavia*' scenario primarily by the high ratio in the economy of exports to Russia (especially food products) and of services to Russian-Western trade and especially transit transportation. Therefore, this scenario was designated as '*ferryman/pipeline operator*'.[3]

The general picture of the axes and scenarios thus obtained the following shape (see Fig. 18.1).

The selected metaphors were of considerable use in better focusing the scenario descriptions and emphasizing the links of primary importance.

18.2.5 The Fourth Stage

This fourth stage involves detailing the calculations based on the matrix's vertical axis and the storylines. While the initial (zero-stage) calculations of economic growth and economic structures were based only on the linear logic of 'markets –> investments –> economic growth', adding the construction of a horizontal axis besides the vertical one enables us to pose the question of the sustainability of the economic growth based solely on markets and investments without the component of technological progress. This was particularly relevant regarding the later part of the period under observation. In the mid-1990s, there was no statistical basis for solving this problem in the conditions of a post-socialist economy. Only the economic growth figures of some EU 'latecomer' countries (e.g. the Irish Republic) were available, while these possible analogies were admittedly somewhat remote. It was also complicated to forecast the long-run rate of inflation in Estonia, while it determined the rate of the existing economic structure losing its competitiveness.

[3] The original Estonian designation ('ülevedaja') is somewhat more abstract and cannot be translated literally.

		Geopolitical situation of Estonia	
Technological modernization (IT-based)		“Bastion”	“Gateway”
	“Leader”	“Military info-oasis”	”Interface”
	“Drag-along”	“Periphery of Scandinavia”	“Ferryman/Pipeline operator”

Fig. 18.1 General picture of the axes and scenarios

Therefore, the economists in the working group carried out the necessary forecasts largely on an intuitive basis. In case of the ‘ferryman’ scenario, a sudden crisis caused by the loss of competitiveness was forecast after a period of quite high economic growth. The ‘periphery of Scandinavia’ scenario, where the economic growth of the first half of the period was forecast as rather low, included a more gradual crisis (or rather the economy remaining in a stagnating state characterized by a permanently high unemployment level).

After the completion of the above calculations, the texts of the scenarios obviously needed updating as well.

18.2.6 The Fifth Stage

This fifth stage involved the first use of the scenarios and feedback. The completed scenarios were used for carrying out economic policy-related discussions with leading officials of several Estonian ministries, leading members of all political parties represented in the Estonian parliament, the economics and environmental committees of the parliament,[4] specialists planning the development of various Estonian counties and the counties’ leaders. Brief versions of the scenarios were also published in the

[4] The discussion with the financial affairs committee took place at a somewhat later stage.

mass media, where they provoked a rather active discussion. Several catchwords used in the scenarios reached broader public debates via the mass media.

The scenarios' general logic did not provoke particular objections during the first stage of their use; most participants in the discussions also expressed support for the need to strengthen the direction of technological modernization on the horizontal axis of the matrix. Specific ideas on which type of innovation or information technology policies Estonia needs differed somewhat, but not to a great extent. The positions of representatives of the various political parties differed more on the issue of how large should be the financial resources that the state should concentrate for the realization of this policy. In some cases, politicians or high-level officials attempted to postpone the implementation of such policies until the period when the EU structural funds would become available to Estonia.

While there were no major differences of opinion concerning the vertical axis of the matrix, this would not apply to the horizontal axis. Here, too, the debate did not concern the basic logic of the scenarios but rather concentrated on the level of risk of 'business with Russia' for Estonia and the extent of the government's support for this business. While the scenarios primarily presented economic arguments, political arguments prevailed in case of politicians. The more extreme positions could be summed up in the expression 'the further and the fastest, the better'. At the same time, a large proportion of the participants in the debates considered business ties with Russia and the CIS secondary to those with the EU but significant enough to attempt to contribute to their success.

It can thus be stated that the debate launched on the basis of the scenarios helped Estonian society to better perceive the development alternatives, their hazards and opportunities and thus comprehend which premises were necessary for the realization of one development version or another.

The logic of the scenario package itself contained an idea that development strategies should be built in accordance with the probable geo-economic environments, and therefore it would be impossible to create a single rigid development strategy which would be applicable to all situations; on the contrary, the foci vital for success should be reinforced.

The idea of what these foci should be was clarified in the course of the debates. Yet the debates did not advance major changes, which the authors should have considered in the scenario texts. As the scenarios were published in a book (Raagmaa and Terk 1997), the texts differed from those used in the debates only in nuances rather than in their principal aspects.

18.2.7 The Sixth Stage

This final stage involved the computation of the scenarios based on the macroeconomic model. This stage of the work was carried out primarily on the basis of the needs of some Estonia's largest infrastructure companies (the main financier was the energy firm 'Eesti Energia') and in consideration of the upcoming address to the Parliamentary Financial Affairs Committee. Specialists of the central bank (the Bank of Estonia) were included (Professor Urmas Sepp et al.), and three out of four scenarios (the 'military info-oasis' was discarded as being of low probability) were computed according to the macroeconomic model used in the central bank.[5] The given ICOR coefficients were entered as a basis in the model, together with the hypotheses on the types of crises caused by the decline of competitiveness in the 'periphery of Scandinavia' and 'ferryman' scenarios, and some additional premises derived from Estonia's integration in the EU. The model calculations resulted in parameters for every scenario concerning indicators such as exports, imports, balance of payments, in- and outflow of capital and the amount of foreign debt.

At the request of 'Eesti Energia', a complementary forecast of Estonia's population was carried out. It took as its basis the demographic model used at the Estonian Institute of Economics, which was complemented by estimations concerning migration. The latter were derived on the basis of forecast economic growth, wages level and unemployment, as for every scenario.

Considering the profiles of the clients and main recipients of this stage of the work, the previous scenario texts were no longer updated with new complementary indicators; instead, new, somewhat shorter and more macroeconomic issues-oriented texts were written. These scenario texts were subsequently discussed with the clients, as well as with the members of the corresponding parliamentary committee.

18.3 Some Conclusions

Viewing this package of scenarios in retrospect, one could claim that it turned out to be quite usable, while the qualitative descriptions of the scenarios appeared somewhat more adequate than the forecast quantitative parameters.[6] Based on the scenarios package, some influence was exerted on the development of the post-stabilization economic policy in Estonia, for example, by explaining to politicians and higher

[5]This was the World Bank Revised Minimum Standard Model, usually known as RMSM, a means of analysis and forecasting, which is used for assessing macroeconomic equilibrium, especially in relation to the cost of serving foreign loans. The model's behaviour formulas are based on the data of the previous development period of 21 selected export-oriented countries.

[6]The situation in Estonia, which developed in reality, contains elements from the 'periphery of Scandinavia' and the 'interface' scenarios and to a lesser extent from the 'ferryman' scenario. Most of the important links between the development parameters, drafted in the scenarios, were validated by reality. At the same time, economic growth in Estonia turned out to be somewhat higher until 2007 (incl.), and unemployment in the second half of the period slightly lower than the assumptions used in the scenario package would have permitted to presume. It can be argued that the mid-1990s' economic theory tended to underestimate the post-socialist countries' rate of convergence. However, definitive conclusions cannot be made until the end of the time period under observation, that is, before the year 2010. Moreover, in 2008, economic growth in Estonia was replaced by decline, and therefore, the adequacy of growth forecasts cannot be judged before the ending of the economic cycle.

officials certain aspects of economic relations with Russia and the need for an innovation policy.

The experience of the 'Estonia 2010' project demonstrated both the significance of calculations based on analogous countries' experience for the subsequent generation of storylines and, vice versa, the importance of qualitative descriptive texts as a basis from which the need for complementary calculations emerges. From the viewpoint of the user (politician or other type of decision maker), the most preferable option is a qualitative description which opens up the meaning of the scenario, whose general adequacy could be demonstrated with calculations concerning some significant parameters. The alternation of qualitative and quantitative logic could be iterative and multilayered. Using the common basic construction, differently focused and stylized scenario texts can be built for various groups of users with special needs, while a need for complementary calculations could emerge in the case of every such variant.

References

Bell, W. (1997). *Foundations of futures studies* (Vol. 1). New Jersey: Transaction.

Maailm ja Eesti tulevikutrendid. (1996). Tallinn/Tartu: EV Keskkonnaministeerium ja ETI.

Raagmaa, G., & Terk, E. (1997). *Eesti tulevikustsenaariumid*. Tallinn/Tartu: EV Keskkonnaministeerium ja ETI.

Terk, E. (1999). Estonia's economic development: Achievements, contradictions, prospects. In: *Estonian human development report*. Tallinn: UNDP.

Wright, G., & Cairns, G. (2011). *Scenario thinking: Practical approaches to the future*. New York: Palgrave Macmillan.

Printed by Publishers' Graphics LLC
AMZ20121206.19.20.14